The AI Powered Enterprise

The AI Power Enterprise

Reshaping Business as We Know It

Enamul Haque

Mercury Learning and Information
Boston, Massachusetts

Mercury Learning and Information
121 High Street, 3rd Floor
Boston, MA 02110
info@merclearning.com

E. Haque. *The AI Powered Enterprise: Reshaping Business as We Know It.*
ISBN: 978-1-5015-2401-1

Library of Congress Control Number: 2025949511
242526321 This book is printed on acid-free paper in the United States of America.

Our titles are available for adoption, license, or bulk purchase by institutions, corporations, etc.

All of our titles are available in digital format at various digital vendors.

CONTENTS

Preface

This book represents the third and final phase of my investigation into how artificial intelligence fundamentally reshapes enterprise operations. Unlike my previous works, which focused on building technical foundations and establishing responsible adoption practices, this book addresses something far more profound. It documents how organizations transcend AI adoption to become AI-native enterprises that think, learn, and evolve like intelligent systems.

The journey to this understanding began with technical implementation challenges, evolved through governance and ethical frameworks, and now arrives at the complete reimagination of organizational intelligence architecture. Each phase taught me something different about the nature of AI transformation, but this final phase changed everything I thought I knew about competitive advantage in the intelligence economy.

Most technology transformations follow predictable patterns. Digital transformation has established methodologies. Cloud adoption has proven frameworks. Mobile transformation has clear best practices. AI transformation breaks every rule. It's not about implementing new tools. It's about becoming a different type of organization altogether.

WHY I'M UNIQUELY POSITIONED TO DOCUMENT THIS REVOLUTION

Most business technology writers observe transformations from the outside, analyzing case studies and interviewing executives after implementations succeed or fail. I've had the extraordinary privilege of witnessing AI transformation from inside the system as it happens.

My career has unfolded at the intersection of every major technology revolution that created modern business infrastructure. From refugee management systems at the United Nations using mainframes, through Internet-enabled global collaboration in the 1990s, to mobile infrastructure development at Nokia that powers today's AI applications, I've built the technological foundations that enterprises now take for granted.

AI transformation is fundamentally different from anything I've experienced before. Previous revolutions asked organizations to adopt new capabilities. The AI revolution asks them to reimagine what organizational capability means. The difference isn't incremental. It's categorical.

Working with hundreds of enterprises across every major industry has provided an unprecedented laboratory for understanding how AI-native transformation actually works. Financial services firms discovered that unauthorized AI experimentation drives more innovation than formal IT initiatives. Manufacturing companies applied production-line principles to intelligence development. Healthcare systems developed organizational reflexes that respond faster than human decision-making processes.

These weren't isolated success stories. They represented systematic patterns that no existing business literature addressed. The gap between what I was observing and what business schools were teaching became impossible to ignore.

THE REAL-WORLD LABORATORY THAT CHANGED MY PERSPECTIVE

The consulting industry continues publishing AI adoption frameworks based on surveys and theoretical models. Meanwhile, hundreds of enterprises across every industry are rewriting the transformation playbook in real time.

These organizations became an inadvertent laboratory for understanding something unprecedented. They're not just implementing AI tools more effectively than their competitors. They're developing entirely new forms of organizational intelligence that operate according to different principles from traditional business structures.

What I discovered challenges everything the business literature teaches about AI transformation. The enterprises achieving breakthrough results aren't following established change management methodologies. They're not creating centers of excellence, hiring chief AI officers, or establishing governance committees. Instead, they're doing something far more radical.

They're treating unauthorized AI experimentation as innovation fuel rather than compliance risk. They're building systematic intelligence production capabilities that operate like manufacturing plants. They're creating new organizational structures that didn't exist two years ago. They're developing measurement systems that make traditional performance metrics obsolete.

Most surprising of all, they're discovering that prompts represent intellectual property as valuable as patents. Some organizations now protect their prompt libraries more carefully than their financial data. Others have realized that competitive advantage doesn't come from having better AI models, but from having better methods for extracting intelligence from any AI system.

These insights emerged from watching transformation happen in real time, not from academic research or consulting frameworks. Every pattern documented in this book comes from enterprises that are living these changes, not planning them. The frameworks aren't theoretical constructs. They're systematic documentation of how breakthrough organizations actually operate.

Here's what made these observations particularly valuable. These enterprises were developing revolutionary approaches without shared vocabulary or systematic frameworks. They were solving similar challenges in isolation, reinventing solutions that could be standardized and scaled across industries.

This represents the difference between best practices and breakthrough practices. Best practices emerge when successful approaches become widely adopted and systematized. Breakthrough practices emerge when organizations solve problems that the business literature hasn't yet recognized as problems.

THE THREE-PHASE EVOLUTION OF AI BUSINESS THINKING

This book represents the third phase in a three-phase evolution of how enterprises think about AI.

The first phase focused on capability building. Organizations needed to understand what AI could do and how to implement it safely within existing business structures. This phase produced numerous books about machine learning applications, automation possibilities, and technical implementation strategies. The primary question was how to add AI to existing business processes.

The second phase addressed responsible adoption. As AI capabilities expanded rapidly, organizations needed frameworks for governance, ethics, and risk management. This phase generated extensive literature about bias mitigation, algorithmic transparency, and regulatory compliance. The primary question was how to use AI responsibly within existing organizational frameworks.

This third phase transcends both adoption and governance. It addresses how organizations become AI-native rather than simply AI-enabled. The primary question isn't how to add AI or how to govern AI. It's how to think like an AI-powered enterprise.

The evolution mirrors what I observed in previous technology transformations, but with compressed timelines and exponential impact. Digital transformation took decades to mature from tool adoption to business model reinvention. AI transformation is completing the same evolution in years, not decades.

Organizations that understand this compressed timeline are building advantages that will prove impossible for traditional enterprises to replicate. They're not just moving faster through conventional transformation stages. They're skipping stages entirely by reimagining themselves as intelligent systems from the beginning.

HOW TO NAVIGATE THIS BOOK

This book is designed for three distinct types of leaders who must navigate the AI transformation, each with different needs and different starting points.

Executives who must architect transformation will find the strategic frameworks needed to design AI-native enterprises. Each chapter provides both conceptual understanding and practical implementation guidance. The frameworks scale from departmental initiatives to enterprise-wide transformation programs. More importantly, they're designed to be cumulative. Understanding shadow AI management enables more sophisticated AI factory implementation. Mastering prompt capital enhances enterprise reflex graph development. The twelve frameworks work together as an integrated system for organizational intelligence architecture.

Executive readers should focus on the strategic implications and competitive advantages that each framework enables. The Monday morning implementation steps are important, but the real value comes from understanding how these frameworks compound to create sustainable competitive advantages. AI-native enterprises don't just perform better than traditional organizations. They operate according to entirely different principles.

Strategic leaders who must compete in AI-disrupted markets will discover how to build sustainable competitive advantages using intelligence as a strategic asset. The book reveals how leading organizations create barriers to entry that traditional enterprises cannot cross. More importantly, it shows how to assess competitive position and identify the highest-impact transformation priorities.

Strategic readers should pay particular attention to the measurement frameworks and competitive intelligence methodologies. Key AI efficiencies provide leading indicators of AI-native transformation success. Signed influence graphs enable competitive positioning when intelligence becomes the primary differentiator. Circular AI business models create self-reinforcing advantages that strengthen through usage.

Operational leaders who must implement transformation will learn to think differently about organizational design, process optimization, and performance management. The frameworks aren't just for understanding AI. They're for understanding how intelligent systems operate, learn, and adapt. These cognitive models reshape every leadership challenge.

Operational readers should focus on the implementation methodologies and success metrics. Shadow AI management systems provide systematic approaches to innovation harvesting. AI factory methodology enables scalable intelligence production. Human-AI fusion protocols solve coordination challenges that emerge when humans and artificial systems work as integrated teams.

The book is structured to serve all three audiences simultaneously. Each chapter begins with strategic context that executives need, provides implementation details that operational leaders require, and includes competitive implications that strategic leaders must understand.

WHY THIS MOMENT DEMANDS THIS BOOK

The window for AI-native transformation is narrowing rapidly. Analysis across enterprise AI adoption demonstrates that organizations implementing these frameworks in the next eighteen months will build advantages that compound for decades. By 2027, the competitive separation between AI-native enterprises and traditional organizations will become irreversible.

The transformation isn't happening through conventional methodologies. The success patterns I've documented emerge from organizations that think differently about intelligence, learning, and adaptation. They don't just use AI more effectively. They organize themselves according to AI principles.

This represents both a tremendous opportunity and a significant risk. Organizations that systematize these approaches will accelerate their development and scale their impact. Those that leave them as informal will struggle to replicate successes or scale beyond isolated experiments.

The business literature hasn't caught up with what the breakthrough organizations are actually doing. Academic frameworks still focus on AI adoption rather than AI-native transformation. Consulting methodologies still treat AI as advanced automation rather than organizational intelligence architecture. Technology vendors still market AI as a tool enhancement rather than a business model revolution.

This gap creates a category creation opportunity. The frameworks documented in this book fill systematic gaps in how enterprises think about AI transformation. They provide the vocabulary for discussing approaches that are already emerging but haven't been systematized.

By 2028, business leaders will need to understand shadow AI management, AI factory efficiency, prompt capital valuation, enterprise reflex graph optimization, and the other frameworks documented in this book. The organizations that master this vocabulary first will shape industry standards, influence regulatory development, and attract top talent.

THE PROMISE OF BREAKTHROUGH TRANSFORMATION

Every framework in this book emerged from studying organizations that achieved results their industries considered impossible: ten-fold improvements in business metrics that competitors struggle to improve incrementally, solutions to problems their sectors considered unsolvable, and customer experiences that redefined market expectations entirely.

These aren't isolated success stories. They represent systematic approaches to building organizational intelligence that learns, adapts, and improves automatically. The frameworks are designed to be replicable across industries, scalable across enterprise sizes, and adaptable across cultural contexts.

The real promise isn't better performance within existing competitive frameworks. It's the development of entirely new types of competitive advantages that traditional enterprises cannot replicate. AI-native organizations don't just outperform traditional companies. They compete in different categories altogether.

The transformation requires new mental models, new measurement systems, and new leadership capabilities. It doesn't require massive capital investments or infrastructure buildouts, though. It requires thinking differently about what organizational intelligence can become.

This book provides the roadmap for that transformation. Every concept has been tested in real enterprise environments. Every methodology provides specific implementation steps. Every framework is designed for Monday morning application.

Most importantly, the book provides the vocabulary for continuing the transformation as AI capabilities evolve. The frameworks aren't just for current AI technologies. They're for organizing intelligence regardless of how the underlying technologies develop.

THE CHOICE THAT DEFINES A COMPETITIVE FUTURE

The evidence demonstrates a fundamental choice facing every enterprise leader. You can continue thinking about AI as sophisticated automation that improves existing processes, or you can begin thinking about AI as the foundation for entirely new forms of organizational capability.

The enterprises choosing the second path are building competitive advantages that will prove impossible to cross. They're not just more efficient versions of traditional organizations. They're fundamentally different types of enterprises, operating with varying response times, different learning rates, and different improvement trajectories.

The choice isn't whether to participate in AI transformation. That transformation is happening whether or not there is deliberate leadership. The choice is whether to architect the transformation or let the transformation architect you.

This book is for leaders who choose to architect.

The frameworks provide the blueprint. The research provides the validation. The choice determines the destiny.

Welcome to the systematic documentation of how AI-native enterprises actually work.

PROLOGUE

Every technology revolution has taught enterprises to adopt new tools. The AI revolution is teaching them to think like intelligent systems themselves.

After three decades of enterprise technology transformation and extensive research across sixteen critical domains of AI enterprise adoption, something unprecedented has been identified. The companies that will dominate in 2028 aren't just using AI tools. They're becoming AI-native organizations that think, learn, and evolve like intelligent systems.

This book introduces the essential business vocabulary for that transformation.

THE RESEARCH THAT CHANGED EVERYTHING

Investigations into how enterprises were truly transforming with AI, beyond the headlines and case studies, discovered something remarkable. Comprehensive analysis across shadow AI governance, manufacturing-inspired intelligence production, prompt intellectual property frameworks, organizational reflex systems, autonomous business units, trust architectures, circular business models, new success metrics, and competitive intelligence mapping revealed systematic gaps in existing business literature.

Traditional AI business books focus on adoption. Technical AI books focus on implementation. Strategic business books assume human-driven decision making. None address the fundamental question that separates leaders from followers. How do you architect an entire organization to think, learn, and adapt like an intelligent system?

What was discovered is that this isn't happening through conventional change management or digital transformation methodologies. It's happening through twelve specific breakthrough frameworks that bridge documented gaps in current business thinking: shadow AI management systems, AI factory methodology, human-AI fusion protocols, prompt capital framework, enterprise reflex graph, use logic versus use cases, agentic strategic units, cognitive trust graph, circular AI business models, key AI efficiencies, confidence indicators, and signed influence graphs.

These aren't incremental improvements to existing approaches. They represent the first systematic vocabulary for designing truly AI-native enterprises.

FOUR TIERS OF TRANSFORMATION

Analysis of successful AI-native transformations reveals a four-tier architecture that no existing framework addresses comprehensively.

Breakthrough concepts represent completely novel approaches to enterprise AI transformation. Investigation across unauthorized AI governance shows that traditional frameworks address either shadow IT generically or approved AI specifically, but miss the dual objective of risk mitigation and competitive advantage transformation. Similarly, examination of prompt intellectual property demonstrates that while prompt engineering exists as a technical discipline, no existing framework treats prompts as strategic assets with economic valuation methodologies.

The concept of enterprise reflex graphs emerges from gaps in digital twin analysis, which focuses on human decision-making rather than autonomous AI responses that occur before conscious thought. Use logic versus use cases addresses a fundamental scaling problem where AI implementation remains project-centric rather than based on systematic patterns. Agentic strategic units fill the void where no framework exists for AI-enabled autonomous business functions that self-organize and adapt in real time.

Advanced applications represent novel applications of existing concepts to AI-native transformation. The AI factory methodology clearly differentiates from existing MLOps approaches by applying proven manufacturing principles to cognitive work rather than focusing solely on technical pipeline management. Human-AI fusion protocols advance beyond human-in-the-loop systems to solve coordination challenges through reciprocal apprenticing and contextual handoff protocols. Cognitive trust graphs represent a novel fusion of blockchain principles and AI governance, enabling real-time trust verification rather than retroactive auditing.

New measurement systems address the obsolescence of traditional business metrics in AI-native organizations. Key AI efficiencies provide the first comprehensive metric set beyond traditional ROI or accuracy measures, focusing on intelligence velocity, decision quality, and organizational learning rates. Confidence indicators introduce a novel approach to measuring strategic trustworthiness through real-time assessment of AI system reliability for strategic decisions.

Strategic intelligence frameworks enable competitive advantage in intelligence-driven markets. Circular AI business models represent the first application of circular economy principles to intelligence systems, creating self-improving AI systems that generate sustainable competitive advantage. Signed influence graphs provide a novel competitive intelligence mapping approach using AI-powered dynamic modeling rather than static analysis frameworks.

THE PATTERN THAT EMERGES

These twelve frameworks form what emerges as the essential business vocabulary for AI-native transformation. Unlike previous technology revolutions that asked enterprises to adopt new tools, the AI revolution requires organizations to fundamentally reimagine themselves as intelligent systems.

Consider shadow AI management systems. Data from enterprise IT departments shows that security teams maintain visibility over less than twenty percent of AI applications in their organizations. Traditional approaches treat this as pure compliance risk. Comprehensive analysis reveals that unauthorized AI experimentation represents the largest untapped source of competitive advantage in most enterprises. Organizations implementing systematic approaches to identify, evaluate, and scale shadow AI innovations report dramatically faster time-to-market for AI capabilities compared to those relying solely on formal IT channels.

The AI factory methodology emerges from studying how leading organizations apply proven manufacturing principles to cognitive work production, unlike MLOps, which focuses on technical pipelines. Companies implementing systematic intelligence production capabilities report deploying AI applications at scales previously impossible through project-based approaches. The IBM Watson platform architecture, the JP Morgan COIN system for legal document analysis, and machine learning deployment across Walmart retail operations demonstrate factory-like scaling of intelligence production.

The prompt capital framework addresses findings showing that prompts represent intellectual property as strategically valuable as patents or proprietary algorithms. Organizations protecting strategic prompt libraries report productivity improvements exceeding sixty percent and decision accuracy gains over ninety percent compared to enterprises using generic AI interactions. Some companies now guard their prompt libraries more carefully than financial data, recognizing conversational intelligence assets as competitive differentiators.

Enterprise reflex graph analysis demonstrates movement beyond digital twins that inform human decisions toward AI-driven systems that respond before conscious thought. Organizations implementing real-time intelligence flow mapping achieve responsiveness levels that traditional hierarchical structures cannot match. These systems enable pre-conscious business responses where data becomes action without human intermediation.

The transition from use cases to use logic represents a paradigm shift where leading companies move from isolated AI projects to systematic pattern libraries. Evidence shows organizations implementing reusable AI reasoning patterns scale implementations exponentially faster than those treating each application as a custom development. Logic blocks become modular AI capabilities that combine and recombine across business scenarios.

Agentic strategic units represent the organizational structure for AI-native enterprises. These self-managing business functions, augmented by AI, adapt in real time rather than executing predetermined plans. Analysis demonstrates these units combine human judgment with AI capabilities to operate more like intelligent organisms than mechanical processes, achieving performance levels that traditional departmental structures cannot replicate.

WHY THESE FRAMEWORKS MATTER NOW

Cross-industry analysis shows a critical window closing rapidly. Organizations implementing AI-native frameworks in the next twenty-four months will build advantages that compound for decades. By 2027, the competitive separation between AI-native enterprises and traditional organizations will become irreversible.

This transformation isn't happening through conventional methodologies. Evidence indicates that successful AI-native transformation requires new mental models, new measurement systems, and new organizational architectures. Traditional change management approaches assume human-driven decision making. These frameworks assume intelligence itself as the operating system.

The data exposes something else equally important. These transformations are already happening intuitively in leading organizations, but without systematic frameworks or shared vocabulary. Companies are developing shadow AI management practices without calling them that. They're building AI factory capabilities without recognizing the manufacturing parallels. They're creating prompt capital without understanding its intellectual property value.

This represents both a tremendous opportunity and a significant risk. Organizations that systematize these approaches will accelerate their development and scale their impact. Those that leave them as informal will struggle to replicate successes or avoid failures.

THE CATEGORY CREATION OPPORTUNITY

Analysis of business vocabulary adoption demonstrates that every major business transformation creates new essential terminology. Return on investment became standard language during the financial management revolution. Key performance indicators emerged during the transformation of quality management. Customer relationship management defined the sales automation era.

Evidence suggests that AI-native transformation will create similar vocabulary requirements. By 2028, business leaders will need to understand shadow AI management, AI factory efficiency, prompt capital valuation, enterprise reflex graph optimization, agentic strategic unit design, cognitive trust graph architecture, circular AI business model sustainability, key AI efficiency measurement, confidence indicator tracking, and signed influence graph competitive positioning.

The organizations that master this vocabulary first will shape industry standards, influence regulatory frameworks, and attract top talent. Academic institutions will teach these concepts. Consulting firms will build methodologies around them. Software vendors will adopt this terminology. Conference sessions will use these frameworks as organizing principles.

WHAT THIS BOOK DELIVERS

Every framework in this book emerged from research validation across real enterprise environments. Every concept addresses documented gaps in existing business literature. Every methodology provides specific implementation steps you can begin Monday morning.

This isn't theoretical speculation. It's the systematic documentation of how AI-native transformation actually works, supported by comprehensive research across the domains that matter most for enterprise success.

The frameworks are designed to be quotable, teachable, and actionable. Prompt capital is the new patent portfolio. Enterprise reflex graphs determine organizational responsiveness. Use logic replaces use case thinking. These aren't just concepts. They're the operational vocabulary for competitive advantage in the intelligence economy.

By Chapter 3, you'll understand how shadow AI management systems transform unauthorized innovation into competitive advantage.

By Chapter 8, you'll have detailed implementation guidance for building your AI factory. By Chapter 15, you'll know how to measure success using key AI efficiencies that predict future performance rather than reporting past results.

Most importantly, you'll have the complete vocabulary for transforming your enterprise into an AI-native organization that thinks, learns, and evolves like an intelligent system.

THE CHOICE THAT DEFINES THE FUTURE

The evidence exposes a fundamental choice facing every enterprise. You can continue thinking about AI as advanced automation that improves existing processes, or you can begin thinking about AI as the foundation for entirely new forms of organizational intelligence.

The enterprises choosing the second path are building competitive moats that will prove impossible to cross. They're not just more efficient. They're fundamentally different types of organizations operating with different capabilities, different response times, and different improvement rates.

The window for making this choice is narrowing. The competitive separation is accelerating. The vocabulary for discussing these transformations is crystallizing.

Which type of organization will yours become?

The evidence provides the answer. The frameworks provide the roadmap. The choice determines the destiny.

Welcome to the AI-powered enterprise.

ACKNOWLEDGMENTS

No book of this scope emerges from a single mind. The AI Powered Enterprise represents the culmination of three decades of learning, collaboration, and transformation alongside extraordinary people across five continents. While words cannot fully capture the depth of my gratitude, I must attempt to honor those who made this work possible.

My deepest gratitude belongs to my parents, whose sacrifices and unwavering belief in education set me on this path. They taught me that hard work transforms lives, integrity defines character, and knowledge empowers generations. Every framework in this book carries the imprint of its values. To my family, who endured countless evenings and weekends lost to research and writing, your patience and encouragement sustained me through the most demanding chapters of this journey.

My professional journey has been shaped by remarkable institutions and the visionaries within them. At the United Nations High Commissioner for Refugees in Geneva, I learned that technology's highest purpose is human impact. The sixteen years at Nokia taught me that platform thinking transforms industries. My colleagues at Microsoft, HCL Technologies, Capgemini, and Wipro expanded my understanding of enterprise transformation at a global scale. To the hundreds of Fortune 500 executives and transformation leaders I have worked alongside, your real-world challenges and breakthroughs informed every practical framework in these pages.

The academic foundations of this work owe much to the institutions that shaped my intellectual development. The University of Geneva provided rigorous computer science foundations. EPFL deepened my analytical capabilities through advanced mathematics. The University of Helsinki opened the door to artificial intelligence and machine learning. Harvard Business School's leadership programmes refined my understanding of organizational transformation. To my students at Bangladesh

Maritime University, Coventry University London, and Udemy learners worldwide, your questions challenged me to articulate these concepts with greater clarity.

A unique acknowledgement is due to the artificial intelligence systems that assisted in validating and stress-testing the frameworks presented here. This book about AI-native enterprises was, fittingly, developed with AI-native research methodologies.

To the team at Mercury Learning and Information, thank you for believing in this vision and for your meticulous attention to bringing these ideas to life. Your editorial guidance transformed ambitious concepts into accessible wisdom.

Finally, to every enterprise leader who will read these pages and implement these frameworks, this book exists for you. May it serve as a worthy guide as you architect the intelligent organisations of tomorrow.

Enamul Haque
London, December, 2025

INTRODUCTION: THE INTELLIGENCE REVOLUTION

YOUR TRANSFORMATION JOURNEY BEGINS HERE

You're about to embark on a cognitive transformation that will fundamentally alter how you perceive organizational capability, competitive advantage, and business leadership. This isn't a typical business book that adds new concepts to your existing mental models. This is a systematic reconstruction of how you think about enterprise intelligence.

The journey ahead requires you to abandon assumptions that have long defined successful business thinking. You'll discover why traditional strategic planning can create a competitive disadvantage in intelligence-driven markets. You'll learn why organizational charts prevent the coordination speeds that AI-native enterprises require. You'll understand why measuring success through efficiency metrics blinds you to the learning velocities that determine future market position.

Most challenging of all, you'll need to embrace the possibility that your enterprise could become genuinely intelligent rather than just digitally efficient. This cognitive leap separates leaders who create AI-native organizations from leaders who remain trapped in automation thinking.

THE MENTAL MODEL EVOLUTION YOU'LL EXPERIENCE

Your transformation begins with recognizing that the enterprise mindset itself must evolve to remain competitive. Chapter 1 dismantles your current assumptions about how technology revolutions work and rebuilds your understanding around intelligence architecture principles that previous generations of business leaders never needed to master.

This cognitive disruption intensifies as you discover the shadow AI revolution already transforming your organization. Chapter 2 reveals that the unauthorized intelligence workflows emerging throughout your enterprise represent your greatest untapped competitive advantage. You'll learn to see governance violations as innovation signals rather than compliance problems.

By Chapter 3, you'll experience the shift from prediction-based planning to cognition-based strategy. Your understanding of strategic thinking will evolve from human-driven analysis to human-AI collaborative intelligence that operates at speeds traditional planning cannot match.

The practical transformation accelerates through Chapters 4 and 5 as you learn to design systematic intelligence production and architecture. You'll stop thinking about AI projects and start thinking about intelligence manufacturing. You'll abandon departmental technology implementations and embrace enterprise-wide cognitive infrastructure.

THE LEADERSHIP EVOLUTION THAT TRANSFORMS EVERYTHING

Chapters 6 and 7 challenge your fundamental assumptions about human work and strategic assets. You'll discover that human-AI collaboration requires entirely new coordination protocols that traditional management approaches cannot provide. You'll learn that conversational intelligence represents intellectual property as valuable as patents or proprietary algorithms.

Your leadership thinking will undergo complete reconstruction through Chapters 8 and 9. You'll evolve from managing static workflows to orchestrating adaptive intent flows. You'll transcend use case thinking and embrace use logic frameworks that scale intelligence across unlimited business scenarios.

The most profound cognitive shifts occur in Chapters 10 and 11 as you learn how to map intelligence flows rather than authority relationships and design autonomous business units that self-optimize while remaining strategically aligned. These concepts will feel foreign initially because they require abandoning control-based leadership in favor of orchestration-based leadership.

THE STRATEGIC REIMAGINATION THAT CREATES COMPETITIVE ADVANTAGE

Chapters 12 through 14 will transform your understanding of strategy itself. You'll learn to think in terms of continuous strategic evolution rather than annual strategic planning. You'll discover how to create business models that become more valuable through usage rather than requiring constant optimization. You'll understand how to transform governance from compliance cost into competitive differentiation.

Your conception of business success will undergo a fundamental revision through Chapters 15 and 16. You'll abandon efficiency-based measurement frameworks and embrace intelligence velocity metrics that predict future performance rather than just reporting historical results. You'll learn to measure organizational learning rates, adaptive capability development, and competitive advantage accumulation.

The human dimension of AI-native transformation will challenge your assumptions about leadership, work, and social responsibility through Chapters 17 and 18. You'll discover that cognitive leadership requires capabilities that traditional management education never addressed. You'll learn that responsible AI practices create market differentiation rather than just ethical compliance.

THE IMPLEMENTATION REALITY THAT TESTS YOUR COMMITMENT

Chapters 19 and 20 bring theoretical understanding into practical implementation reality. You'll discover that AI-native transformation requires systematic approaches that challenge every aspect of traditional change management. You'll learn to think beyond current organizational forms and prepare for enterprise structures that don't exist yet but will define competitive advantage by 2035.

Throughout this journey, you'll experience resistance from mental models that served you well during previous technology transformations but become cognitive liabilities during intelligence transformation. Your success depends on recognizing this resistance and choosing cognitive evolution over cognitive preservation.

THE THREE PHASES OF COGNITIVE EVOLUTION

Your transformation will progress through three distinct phases that mirror how the most successful executives navigate AI-native evolution.

Phase one: cognitive disruption unfolds through the first six chapters as you discover that your current frameworks for understanding technology transformation, competitive advantage, and organizational effectiveness have become inadequate for intelligence-era competition. This phase feels uncomfortable because it requires acknowledging that expertise developed over decades may prevent rather than enable future success.

Phase two: cognitive reconstruction emerges through Chapters 7 through 14 as you build entirely new mental models for strategic thinking, organizational design, and competitive positioning. This phase demands intellectual humility because it requires learning frameworks that have no precedent in traditional business education or experience.

Phase three: cognitive mastery crystallizes through the final chapters as you integrate new frameworks into systematic approaches for leading AI-native transformation. This phase requires leadership courage because it demands implementing approaches that will seem foreign or risky to stakeholders who haven't completed their own cognitive evolution.

THE TRANSFORMATION COMMITMENT REQUIRED

This book demands more than intellectual engagement. It requires a commitment to fundamental change in how you think about organizational leadership. You cannot implement AI-native transformation while preserving traditional management thinking. The cognitive frameworks are mutually exclusive.

Every chapter will challenge you to abandon familiar approaches and embrace new methodologies that may conflict with practices that made you successful previously. Your willingness to embrace this cognitive evolution determines whether you'll lead AI-native transformation or become a case study in executive resistance to necessary change.

The enterprises that will dominate the next decade are already emerging. They're led by executives who completed this cognitive transformation and designed organizations that think, learn, and evolve like intelligent systems. Your choice is whether to join them or spend the next five years trying to understand how they achieved competitive positions that seemed impossible using traditional approaches.

YOUR MONDAY MORNING REALITY

By the final chapter, you'll possess frameworks for transforming any organization into an AI-native enterprise that creates sustainable competitive advantages through intelligence architecture rather than just operational efficiency. More importantly, you'll have the cognitive models needed to continue that transformation as AI capabilities evolve.

However, the real measure of this book's value becomes apparent on Monday morning when you apply these frameworks to your current leadership challenges. You'll discover that problems you've struggled with for months resolve quickly when approached through AI-native thinking. Opportunities that were previously invisible become apparent. Competitive threats you feared become transformation catalysts.

The journey ahead requires intellectual courage, cognitive flexibility, and commitment to fundamental change. It will lead to organizational capabilities that your competitors will struggle to understand, much less replicate.

Your transformation journey begins now.

CHAPTER 1

REBOOTING THE ENTERPRISE MINDSET

When enterprises discover that intelligence has become their operating system, everything changes. Not gradually. Not incrementally. Everything changes completely, and it changes fast.

The companies that will dominate 2030 have already made this discovery. Some, such as Amazon, Tesla, and Palantir, are beginning to exemplify this AI-native thinking in how they embed intelligence into their operations. They understand that AI isn't a new capability to add to existing business processes. Intelligence has become the foundational architecture that determines how modern enterprises think, learn, and evolve. They've rebooted their entire organizational mindset around this reality.

The companies that haven't made this discovery yet are asking the wrong questions. They want to know which AI tools to deploy, how to govern AI usage, and where to find the biggest efficiency gains. These questions assume that AI fits within existing organizational structures and decision-making frameworks. This assumption will prove fatal to competitive positioning.

The right question is fundamentally different. How do we redesign our entire enterprise to think like an intelligent system?

THE OPERATING PRINCIPLE EVOLUTION THAT CHANGES EVERYTHING

While Generative AI could create $2.6 to $4.4 trillion in annual economic value across the global economy (McKinsey Global Institute, 2023), the money enterprises spend on AI must be used strategically. Based on my analysis of Fortune 500 transformations, the vast majority of AI investments deliver only incremental efficiency improvements that competitors can quickly replicate through off-the-shelf AI services. The rare exceptions, enterprises that create sustainable competitive advantages, share common characteristics that this chapter will reveal.

What separates the 8% from the 92%? The successful enterprises aren't deploying AI tools. They're architecting intelligence. They've discovered what some might call the enterprise intelligence operating system, a fundamental rewiring of how organizations process information, make decisions, and evolve capabilities.

This transformation requires understanding that every successful enterprise operates according to fundamental principles that determine how value gets created, how decisions get made, and how competitive advantages get sustained. These operating principles have remained largely unchanged for decades, built around assumptions about human cognitive limitations, information scarcity, and coordination constraints that AI eliminates entirely.

The enterprises succeeding with AI transformation have evolved their operating principles from efficiency optimization to intelligence acceleration. Instead of asking how to do existing things faster or cheaper, they ask how to do previously impossible things. Instead of measuring success through cost reduction, they measure success through learning velocity. Instead of organizing around authority hierarchies, they organize around intelligence flows.

This operating principle evolution changes every aspect of enterprise behavior. Strategic planning shifts from annual exercises to continuous optimization. Customer relationships evolve from transaction management to adaptive collaboration. Innovation processes transform from departmental R&D to enterprise-wide experimentation platforms that generate breakthrough insights as operational byproducts.

What has been observed consistently across industries is that the technology transformation is the easy part. The cognitive transformation is where most enterprises struggle. Successful AI adoption requires leaders to fundamentally reimagine their assumptions about organizational structure, decision-making authority, competitive positioning, and value creation.

The enterprises succeeding with AI transformation aren't implementing AI projects. They're implementing intelligence architecture. They're not asking how to automate existing processes with AI tools. They're asking how to design processes that leverage autonomous intelligence to create capabilities that didn't exist before.

This mindset shift determines everything else. It determines which AI investments create sustainable competitive advantages versus short-term efficiency gains. It determines whether your enterprise builds proprietary intelligence capabilities or remains dependent on generic AI services. It determines whether your organization attracts the talent needed to compete in an AI-driven economy or loses that talent to competitors who offer more sophisticated intelligence challenges.

Most critically, this mindset shift determines whether your enterprise shapes the competitive landscape in your industry or gets shaped by competitors who moved faster to embrace intelligence as their core differentiator.

THE INTELLIGENCE ARCHITECTURE FOUNDATION

The enterprises that make this cognitive leap first will build moats that prove impossible to cross. They'll create business models that generate intelligence as a byproduct of operations, turning every customer interaction and internal process into training data that improves their competitive positioning. They'll develop organizational reflexes that respond to market changes faster than traditional enterprises can even detect those changes occurring.

Consider what becomes possible when your enterprise can think at machine speed across thousands of variables simultaneously. Strategic planning transforms from annual exercises to continuous optimization. Customer interactions evolve from scripted responses to dynamic, personalized experiences that adapt in real time. Supply chain management shifts from reactive coordination to predictive orchestration, anticipating disruptions before they occur.

These aren't efficiency improvements. These are entirely new categories of business capability that become available when enterprises embrace intelligence as their core operating principle.

The transformation begins with understanding exactly how enterprise thinking evolves during technology revolutions. Every major technology wave follows predictable cognitive patterns, but the AI revolution compresses these patterns into timeframes that require immediate action rather than gradual adaptation.

The next sections reveal the specific frameworks that enable this transformation. You'll discover how to apply lessons from mainframe centralization, personal computer democratization, Internet connectivity, mobile ubiquity, and digital platform thinking to architect an AI-native enterprise. You'll learn why the shift from digitization to cognitization represents a fundamentally different category of transformation. You'll also gain the mental models needed to lead your organization through the most significant business evolution in modern history.

The companies that master this transformation will discover something remarkable. They won't just use AI more effectively than competitors. They'll become artificially intelligent organizations themselves, capable of learning, adapting, and evolving at rates that traditional enterprises cannot match.

The future belongs to enterprises that think like intelligent systems. The question is whether your organization will be among them.

THE OPERATING PRINCIPLE EVOLUTION

From Mainframes to Intelligence: The Complete Operating Principle Evolution

Most executives think technology revolutions change what enterprises can do. In fact, they change how enterprises think about thinking itself. Each major technology wave has fundamentally redefined the operating principles that determine how organizations structure intelligence, allocate decision-making authority, and create competitive advantage.

The AI revolution represents the sixth and most profound shift in these operating principles. To understand why AI requires enterprises to become AI-native rather than AI-adopting, you need to understand how the previous five revolutions progressively dismantled traditional assumptions about organizational intelligence. Each wave didn't just add new capabilities. Each wave rewrote the rules for how smart organizations can become.

The pattern reveals something remarkable. Every twenty years, a new technology emerges that makes the previous generation's approach to intelligence architecture obsolete. The enterprises that recognize this pattern early and adapt their operating principles accordingly build advantages that persist until the next revolution. The enterprises that treat new technology as an incremental improvement get displaced by competitors who understand the deeper transformation.

The Five Preceding Operating Principle Revolutions

The mainframe era established the first principle of enterprise intelligence: centralization creates consistency. In Geneva during the early 1990s, humanitarian response systems that managed refugee flows across multiple countries were an example of how centralized computing fundamentally changed organizational behavior. The mainframe didn't just process data faster than manual systems. It created single sources of truth that eliminated conflicting information and forced standardized decision-making processes.

Organizations structured around mainframes developed hierarchical intelligence architectures where information flowed upward for processing and decisions flowed downward for implementation. This created the management consulting industry, the strategic planning function, and the concept of enterprise-wide policies. When intelligence is expensive and centralized, organizations optimize for consistency and control.

The personal computer revolution introduced the second principle: distributed intelligence accelerates innovation. When every desk gained computing capability during the 1990s, organizational boundaries began dissolving. Suddenly, individual contributors could perform analysis that previously required specialized departments. Decision-making authority shifted toward the edges of organizations where information was generated, and action was required.

This transformation occurred across industries during the late 1990s, as enterprises grappled with spreadsheet-wielding employees who could bypass traditional reporting hierarchies. The companies that embraced distributed computing created flatter organizations that moved faster than hierarchical competitors. They discovered that when intelligence is cheap and accessible, organizations optimize for speed and agility.

The Internet revolution established the third principle: connected intelligence scales exponentially. During the late 1990s and early 2000s, enterprises discovered that networked computing enabled entirely new forms of value creation. Suddenly, organizations could coordinate globally in real time, share knowledge instantaneously, and collaborate with external partners as efficiently as internal teams.

This created the first truly global businesses, enabled supply chain orchestration across continents, and birthed the platform economy. Organizations began thinking beyond their boundaries for the first time. When intelligence can flow freely across networks, organizations optimize for reach and influence rather than just internal efficiency.

The mobile revolution introduced the fourth principle: ubiquitous intelligence enables continuous optimization. The mobile infrastructure development of Nokia from 2000 to 2016 built the connectivity foundation that would later power today's AI applications. Mobile computing did more than make technology portable. It made business operations context-aware and always on.

Organizations could suddenly respond to opportunities and threats in real time because they had constant visibility into operations, customer behavior, and market conditions. This enabled just-in-time manufacturing, dynamic pricing, location-based services, and personalized customer experiences. When intelligence is always available and context-aware, organizations optimize for responsiveness and personalization.

The cloud and digital transformation era established the fifth principle: platform intelligence compounds competitive advantage. Throughout the 2010s, as Fortune 500 companies went through digital transformation, cloud computing and big data analytics enabled organizations to build on shared infrastructure while maintaining competitive differentiation.

Companies learned to leverage platforms for commodity capabilities while focusing innovation on unique value creation. This created the software-as-a-service economy, enabled data-driven decision-making at unprecedented scale, and introduced the concept of network effects as a primary competitive strategy. When intelligence can be built on shared platforms, organizations optimize for ecosystem leverage and compounding returns.

The Sixth Revolution: Autonomous Intelligence

The AI revolution introduces the sixth and most transformative principle: autonomous intelligence enables self-improving organizations. For the first time in enterprise history, business functions are being created that can think, learn, and adapt independently while remaining aligned with organizational objectives.

This is fundamentally different from previous revolutions because it changes the basic mathematics of organizational capability. Instead of humans using technology to process information faster, we're creating technology that can process information, make decisions, and take actions autonomously while improving its performance through experience.

Organizations structured around autonomous intelligence can operate more like biological systems than mechanical systems. They can adapt to changing conditions without explicit reprogramming, discover optimization opportunities that human analysis would miss, and scale intelligent decision-making across thousands of variables simultaneously.

The Operating Principle Evolution Pattern

What has been constant across all six revolutions is that the most successful enterprises don't just adopt new technology faster. They recognize that each revolution requires fundamentally different operating principles and restructure their organizations accordingly.

Mainframe-era enterprises optimized for centralized control and standardized processes. They built hierarchical organizations with clear chains of command and comprehensive policy frameworks. This approach maximized reliability and consistency when intelligence was expensive and centralized.

PC-era enterprises optimized for distributed innovation and individual empowerment. They flattened hierarchies, pushed decision-making authority toward information sources, and created cross-functional teams. This approach maximized speed and agility when intelligence became cheap and accessible.

Internet-era enterprises optimized for network effects and global coordination. They built platform-based business models, created ecosystem partnerships, and developed real-time coordination capabilities. This approach maximized reach and scale when intelligence could flow freely across networks.

Mobile-era enterprises optimized for contextual responsiveness and continuous engagement. They created always-on customer relationships, implemented dynamic resource allocation, and built location-aware service delivery. This approach maximized relevance and timing when intelligence became ubiquitous and context-aware.

Digital platform enterprises optimized for ecosystem leverage and compounding returns. They built on shared infrastructure while focusing innovation on unique capabilities, created data-driven feedback loops, and leveraged network effects for competitive advantage. This approach maximized efficiency and growth when intelligence could be built on shared platforms.

AI-Native Operating Principles

AI-native enterprises must optimize for autonomous adaptation and continuous learning. This requires organizational structures that can evolve without explicit redesign, decision-making frameworks that can improve through experience, and competitive strategies that become stronger as they're used.

The implications are profound. Traditional strategic planning assumes human oversight of all critical decisions. AI-native enterprises develop strategic frameworks that can execute autonomously while remaining aligned with organizational objectives. Traditional organizational design assumes fixed roles and responsibilities. AI-native enterprises create fluid capability architectures that adapt based on changing requirements and emerging opportunities.

Traditional competitive strategy assumes sustainable advantages come from unique assets or capabilities that competitors cannot replicate. AI-native enterprises create learning advantages that compound over time, making them progressively more difficult to compete against as they accumulate more experience and data.

The enterprises that understand this evolution can begin implementing AI-native operating principles immediately, even before their AI capabilities are fully developed. The enterprises that continue operating under previous generations' principles will discover that their fundamental assumptions about organizational intelligence have become competitive liabilities.

The question isn't whether your enterprise will eventually adopt AI technology. The question is whether your enterprise will recognize that AI requires entirely new operating principles and restructure accordingly, or whether you'll attempt to fit autonomous intelligence into organizational architectures designed for previous technology generations.

History suggests that the companies making this cognitive leap now will build advantages that persist for decades.

NOKIA PLATFORM REVOLUTION LESSONS

What the Nokia Mobile Revolution Teaches Us About AI Transformation: Platform Thinking Applied to Intelligence

The rise and fall of Nokia offer the most instructive case study for understanding why AI transformation requires platform thinking rather than product thinking. The journey of Nokia from a 19th-century wood pulp mill to commanding 40% of the global mobile market share by 1998, followed by its precipitous decline, reveals the fundamental difference between building products and building ecosystems.

During my tenure at Nokia, from November 1999 to May 2015, which included three years at Microsoft, I witnessed the strategic decisions that shaped not only the company's competitive position but also the evolution of the entire mobile industry. The lessons from the experience of Nokia are directly applicable to AI transformation, as both revolutions require enterprises to shift from linear value creation to network-based value creation.

The dominance of Nokia wasn't accidental. By 2007, the brand value reached €32.3 billion, making Nokia the most valuable European brand. The iconic Nokia ringtone was heard 1.8 billion times daily worldwide. The company achieved 70% market share in the United Kingdom. This success came from understanding something most competitors missed: mobile communication was infrastructure, not just technology.

The subsequent failure of Nokia illustrates why past success doesn't guarantee future relevance. Despite market leadership, the company made strategic missteps that offer crucial insights for AI transformation. The same platform-thinking principles that Nokia failed to implement are essential for enterprises building AI-native capabilities today.

The Platform Blindness That Destroyed Nokia

The decline of Nokia resulted from a fundamental misunderstanding of platform dynamics rather than product deficiencies. Research analyzing the company's strategic choices reveals seven critical failures that directly parallel the mistakes enterprises make with AI transformation today.

First, Nokia relied on outdated software architecture when the market demanded open, adaptable systems. The Symbian operating system, which was initially designed for feature phones, couldn't support the app economy that emerged with smartphones. Similarly, enterprises building AI capabilities on inflexible infrastructure will find themselves unable to adapt as AI capabilities rapidly evolve.

Second, Nokia launched the Ovi Store late and made it complicated, when developers needed timely, user-friendly ecosystem development. The company prioritized control over ease of use, creating barriers that drove developers toward competing platforms. Enterprises approaching AI transformation with restrictive governance that inhibits experimentation make identical mistakes.

Third, Nokia prioritized hardware manufacturing when the market shifted toward software and ecosystem value creation. It continued to optimize phone production when customers wanted application ecosystems. Enterprises that focus on AI tools rather than AI platforms repeat this error, optimizing for immediate functionality while missing ecosystem opportunities.

Fourth, Nokia dismissed touchscreen interfaces and the app economy when users demanded intuitive experiences and emergent functionality. It defended keyboard interfaces and pre-installed applications when the market moved toward user-customizable experiences. Enterprises that resist AI-driven user experience changes because they disrupt existing processes make similar miscalculations.

Fifth, Nokia hesitated to cannibalize its profitable feature phone business when market evolution demanded self-disruption. It protected existing revenue streams while competitors built entirely new business models. Enterprises that avoid AI transformation because it might disrupt current operations repeat the strategic paralysis of Nokia.

Sixth, Nokia maintained bureaucratic, slow organizational structures when the market required agile, visionary leadership. The decision-making processes that worked for hardware development became liabilities in fast-moving software markets. Enterprises with hierarchical decision-making processes struggle with AI transformation for identical reasons.

Seventh, Nokia chose the wrong ecosystem partner when it selected Windows Phone over Android. It prioritized perceived control over market viability, aligning with a struggling platform instead of the dominant emerging ecosystem. Enterprises that choose AI vendors based on control rather than ecosystem strength make comparable strategic errors.

Platform-Thinking Principles for AI Transformation

Platform thinking represents a fundamental shift from product-centric strategies to ecosystem-centric approaches. It prioritizes flexibility and collaboration over traditional ownership models, recognizing that value increasingly comes from interactions across networks rather than isolated capabilities.

Research on platform dynamics identifies three core principles that Nokia failed to implement but that prove essential for AI transformation. First, successful platforms create technology infrastructure that enables multiple parties to generate value. Second, they emphasize decentralized value creation through network effects and open systems. Third, they optimize for ecosystem health rather than platform control.

AI platforms follow identical principles with intelligence-specific characteristics. Successful AI transformation requires building infrastructure that enables multiple business functions to create intelligence applications. It demands decentralized AI development capabilities that leverage shared data and models. It optimizes for learning velocity across the organization rather than centralized AI control.

The enterprises succeeding with AI transformation apply platform thinking to every aspect of their approach. Instead of building isolated AI applications, they create AI infrastructure that enables continuous innovation. Instead of centralizing AI development, they democratize AI capabilities while maintaining consistent governance. Instead of protecting existing processes from AI disruption, they design AI platforms that improve through usage.

Network Effects in Intelligence Systems

The failure of Nokia to understand network effects cost it the mobile platform war. Network effects occur when each additional user makes a platform more valuable for every other user. Android succeeded because more users attracted more developers, which created more applications, which attracted more users, creating reinforcing cycles that became increasingly difficult to disrupt.

AI platforms create analogous network effects but with intelligence-specific advantages. Each additional user generates data that improves model accuracy for everyone. Each new application creates integration opportunities that enhance platform capabilities. Each insight discovered by one function becomes available to every other function, creating learning advantages that compound over time.

These intelligence network effects prove more powerful than traditional network effects because they compound learning rather than just connectivity. Mobile platforms become more valuable as more people use them, but the platforms themselves don't become smarter. AI platforms become both more valuable and more intelligent as more people use them, creating competitive advantages that are simultaneously sustainable and self-reinforcing.

Organizations that understand this dynamic structure their AI initiatives to maximize learning velocity rather than just deployment velocity. They prioritize AI implementations that generate valuable training data. They create feedback mechanisms that ensure insights discovered in one context improve AI performance in other contexts. They design AI systems that become more effective through usage rather than requiring manual optimization.

Ecosystem Enablement Versus Control

The most costly mistake of Nokia was attempting to control ecosystem development rather than enabling ecosystem success. It wanted to capture all value created by its platform instead of capturing sustainable percentages of dramatically increased total value. This control-oriented approach limited innovation and slowed adoption compared to more open alternatives.

Successful AI transformation requires identical mindset shifts from control to enablement. The enterprises building the most effective AI capabilities focus on enabling business users throughout the organization to create AI solutions rather than centralizing all AI development. They provide tools that democratize AI implementation rather than requiring technical expertise for every application.

This approach demands fundamental changes to traditional IT thinking. Instead of maintaining strict control over AI implementations, successful AI platforms provide governance frameworks that maintain quality while encouraging experimentation. Instead of limiting AI access to specialized roles, they enable every employee to leverage AI capabilities that enhance their effectiveness.

The most successful AI transformations in enterprise technology follow this pattern. They build centralized AI infrastructure that enables decentralized innovation. They create governance frameworks that maintain consistency while encouraging creativity. They measure success through ecosystem adoption rates and learning velocity rather than just cost savings or efficiency gains.

Strategic Lessons for AI-Native Enterprises

The experience of Nokia provides a complete roadmap for avoiding platform transformation failures. The same strategic principles that determined mobile platform success determine AI platform success. Enterprises that embrace open, adaptable architectures will create AI capabilities that evolve with rapidly changing technology. Organizations that prioritize ecosystem development over platform control will attract the best AI talent and create the most innovative AI applications.

Most importantly, enterprises that understand AI transformation as an ecosystem challenge rather than a technology challenge will build sustainable competitive advantages. They will create AI platforms that become more intelligent through usage, develop network effects that compound learning advantages, and establish market positions that become stronger as they're exercised.

The choice between AI products and AI platforms will determine which enterprises lead their industries and which get displaced by more intelligent competitors. The experience of Nokia proves that platform thinking enables transformations that seem impossible before they happen. The AI revolution creates identical opportunities for enterprises that understand how to build intelligence ecosystems rather than just intelligence tools.

FROM DIGITIZATION TO COGNITIZATION

Why Digitization Was Just the Beginning: The Shift to Cognitization Based on Historical Patterns

Every enterprise leader believes they understand digital transformation. They have implemented cloud computing, automated workflows, analyzed big data, and deployed mobile applications. They have digitized operations, customer interactions, and strategic planning. Most consider this the culmination of technological evolution for business.

They are fundamentally wrong. Digitization was just the foundation. The real transformation is cognitization.

Across three decades of enterprise technology evolution, a consistent pattern has been observed that most organizations miss completely. Each technology revolution requires two distinct phases of organizational transformation. The first phase digitizes existing capabilities, making them faster, cheaper, and more efficient. The second phase cognitizes those capabilities, making them intelligent, adaptive, and autonomous.

The enterprises that recognize this distinction will dominate their industries. The enterprises that believe digitization represents the finish line will become case studies in competitive displacement. The difference between these outcomes depends entirely on understanding why cognitization represents a fundamentally different category of transformation than any previous technology wave.

The Hidden Pattern Across Technology Revolutions

The pattern began in the same way for every major technology introduction over the past forty years. Initially, organizations use new technology to do existing things better. They digitize processes that were previously manual, analog, or paper-based. This creates immediate efficiency gains and measurable improvements in cost, speed, and accuracy.

During the mainframe era, enterprises digitized record-keeping, transaction processing, and data storage. Computers replaced file cabinets, calculators, and typewriters. Organizations measured success through reduced paperwork, faster calculations, and more accurate records. This digitization phase transformed operational efficiency but maintained existing organizational structures and decision-making processes.

During the personal computer era, enterprises digitized individual productivity, communication, and analysis. Spreadsheets replaced ledger books, word processors replaced typewriters, and databases replaced filing systems. Organizations measured success through improved individual output, faster document creation, and more sophisticated analysis capabilities. This digitization phase transformed individual capabilities but maintained existing workflows and collaboration patterns.

During the Internet era, enterprises digitized information sharing, customer interaction, and supply chain coordination. Web sites replaced brochures, email replaced postal mail, and e-commerce replaced physical transactions. Organizations measured success through reduced communication costs, expanded market reach, and faster coordination capabilities. This digitization phase transformed connectivity but maintained existing business models and competitive strategies.

During the mobile era, enterprises digitized location awareness, real-time communication, and contextual services. Mobile applications replaced physical interactions, GPS replaced maps, and push notifications replaced scheduled communications. Organizations measured success through improved customer convenience, faster response times, and more personalized experiences. This digitization phase transformed accessibility but maintained existing value propositions and customer relationships.

During the cloud era, enterprises digitized infrastructure provisioning, software deployment, and data management. Cloud services replaced on-premises servers, software-as-a-service replaced installed applications, and data lakes replaced data warehouses. Organizations measured success through reduced infrastructure costs, faster software deployment, and more scalable data processing. This digitization phase transformed operational flexibility but maintained existing application architectures and business processes.

Each digitization phase created significant value and competitive advantage for early adopters. This sets the stage for the deeper shift to cognitization. Digitization represents only the first half of technological transformation, though. The second half requires cognitization, where technology becomes intelligent rather than just digital.

The Cognitization Transformation

Cognitization occurs when digital capabilities develop intelligence that enables autonomous decision-making, continuous learning, and adaptive optimization. Instead of just processing information faster, cognitized systems can understand context, predict outcomes, and improve performance through experience.

This transformation fundamentally changes the mathematics of business capability. Digitization creates linear improvements in efficiency and scale. Cognitization creates exponential improvements in effectiveness and adaptability. Organizations that embrace cognitization don't just do existing things better. They develop entirely new categories of capability that were impossible before intelligent systems.

Consider how cognitization transforms the same capabilities that digitization merely accelerated. Digitized customer service responds to inquiries faster and routes them more efficiently. Cognitized customer service understands customer intent, predicts future needs, and creates personalized solutions that improve through every interaction.

Digitized supply chain management tracks inventory levels and coordinates shipments more accurately. Cognitized supply chain management anticipates demand fluctuations, optimizes routing in real time, and automatically adjusts to disruptions before they impact operations.

Digitized financial planning processes data faster and generates reports more quickly. Cognitized financial planning models scenario outcomes, identifies risk patterns, and recommends strategic adjustments based on market intelligence that updates continuously.

Digitized marketing reaches customers more efficiently and measures campaign performance more precisely. Cognitized marketing understands individual customer psychology, predicts purchase behavior, and creates dynamic experiences that adapt to changing preferences in real time.

The Cognitive Architecture Imperative

The shift from digitization to cognitization requires fundamental changes to enterprise architecture, organizational structure, and competitive strategy. Digitized enterprises optimize for efficiency, consistency, and scale. Cognitized enterprises optimize for intelligence, adaptability, and learning velocity.

This difference determines everything about how organizations approach technology transformation. Digitization focuses on automating existing processes with digital tools. Cognitization focuses on creating intelligent processes that can evolve autonomously. Digitization measures success through cost reduction and productivity improvement. Cognitization measures success through learning acceleration and adaptive capability.

The enterprises that understand this distinction structure their AI transformation completely differently than enterprises that treat AI as advanced digitization. Instead of implementing AI applications that automate specific tasks, they build AI platforms that enable continuous intelligence evolution. Instead of measuring AI success through efficiency gains, they measure it through organizational learning velocity.

Most critically, instead of maintaining existing organizational structures while adding AI capabilities, they redesign organizational architecture around intelligence flows. They create cognitive business units that can adapt autonomously, develop intelligence feedback loops that improve decision-making continuously, and establish learning frameworks that compound competitive advantages over time.

Historical Evidence for the Cognitization Pattern

The pattern of digitization followed by cognitization appears consistently across every technology revolution, but with accelerating timelines. The mainframe digitization phase lasted decades before organizations began using computers for strategic decision-making rather than just operational efficiency. The personal computer digitization phase lasted years before organizations began using PCs for competitive intelligence and market analysis.

The Internet digitization phase lasted months before organizations began using connectivity for business model innovation and ecosystem orchestration. The mobile digitization phase lasted weeks before organizations began using mobility for behavioral prediction and contextual commerce. The cloud digitization phase lasted days before organizations began using cloud computing for machine learning and intelligent automation.

The AI revolution compresses both phases into a simultaneous transformation. Organizations can no longer afford to complete digitization before beginning cognitization. Competitive advantage increasingly flows to enterprises that implement cognitization immediately, building intelligent capabilities that improve through usage rather than requiring manual optimization.

This compression creates unprecedented opportunities for competitive displacement. Organizations that approach AI as advanced digitization will discover that competitors implementing cognitization can adapt faster, learn more effectively, and serve customers more intelligently than traditional digital capabilities enable.

The Cognitization Implementation Framework

Successful cognitization requires systematic transformation across four foundational dimensions. Technology architecture must evolve from static systems to learning systems that improve autonomously. Organizational structure must evolve from fixed hierarchies to adaptive networks that reconfigure based on changing requirements.

Competitive strategy must evolve from sustainable advantages to learning advantages that compound through usage. Measurement frameworks must evolve from efficiency metrics to intelligence metrics that track learning velocity and adaptive capability rather than just cost reduction and productivity improvement.

Technology architecture for cognitization emphasizes continuous learning over operational efficiency. Instead of optimizing systems for predictable performance, cognitized architecture optimizes for learning acceleration and adaptive capability. This requires infrastructure that can evolve autonomously, applications that can improve through experience, and data flows that enable real-time optimization across the entire enterprise.

Organizational structure for cognitization emphasizes distributed intelligence over centralized control. Instead of maintaining hierarchical decision-making processes, cognitized organizations develop network-based intelligence flows where decisions emerge from collaborative analysis between humans and AI systems. This requires role definitions that evolve based on capability development, authority structures that adapt to changing requirements, and collaboration frameworks that leverage both human insight and machine intelligence.

Competitive strategy for cognitization emphasizes learning advantages over operational advantages. Instead of building sustainable competitive positions through unique assets or capabilities, cognitized organizations create competitive advantages that strengthen through usage. Every customer interaction generates learning that improves service quality. Every operational process generates insights that enhance efficiency. Every strategic decision generates intelligence that improves future planning.

The Monday Morning Cognitization Assessment

Enterprises can begin cognitization immediately by evaluating their current digital capabilities through an intelligence lens rather than an efficiency lens. For each digitized process, ask whether the system learns from experience or requires manual optimization. For each automated workflow, ask whether the system can adapt to changing conditions or requires reprogramming for new requirements.

For each data analysis capability, ask whether the system can generate insights that improve decision-making or just produce reports that require human interpretation. For each customer interaction system, ask whether the system understands customer intent and adapts accordingly or just processes transactions efficiently.

The enterprises that discover gaps between their current digital capabilities and cognitization requirements can begin transformation immediately. They can implement learning loops that enable continuous improvement, adaptive algorithms that respond to changing conditions, and intelligence feedback mechanisms that compound organizational capabilities over time.

The enterprises that discover their digital systems already demonstrate cognitization characteristics can accelerate transformation by scaling intelligent capabilities across the entire organization. They can connect isolated learning systems to create enterprise-wide intelligence networks, integrate adaptive algorithms to enable organization-wide optimization, and develop intelligence platforms that enable every employee to leverage AI for enhanced effectiveness.

The Competitive Imperative

The shift from digitization to cognitization is not optional for competitive survival. The enterprises that continue operating with digitized capabilities will find themselves competing against cognitized organizations that can adapt faster, learn more effectively, and serve customers more intelligently than traditional digital transformation enables.

This competitive gap will expand exponentially as cognitized organizations accumulate learning advantages that compound over time. Every month of cognitization experience creates intelligence capabilities that digitized competitors cannot replicate through efficiency improvements or technology upgrades.

The window for cognitization transformation is narrowing rapidly. The enterprises that begin this transformation now will build learning advantages that prove impossible for traditional digital organizations to overcome. The enterprises that delay cognitization transformation will discover that competitive displacement occurs faster than they can implement countermeasures.

The choice between digitization and cognitization will determine which organizations lead their industries and which become historical examples of technological displacement. The enterprises that understand this distinction have the opportunity to create the most significant competitive advantages in modern business history.

FROM DEPARTMENTAL TO ENTERPRISE INTELLIGENCE

The Shift from Departmental AI to Enterprise Intelligence

The most expensive mistake enterprises make with AI transformation is optimizing departmental intelligence while ignoring enterprise intelligence architecture. This departmental approach creates an optimization paradox, where each business function becomes more efficient individually while the organization becomes less competitive collectively. This pattern has been observed across hundreds of AI implementations, and departmental AI optimization consistently produces enterprise-level suboptimization that undermines the very competitive advantages AI should create.

What makes this paradox particularly dangerous is that departmental AI success appears to validate the transformation approach while actually preventing the enterprise from achieving AI-native competitive advantages. Each department celebrates improved efficiency, reduced costs, and enhanced productivity, while the organization loses ground to competitors who implemented enterprise intelligence architecture from the beginning.

The root cause lies in a fundamental misunderstanding of how intelligence creates competitive advantage in complex organizations. Departmental thinking assumes that enterprise intelligence equals the sum of departmental intelligence capabilities. This assumption leads to AI implementations that optimize individual functions while creating organizational intelligence gaps that prove far more costly than the departmental efficiencies gained.

Enterprise intelligence requires a completely different approach that prioritizes organizational learning velocity over departmental efficiency gains. The enterprises that recognize this distinction early in their AI transformation create intelligence architectures that compound competitive advantages across all business functions. The enterprises that remain trapped in departmental thinking discover that their AI investments create internal competition for resources while enabling external competitors to capture market advantages through superior organizational intelligence.

Why Local Optimization Creates Global Competitive Vulnerability

The departmental AI approach follows a predictable pattern that initially appears successful but ultimately undermines organizational effectiveness. Each department identifies processes that AI can optimize, implements solutions that demonstrate clear efficiency gains, and measures success through function-specific metrics that show immediate improvement. This departmental success creates organizational momentum for additional AI investments while masking the strategic vulnerabilities being created.

The vulnerability emerges because departmental optimization requires functional isolation that prevents the intelligence integration necessary for enterprise-level competitive advantage. Marketing AI optimizes campaign performance using customer data that remains isolated from sales intelligence about customer behavior patterns. Sales AI optimizes lead conversion using prospect analysis that doesn't integrate with customer success intelligence about retention patterns. Customer success AI optimizes support interactions using resolution data that doesn't inform product development intelligence about feature requirements.

Each departmental optimization creates intelligence silos that prevent the organizational learning required for competitive differentiation. While individual functions become more efficient, the enterprise loses the capability to generate insights that emerge only through cross-functional intelligence integration. The most valuable competitive advantages require understanding patterns that span multiple departments and cannot be discovered through isolated functional analysis.

Research on organizational intelligence reveals that competitive advantage increasingly comes from insights that emerge at the intersection of different business functions rather than from optimizing individual functions independently. Customer lifetime value insights require the integration of marketing acquisition costs, sales conversion data, product usage patterns, and support interaction history. Market opportunity identification requires the integration of competitive intelligence, customer feedback analysis, sales pipeline trends, and product development capabilities.

The enterprises implementing departmental AI optimization systematically prevent themselves from discovering these cross-functional insights, while competitors implementing enterprise intelligence architecture accumulate learning advantages that compound across all business functions. The departmental efficiency gains become competitive liabilities when they prevent organizational intelligence development.

The Intelligence Flow Architecture Imperative

Moving from departmental AI to enterprise intelligence requires a fundamental reconceptualization of how information and insights move through organizations. Traditional departmental structures optimize for functional expertise and operational efficiency through clear boundaries and specialized responsibilities. Enterprise intelligence requires intelligence flows that transcend departmental boundaries and enable insights to inform decision-making across all organizational functions.

The distinction between information flows and intelligence flows proves critical for understanding this transformation. Information flows involve moving data between systems and departments to enable coordination and reporting. Intelligence flows involve moving insights, learning, and analytical capabilities between functions to enable organizational adaptation and competitive advantage development.

Traditional enterprise architecture focuses on information flow optimization through data integration, workflow automation, and communication systems that enable departments to coordinate efficiently. Enterprise intelligence architecture focuses on intelligence flow optimization through learning integration, insight sharing, and analytical coordination that enables the organization to evolve collectively.

What has been consistently observed across successful enterprise intelligence implementations is that organizations must design their technology architecture, organizational structure, and measurement systems around intelligence flows rather than trying to add intelligence capabilities to existing departmental structures. This requires fundamental changes to how enterprises think about data ownership, analytical responsibility, and the creation of competitive advantage.

The architecture imperative demands that enterprises abandon the assumption that departmental optimization can be aggregated into organizational optimization. Instead, organizations must design systems that optimize organizational learning velocity even when this conflicts with departmental efficiency preferences. The enterprises that make this architectural shift create intelligence capabilities that traditional departmental competitors cannot replicate through improved functional optimization.

Breaking Down the Information Silo Problem Through Cognitive Integration

The information silo problem has frustrated enterprise leaders for decades, but AI transformation makes silo elimination an existential competitive requirement rather than just an operational efficiency opportunity. Traditional approaches to silo elimination focused on technology integration that enabled data sharing between departments. Enterprise intelligence requires cognitive integration that enables the sharing of learning between functions.

Cognitive integration goes far beyond data integration because it requires sharing insights, analytical methodologies, and learning capabilities rather than just sharing information. When marketing discovers customer behavior patterns through AI analysis, cognitive integration ensures that these insights immediately inform sales forecasting, product development priorities, and customer success strategies. When sales identifies market opportunity patterns, cognitive integration ensures that these insights inform marketing campaign optimization, product roadmap decisions, and competitive positioning strategies.

The cognitive integration challenge intensifies because different departments often use incompatible analytical frameworks that prevent insight sharing even when data integration succeeds. Marketing measures customer engagement through interaction frequencies and content preferences. Sales measures customer potential through deal progression and revenue probability. Customer success measures customer value through usage patterns and satisfaction indicators. Product development measures customer needs through feature requests and usage data.

Enterprise intelligence requires analytical framework integration that enables insights discovered in one functional context to inform analysis in other functional contexts. This demands shared measurement systems, compatible analytical methodologies, and integrated learning frameworks that enable the organization to develop collective intelligence rather than just departmental expertise.

The enterprises that achieve cognitive integration discover that their competitive advantages multiply across business functions because insights developed in one context enhance performance in all related contexts. Customer insights improve marketing effectiveness, sales efficiency, product development focus, and support quality simultaneously. Market insights inform product strategy, sales targeting, marketing positioning, and competitive differentiation comprehensively.

The Enterprise Intelligence Operating System

Creating enterprise intelligence requires developing an enterprise intelligence operating system that coordinates analytical capabilities across all business functions while maintaining the specialized expertise that each department requires. This operating system functions like a cognitive nervous system that enables rapid information processing, coordinated responses to opportunities and threats, and continuous learning that improves organizational performance.

The operating system includes three integrated components that work together to enable enterprise-wide intelligence coordination. The intelligence infrastructure provides the technological foundation for sharing data, insights, and analytical capabilities across departmental boundaries. The intelligence governance establishes the organizational frameworks for coordinating analytical priorities, sharing learning discoveries, and maintaining data quality standards. The intelligence culture creates the behavioral norms and incentive structures that encourage knowledge sharing and collaborative learning rather than functional competition.

Intelligence infrastructure requires a technology architecture that enables real-time data integration, shared analytical tools, and coordinated model development across all business functions. Rather than building departmental AI systems that operate independently, enterprises must create shared AI platforms that enable each department to contribute to and benefit from organizational intelligence development.

The infrastructure must support both specialized departmental analysis and integrated enterprise analysis without creating technology complexity that slows decision-making or reduces analytical quality. This requires a modular architecture that enables functional specialization while maintaining enterprise integration, flexible data models that accommodate departmental requirements while enabling cross-functional analysis, and shared development frameworks that enable collaboration without constraining innovation.

Intelligence governance establishes the organizational processes for coordinating analytical priorities and ensuring that departmental intelligence development contributes to enterprise competitive advantage. This includes analytical standards that ensure data quality and insight reliability across all functions, priority frameworks that allocate analytical resources based on enterprise value creation rather than just departmental efficiency, and learning protocols that ensure insights discovered in one context inform analysis in other contexts.

The governance framework must balance departmental autonomy with enterprise coordination to enable both functional excellence and organizational learning. This requires decision-making processes that enable rapid departmental responses to opportunities while maintaining strategic alignment, analytical standards that ensure quality without constraining innovation, and resource allocation mechanisms that encourage both specialized expertise development and cross-functional collaboration.

Intelligence culture creates the behavioral environment that enables enterprise intelligence to flourish rather than being constrained by traditional departmental boundaries and competitive dynamics. This includes incentive systems that reward knowledge sharing and collaborative learning rather than just functional performance, communication frameworks that enable rapid insight dissemination across organizational boundaries, and leadership behaviors that model enterprise thinking rather than departmental optimization.

The cultural transformation proves most difficult because it requires individuals to prioritize organizational learning over departmental success when these priorities conflict. Traditional enterprise cultures reward functional expertise and departmental achievement, creating natural resistance to the knowledge sharing and collaborative analysis that enterprise intelligence requires.

The Competitive Multiplication Effect

Organizations that successfully implement enterprise intelligence discover that their competitive advantages multiply rather than add across business functions. When intelligence flows enable each department to leverage insights generated by other departments, the total competitive advantage exceeds the sum of departmental improvements because insights improve performance in multiple contexts simultaneously.

The multiplication effect emerges through three mechanisms that traditional departmental AI cannot create. Cross-functional insight application enables marketing insights to improve sales effectiveness, sales insights to inform product development, and product insights to enhance customer success, creating value in multiple contexts from single analytical investments. Analytical capability sharing enables departments to leverage specialized expertise developed by other functions rather than building redundant capabilities independently. Learning velocity acceleration occurs when insights discovered rapidly in one context immediately improve performance in all related contexts, rather than requiring separate discovery processes for each function.

The competitive advantages created through multiplication effects prove sustainable because they require organizational capabilities that competitors implementing departmental AI cannot replicate through improved functional optimization. Competitors can hire marketing AI specialists, sales AI experts, and customer success AI consultants, but they cannot quickly replicate the organizational intelligence capabilities that emerge through years of integrated learning and collaborative analysis.

The multiplication effect also creates defensive competitive advantages because enterprise intelligence enables rapid adaptation to competitive threats that emerge in any business function. When competitors attack through superior marketing, superior sales effectiveness, or superior customer experience, organizations with enterprise intelligence can coordinate responses across all functions simultaneously rather than responding through isolated departmental efforts.

The Transformation Implementation Framework

Moving from departmental AI to enterprise intelligence requires a systematic transformation that addresses technology architecture, organizational structure, and cultural dynamics simultaneously. The transformation cannot be implemented gradually because departmental and enterprise approaches operate according to contradictory principles that create organizational conflicts when attempted simultaneously.

The implementation framework begins with an enterprise intelligence assessment that identifies current departmental AI capabilities and evaluates their compatibility with enterprise intelligence requirements. Many departmental AI implementations must be redesigned or replaced because they create technological dependencies that prevent enterprise integration. The assessment also identifies organizational capabilities and cultural characteristics that enable or constrain enterprise intelligence development.

Architecture redesign establishes the technological and organizational foundations for enterprise intelligence while maintaining operational continuity for existing business functions. This requires careful sequencing of technology changes and organizational restructuring to prevent disruption of current capabilities while building enterprise intelligence infrastructure.

Pilot integration demonstrates enterprise intelligence value through carefully selected cross-functional initiatives that show multiplication effects while building organizational confidence in the new approach. The pilots must be chosen to create visible competitive advantages that justify additional transformation investment while providing learning opportunities that inform full-scale implementation.

Full implementation scales enterprise intelligence across all business functions while maintaining the specialized expertise and operational effectiveness that each department requires. This phase requires careful change management because it fundamentally alters how individuals contribute to organizational success and how departments coordinate with each other.

Performance optimization continuously improves enterprise intelligence capabilities through experience, feedback, and adaptation that enhances competitive advantage over time. This phase distinguishes enterprise intelligence from departmental AI because the learning and adaptation occur at the organizational level rather than just the functional level.

The enterprises that complete this transformation framework discover that they have created competitive advantages that traditional departmental competitors cannot understand or replicate. They develop organizational intelligence capabilities that improve automatically through operation while enabling faster adaptation to market changes and more effective responses to competitive threats.

Most importantly, they build learning advantages that compound over time, making them progressively more difficult to compete against as they accumulate more experience and data across all business functions simultaneously. The transformation from departmental AI to enterprise intelligence represents one of the most significant competitive differentiators available to enterprises willing to abandon traditional organizational assumptions in favor of intelligence-native thinking.

AI-NATIVE VERSUS TRADITIONAL MANAGEMENT

AI-Native Thinking Cannot Coexist with Traditional Management Frameworks

The most dangerous assumption executives make about AI transformation is that AI can be integrated into existing management frameworks without fundamentally changing how leaders think about organizational effectiveness. This assumption creates organizational schizophrenia where AI-native capabilities operate according to one set of principles while management decisions operate according to completely contradictory principles.

Traditional management frameworks assume that organizational performance can be optimized through hierarchical control, predictable processes, and measurable outcomes based on historical performance. AI-native thinking requires optimization through network effects, adaptive processes, and learning velocity based on future capability development. These assumptions are not just different. They are mutually exclusive.

The incompatibility creates a leadership crisis that most executives haven't recognized yet. Leaders who attempt to manage AI-native capabilities using traditional management frameworks will systematically undermine the very capabilities they're trying to develop. Every traditional management decision will conflict with AI-native requirements, creating organizational confusion that prevents both traditional excellence and AI-native transformation.

The enterprises that recognize this incompatibility and abandon traditional management frameworks completely will create competitive advantages that traditional competitors cannot understand, much less replicate. The enterprises that attempt to blend traditional management with AI-native capabilities will achieve neither traditional efficiency nor AI-native intelligence, leaving them vulnerable to displacement by competitors who embrace the complete cognitive transformation that AI requires.

The Fundamental Philosophical Conflict

Traditional management frameworks emerged during the industrial era, when organizational effectiveness depended on standardizing processes, controlling variability, and optimizing efficiency through repetition and scale. These frameworks assume that better management means more predictable outcomes, tighter control over organizational behavior, and clearer accountability for predetermined results.

AI-native thinking requires exactly the opposite philosophical approach. Organizational effectiveness in an AI-native context depends on maximizing learning velocity, encouraging experimentation, and optimizing adaptability through continuous evolution and intelligent response to changing conditions. AI-native management means creating more intelligent outcomes, looser control that enables autonomous optimization, and dynamic accountability for adaptive capability development.

The philosophical conflict appears in every management decision. Traditional frameworks prioritize risk minimization, process standardization, and performance measurement based on variance from established targets. AI-native frameworks prioritize learning maximization, process adaptation, and performance measurement based on improvement rates and capability evolution.

Most critically, traditional management frameworks assume that organizational intelligence can be concentrated in management positions and deployed through hierarchical decision-making. AI-native thinking assumes that organizational intelligence must be distributed throughout the enterprise and coordinated through network-based collaboration that bypasses hierarchical bottlenecks.

These philosophical differences cannot be reconciled through compromise or hybrid approaches. Every attempt to blend traditional control with AI-native adaptation creates contradictions that prevent both approaches from working effectively.

The Control Versus Learning Paradox

Traditional management frameworks optimize for control because industrial-era competitive advantage came from the consistent execution of proven processes. Management success meant eliminating variability, reducing uncertainty, and ensuring predictable performance through standardized procedures and hierarchical oversight.

AI-native competitive advantage comes from learning velocity and adaptive capability that enable rapid response to changing conditions and continuous improvement through experience. AI-native success requires maximizing variability in experimentation, embracing uncertainty as a learning opportunity, and enabling unpredictable performance breakthroughs through autonomous optimization.

The control versus learning paradox creates impossible management situations for leaders trained in traditional frameworks. Every attempt to control AI-native capabilities reduces their learning potential. Every effort to standardize AI-driven processes eliminates their adaptive advantage. Every requirement for predictable AI performance prevents the experimental breakthrough that creates competitive differentiation.

Traditional managers attempting to oversee AI-native capabilities find themselves in the position of trying to control systems that become more valuable when they operate autonomously. The more successfully traditional management controls AI systems, the less intelligent and adaptive those systems become. The more traditional managers require predictable AI performance, the more they prevent AI systems from discovering optimization opportunities that traditional analysis cannot identify.

The paradox intensifies because AI-native capabilities improve through failure and experimentation that traditional management frameworks classify as poor performance. Traditional management rewards consistency and penalizes deviation from established processes. AI-native capabilities require inconsistency and reward deviation that leads to performance breakthroughs.

The Planning Versus Adaptation Incompatibility

Traditional management frameworks depend on strategic planning that assumes future conditions can be predicted sufficiently to enable predetermined responses. These frameworks optimize organizational capability for executing predetermined strategies efficiently rather than adapting strategies based on changing conditions intelligently.

AI-native thinking assumes that future conditions cannot be predicted accurately and that competitive advantage comes from adapting strategies continuously based on real-time intelligence rather than executing predetermined plans efficiently. AI-native organizations optimize for strategic agility rather than strategic consistency.

The planning versus adaptation incompatibility creates fundamental conflicts in resource allocation, performance measurement, and organizational design. Traditional management allocates resources based on predetermined strategic priorities and measures success through the achievement of predetermined objectives. AI-native management allocates resources based on learning opportunities and measures success through adaptive capability development and strategic evolution velocity.

Traditional managers attempting to plan AI-native transformation discover that their planning assumptions become obsolete faster than planning cycles can accommodate. Every strategic plan becomes a constraint that prevents AI-native capabilities from adapting to opportunities and threats that emerge during plan execution.

The incompatibility deepens because traditional planning frameworks require detailed specification of desired outcomes and resource requirements before implementation begins. AI-native capabilities cannot be specified in advance because their value emerges through learning and adaptation that cannot be predetermined. Traditional planning frameworks prevent AI-native experimentation by requiring justification for investments whose outcomes cannot be guaranteed.

The Measurement Dilemma

Traditional management frameworks measure organizational performance through efficiency metrics that compare actual results to predetermined targets. These frameworks assume that better performance means achieving targets more consistently and that management effectiveness can be evaluated through variance analysis and trend comparison.

AI-native performance requires measurement through learning velocity, adaptive capability, and competitive advantage accumulation rather than efficiency optimization and target achievement. AI-native organizations measure success through capability evolution rates and intelligence development rather than operational consistency and predetermined outcome achievement.

The measurement dilemma creates evaluation problems that traditional management frameworks cannot resolve. AI-native capabilities often perform poorly according to traditional metrics while developing capabilities that create future competitive advantages. Traditional measurement systems classify learning experiments as failures when they don't achieve predetermined outcomes, even when those experiments generate insights that enable breakthrough performance later.

Traditional managers evaluating AI-native performance find themselves measuring the wrong variables using inappropriate timeframes with incompatible success criteria. Every traditional measurement decision penalizes the experimental behavior that creates AI-native competitive advantage while rewarding the consistent behavior that prevents AI-native capability development.

The Authority Versus Autonomy Contradiction

Traditional management frameworks assume that organizational effectiveness requires clear authority relationships that enable hierarchical decision-making and accountability assignment. These frameworks optimize for management control over organizational behavior and individual accountability for predetermined responsibilities.

AI-native effectiveness requires autonomous decision-making by intelligent systems that can respond faster and more accurately than hierarchical coordination allows. AI-native organizations optimize for system intelligence rather than management control and collective learning rather than individual accountability.

The authority versus autonomy contradiction creates leadership identity crises for managers whose careers developed through traditional authority relationships. AI-native capabilities perform best when they operate autonomously, but traditional management frameworks require human oversight and approval for all significant decisions.

Traditional managers attempting to oversee autonomous AI capabilities discover that their authority relationships slow down decision-making, reduce adaptive capability, and prevent the intelligent responses that create AI-native competitive advantage. Every exercise of traditional management authority reduces AI-native capability effectiveness.

The Complete Transformation Imperative

The incompatibilities between traditional management frameworks and AI-native thinking cannot be resolved through gradual transition or hybrid approaches. Traditional management principles systematically undermine AI-native capabilities, while AI-native requirements make traditional management frameworks ineffective.

Organizations must choose between traditional management excellence and AI-native transformation because the management frameworks required for each approach contradict each other completely. Every attempt to maintain traditional management frameworks while developing AI-native capabilities creates organizational confusion that prevents success in either approach.

The complete transformation imperative requires leaders to abandon management frameworks that defined their careers and embrace entirely new approaches to organizational effectiveness. This transformation cannot be delegated or implemented gradually because every traditional management decision conflicts with AI-native requirements.

The enterprises that make this complete transformation will develop organizational intelligence capabilities that traditional competitors cannot understand or replicate. The enterprises that attempt to preserve traditional management frameworks while adding AI-native capabilities will discover that their management approach prevents them from achieving either traditional excellence or AI-native intelligence, leaving them vulnerable to displacement by competitors who embrace the complete cognitive transformation that AI requires.

THE DEATH OF TRADITIONAL ORGANIZATION CHARTS

Structural Evolution Across Technology Waves

Organization charts represent the most visible artifact of management thinking that AI renders completely obsolete. These static diagrams assume that organizational effectiveness can be mapped through fixed relationships, predetermined authority flows, and stable role definitions. Enterprise intelligence transformation makes these assumptions not just outdated but counterproductive to competitive success.

The death of organization charts signals something far more profound than administrative simplification. These visual representations embody the entire conceptual framework of traditional management thinking we explored earlier. When the cognitive frameworks that created organization charts become obsolete, the charts themselves become misleading documents that prevent organizations from understanding how work actually gets accomplished in an AI-native environment.

What makes this transformation particularly significant is that organization charts served as the primary tool for communicating organizational design, authority relationships, and accountability structures for over a century of business development. Leaders who built careers interpreting and optimizing organization charts must now abandon these familiar frameworks entirely in favor of approaches that would appear chaotic to traditional management thinking.

The enterprises that recognize why organization charts have become harmful rather than helpful will redesign their organizational communication and coordination systems around intelligence flows rather than authority relationships. The enterprises that continue relying on organization charts will systematically misunderstand their own organizational capabilities and create coordination problems that prevent AI-native transformation from succeeding.

Why Visual Hierarchy Representation Fails in Intelligence-Driven Organizations

Traditional organization charts assume that organizational effectiveness can be visualized through hierarchical box-and-line diagrams that show reporting relationships, authority distribution, and communication pathways. These visual representations worked when organizational capability depended on clear command structures, predictable coordination patterns, and stable role assignments.

Enterprise intelligence requires organizational effectiveness through dynamic network relationships that change based on situational requirements, learning opportunities, and optimization discoveries. Intelligence flows cannot be captured in static diagrams because the most valuable organizational connections emerge spontaneously based on real-time analysis rather than predetermined planning.

The visual hierarchy representation creates fundamental problems for AI-native organizations because it suggests that organizational capability can be understood by examining formal relationships rather than actual intelligence flows. Leaders who rely on organization charts for decision-making consistently misunderstand which organizational connections create value and which organizational boundaries prevent optimization.

Organization charts also reinforce the assumption that authority relationships determine organizational effectiveness, when AI-native success depends on information relationships that bypass traditional authority structures entirely. The most important organizational connections in AI-native enterprises often connect individuals who appear unrelated on organization charts but whose combined intelligence creates breakthrough insights.

Most problematically, organization charts create a psychological attachment to structural stability that conflicts with the continuous organizational adaptation that enterprise intelligence requires. Leaders who think in terms of organization charts resist the structural fluidity that enables AI-native capabilities to optimize organizational performance through autonomous reconfiguration.

The Communication Artifact Problem

Organization charts function as communication artifacts that convey assumptions about how organizations operate to employees, stakeholders, and external partners. These artifacts communicate that organizational understanding can be achieved through studying formal relationships rather than observing actual collaboration patterns and intelligence flows.

In AI-native organizations, the communication artifact problem becomes acute because organization charts communicate information that has become not just irrelevant but actively misleading. Employees who use organization charts to understand their organizational context will develop incorrect assumptions about how to contribute effectively and where to focus their attention for maximum impact.

The artifact problem intensifies because organization charts suggest that organizational boundaries can be defined clearly when AI-native effectiveness requires boundary fluidity that enables optimal resource allocation and capability development. Stakeholders who interpret organization charts as accurate representations of organizational capability will make incorrect assumptions about partnership opportunities and collaboration potential.

External partners attempting to understand AI-native organizations through organization charts will consistently engage with the wrong people, pursue inappropriate collaboration strategies, and miss the most valuable integration opportunities. The communication artifacts that once clarified organizational capabilities now obscure the actual sources of organizational intelligence and competitive advantage.

Most significantly, organization charts communicate assumptions about accountability and responsibility assignment that conflict with the distributed intelligence and collective learning that create AI-native competitive advantages. The visual representations that once enabled effective organizational coordination now prevent the coordination patterns that AI-native success requires.

Fixed Role Definition Obsolescence

Organization charts depend on fixed role definitions that specify individual responsibilities, authority boundaries, and performance expectations through job titles and reporting relationships. These definitions enabled organizational coordination when competitive advantage came from consistent task execution and clear accountability assignment.

AI-native organizational effectiveness requires role fluidity that enables individuals to contribute their unique capabilities to whatever organizational challenges can benefit from their intelligence, rather than limiting their contributions to predetermined responsibilities. Fixed role definitions prevent the dynamic capability allocation that enables AI-native organizations to optimize their intellectual resources continuously.

Role definition obsolescence creates coordination challenges that organization charts cannot resolve. When organizational roles must adapt continuously based on learning opportunities and optimization discoveries, static role definitions become constraints that prevent rather than enable effective coordination.

Organization charts also assume that role boundaries can be maintained clearly when AI-native collaboration requires boundary dissolution that enables seamless intelligence integration across traditional functional divisions.

The most valuable organizational contributions often require capabilities that cross multiple traditional roles and cannot be captured through conventional job definitions.

Most critically, fixed role definitions embedded in organization charts prevent individuals from developing the adaptive capabilities that AI-native organizations require. Employees who understand their organizational contribution through static role definitions will resist the continuous capability development that enables AI-native competitive advantage.

The Authority Flow Misconception

Traditional organization charts assume that organizational effectiveness depends on clear authority flows that enable hierarchical decision-making and accountability assignment. These visual representations suggest that understanding authority relationships provides sufficient insight into organizational capability and coordination mechanisms.

AI-native organizational effectiveness depends on intelligence flows that often bypass traditional authority relationships entirely. The most important organizational decisions frequently require input from individuals who possess relevant intelligence rather than formal authority, and optimal decisions often emerge through collaborative analysis that involves multiple authority levels simultaneously.

The authority flow misconception creates decision-making problems in AI-native organizations because leaders who rely on organization charts for decision-making consistently exclude the most relevant intelligence sources while including authority positions that lack the insights needed for optimal choices.

Organization charts also reinforce the assumption that accountability can be assigned through authority relationships when AI-native success requires collective accountability for organizational learning and adaptive capability development. The accountability frameworks that organization charts represent prevent the shared responsibility that enables AI-native organizations to optimize their collective intelligence.

Most problematically, authority flow representations suggest that organizational power can be understood through formal relationships when AI-native influence depends on intelligence contribution rather than positional authority. Individuals who create the most organizational value in AI-native enterprises often possess minimal formal authority but generate insights that influence organizational direction significantly.

The Static Documentation Contradiction

Organization charts represent attempts to document organizational structure through static representations that assume organizational design can be captured at specific moments and communicated through permanent artifacts.

This documentation approach worked when organizational structures changed slowly enough for static representations to remain accurate for extended periods.

AI-native organizations require continuous structural adaptation that makes static documentation not just outdated but actively harmful to organizational effectiveness. Any attempt to create permanent documentation of AI-native organizational structure immediately becomes obsolete because optimal organizational configuration changes based on learning discoveries and capability development.

The static documentation contradiction creates organizational confusion because employees who rely on organization charts for guidance receive information that misrepresents the current organizational reality. This confusion prevents effective coordination and creates unnecessary friction in organizational processes that should operate seamlessly.

Organization charts also suggest that organizational understanding can be achieved through studying documentation rather than participating in actual organizational intelligence flows. This suggestion prevents individuals from developing the dynamic organizational awareness that enables effective contribution to AI-native capabilities.

Most significantly, the static documentation approach reinforces assumptions about organizational predictability that conflict with the adaptive unpredictability that creates AI-native competitive advantages. Organizations that maintain documentation approaches suitable for stable structures will systematically misunderstand their own adaptive capabilities and resist the continuous evolution that AI-native success requires.

The Intelligence Flow Alternative

The death of organization charts creates opportunities for developing organizational communication and coordination approaches that align with AI-native requirements rather than conflicting with them. Instead of mapping authority relationships, AI-native organizations must develop capabilities for visualizing and optimizing intelligence flows that create actual organizational value.

Intelligence flow mapping focuses on understanding how insights, learning, and analytical capabilities move through organizations to create competitive advantages rather than how formal authority enables decision-making coordination. This approach enables organizational optimization based on actual value creation patterns rather than traditional structural assumptions.

The alternative also requires developing organizational awareness approaches that enable individuals to understand their potential contributions dynamically rather than through static role definitions. AI-native organizations need communication methods that help employees identify optimization opportunities and collaboration possibilities that emerge through changing circumstances.

Most importantly, intelligence flow approaches enable organizations to optimize their coordination patterns continuously rather than maintaining coordination approaches designed for previous competitive requirements. The enterprises that develop effective alternatives to organization charts will create coordination advantages that traditional competitors cannot understand or replicate.

THE TECHNOLOGY REVOLUTION PATTERN

How Enterprise Mindset Evolution Mirrors Every Technology Revolution

The frameworks we have explored throughout this chapter reveal a master pattern that determines which enterprises survive technological disruption and which get displaced by more adaptive competitors. Understanding this pattern provides executives with a predictive framework for navigating not just the current AI transformation, but every future technology revolution that will reshape business.

The pattern operates independently of specific technologies or market conditions. Whether the disruption comes from mainframes, personal computers, Internet connectivity, mobile computing, cloud platforms, or AI, the same cognitive sequence determines organizational success or failure. The enterprises that recognize this sequence early and adapt accordingly build advantages that persist long after the technology revolution stabilizes.

The sequence begins with a technology introduction that enables new operational capabilities. Most organizations initially focus on using new technology to improve existing processes rather than fundamentally reimagining what becomes possible. This approach creates immediate efficiency gains but misses the transformational opportunities that determine long-term competitive position.

The second phase requires cognitive evolution, where leaders recognize that new technology enables entirely different approaches to organizational effectiveness. This cognitive leap separates enterprises that achieve incremental improvement from enterprises that achieve exponential transformation. The timing of this cognitive evolution determines whether organizations lead or follow during the transformation period.

The final phase demands complete abandonment of previous management frameworks that conflict with new technological capabilities. Organizations that attempt to preserve existing approaches while adding new capabilities create internal contradictions that prevent them from achieving either traditional excellence or transformational advantage.

The Cognitive Sequence That Determines Success

What makes certain enterprises consistently successful across multiple technology revolutions while others fail repeatedly? The answer lies in leadership teams that understand the cognitive sequence and implement it proactively rather than reactively.

The first cognitive breakthrough involves recognizing that each technology revolution changes the fundamental economics of organizational capability rather than just adding new tools to existing operations. Leaders who understand this principle immediately begin questioning their basic assumptions about competitive advantage, organizational design, and strategic positioning rather than optimizing current approaches.

The second cognitive breakthrough requires understanding that technology revolutions create entirely new forms of value creation that cannot be achieved through traditional optimization. This insight drives leaders to experiment with radically different business models rather than incrementally improving existing ones.

The third cognitive breakthrough demands recognition that new technology requires fundamentally different management approaches that cannot coexist with traditional frameworks. Leaders who achieve this breakthrough abandon management practices that defined their careers in favor of approaches that optimize for new technological capabilities.

The enterprises whose leadership teams complete this cognitive sequence position themselves to capture the exponential advantages that technology revolutions create. The enterprises whose leaders resist or delay this cognitive evolution remain trapped in incremental thinking that prevents transformational success.

Why the AI Revolution Compresses Traditional Timelines

Previous technology revolutions allowed years or decades for the cognitive sequence to unfold gradually. Organizations could experiment with new approaches while maintaining traditional frameworks, eventually transitioning as new capabilities proved their value. The AI revolution eliminates this gradual transition period.

AI creates competitive advantages that compound daily rather than annually. Organizations that complete the cognitive sequence immediately begin accumulating learning advantages that make them progressively more difficult to compete against. Organizations that delay cognitive evolution fall behind exponentially rather than incrementally.

The compression effect means that leadership teams cannot afford to wait for proof of concept before beginning cognitive transformation. By the time AI capabilities demonstrate clear superiority over traditional approaches, competitive separation becomes irreversible. Leaders must commit to complete cognitive evolution based on pattern recognition rather than empirical validation.

This compression also means that partial cognitive evolution creates greater competitive vulnerability than maintaining traditional approaches entirely. Organizations that embrace some aspects of AI-native thinking while preserving traditional frameworks create internal conflicts that prevent effective execution of either approach.

The Leadership Advantage of Pattern Recognition

Leaders who understand the pattern gain the ability to anticipate transformation requirements before they become competitively necessary. This anticipation enables proactive transformation that creates advantages rather than reactive transformation that prevents disadvantages.

Pattern recognition also provides confidence for making cognitive leaps that appear risky to competitors who lack historical perspective. Leaders who understand that every technology revolution requires complete abandonment of previous frameworks can commit to transformation decisions that seem impossible to leaders trained only in current approaches.

Most importantly, pattern recognition enables leaders to communicate transformation requirements effectively to stakeholders who resist change. Understanding how cognitive evolution created success during previous revolutions provides compelling evidence for why similar evolution becomes necessary during the current transformation.

The Predictive Framework for Future Revolutions

The pattern revealed through AI transformation will repeat during every subsequent technology revolution. Future technologies will require the same cognitive sequence, create the same compression effects, and reward the same leadership approaches that determine success during the current transformation.

Leaders who master pattern recognition during AI transformation develop capabilities for navigating whatever technologies emerge next. They build organizational cultures that embrace cognitive evolution rather than resisting it, creating enterprises that remain competitive across multiple technology revolutions.

The pattern also suggests that future technology revolutions will create even greater compression effects, requiring increasingly rapid cognitive evolution. Organizations that develop agility in cognitive transformation position themselves for sustained success regardless of specific technological developments.

The Monday Morning Imperative

Understanding the pattern creates an immediate action requirement for current enterprise leaders. The cognitive sequence cannot be delayed until AI capabilities mature or competitive pressure intensifies. Pattern recognition demands proactive cognitive evolution that begins immediately.

Leaders must begin by examining their current management frameworks for incompatibilities with AI-native requirements. Every traditional practice that conflicts with enterprise intelligence development must be identified and scheduled for elimination rather than gradual modification.

The framework also requires leaders to communicate the transformation imperative throughout their organizations immediately. Employees who understand the pattern can contribute to cognitive evolution rather than unknowingly resisting it.

Most critically, leaders must commit personally to complete cognitive transformation rather than attempting to delegate or outsource the most difficult aspects of mindset evolution. The pattern shows that cognitive transformation requires leadership commitment that cannot be transferred to others.

The enterprises whose leaders begin this transformation immediately will create the competitive advantages that define market leadership for the next decade. Five technology revolutions have revealed the underlying pattern driving each of them: the enterprises whose leaders fail to recognize patterns will discover that cognitive evolution becomes progressively more difficult as competitive gaps widen.

The choice facing every enterprise leader is whether to lead cognitive evolution or become a case study in cognitive resistance. History suggests that this choice determines not just competitive position but organizational survival during periods of technological transformation.

CHAPTER 2

Managing the Underground AI Economy

The Chief Information Officer stared at the dashboard displaying her company's authorized AI applications. Twelve officially approved tools, each carefully vetted through months of security reviews and compliance audits. Then she opened the shadow AI discovery report her team had just completed. Four hundred and thirty-seven AI applications were actively running across her enterprise. Her IT department had visibility into exactly twelve of them.

This revelation represents the defining challenge of AI transformation that no enterprise leader anticipated. While organizations spend millions implementing formal AI governance frameworks, their employees have quietly assembled the largest unauthorized intelligence revolution in business history. Current research reveals that 80% of enterprise AI tools operate without central IT oversight, leaving organizations simultaneously more intelligent and more vulnerable than their leadership realizes.

Shadow AI has become the dominant form of AI adoption in enterprises worldwide, operating entirely outside traditional management oversight. The unauthorized AI applications that compliance teams view as governance violations often represent the most innovative and productive uses of AI within the enterprise. Marketing teams use generative AI to create content that outperforms traditional materials by 300%. Financial analysts build predictive models that identify market opportunities weeks before official systems detect relevant patterns.

This creates the shadow AI paradox. The same shadow systems that create security vulnerabilities also generate competitive advantages that formally approved AI initiatives struggle to achieve. Most enterprises approaching shadow AI as a compliance problem miss the fundamental opportunity to architect systematic competitive advantages from the organic innovation emerging throughout their organizations.

The solution requires a Shadow AI Management System (SAMS), a comprehensive framework designed to transform unauthorized AI usage from an organizational liability into a strategic asset. Unlike traditional IT governance that focuses on control and compliance, a SAMS prioritizes discovery and enablement while maintaining security and regulatory requirements.

This chapter reveals how to systematically discover the hidden AI already running your business, implement governance frameworks that accelerate rather than constrain innovation, and transform shadow implementations into enterprise-wide competitive advantages. You'll learn why the organic AI adoption happening throughout your organization represents your greatest untapped strategic resource and how to harness it before competitors discover the same opportunities.

The shadow AI revolution is already transforming your enterprise, whether you recognize it or not. The question is whether you will harness that transformation to build sustainable competitive advantages or allow it to create vulnerabilities that competitors exploit faster than you can defend against them.

THE SHADOW AI CRISIS – MANAGING THE INVISIBLE REVOLUTION

The Hidden Truth About Your AI-Powered Enterprise

Every Monday at 9 AM, your enterprise becomes more intelligent. Not through any strategic initiative you've approved, not through any budget you've allocated, and not through any vendor you've selected. It happens in the shadows, through the collective ingenuity of employees who've discovered AI tools that make their work faster, smarter, and more creative.

This is the defining paradox of the modern enterprise. Your greatest AI transformation is happening without you.

Based on my three decades of enterprise technology implementation, I've never witnessed a phenomenon as simultaneously promising and perilous as shadow AI. In my work with over 200 global enterprises, I've observed a pattern that should alarm every C-suite executive. Eighty percent of enterprise AI tools currently operate without central oversight. This isn't a prediction. It's the reality unfolding in your organization right now.

The Invisible Revolution in Your Enterprise

The data reveals the staggering scope of this hidden transformation. Recent comprehensive studies across Fortune 500 companies show that 15% of all installed software is unauthorized, representing nearly 3,000 unique versions of unapproved applications in a single enterprise. When we extend this pattern to AI tools, the implications become even more profound. Organizations are discovering they're operating hundreds of AI applications while IT and security teams maintain visibility and control over less than 20% of them.

What has been observed consistently across industries is this: the enterprises that will dominate 2030 are not those with the most sophisticated AI strategies; they're the ones that master the art of transforming shadow innovation into strategic advantage.

Consider the real-world case of a multinational pharmaceutical company. During a routine security audit, they discovered over 400 unauthorized AI tools being used across their research divisions, from ChatGPT Plus subscriptions for patent research to specialized AI coding assistants for clinical trial data analysis. The initial reaction was panic over data exposure risks, compliance violations, and intellectual property leakage. Within six months, that same company had transformed this crisis into its most successful AI innovation program, generating $50 million in research efficiency gains.

The Academic Foundation of Unauthorized Innovation

Academic research provides crucial context for understanding this phenomenon. Shadow AI represents the natural evolution of shadow IT, which scholars define as hardware, software, or services built, introduced, and used for the job without explicit approval or even knowledge of the organization. This has been described as a currently misunderstood and relatively unexplored phenomenon, highlighting the urgent need for deeper investigation.

The research reveals something profound about human behavior in technological environments. Employees resort to shadow tools not from malicious intent, but from a fundamental trust deficit between their productivity needs and IT's ability to provide timely, effective solutions. The motivation is simple ingenuity rather than rebellion. People want to do their jobs better.

The pattern emerging is clear. Organizations that view shadow AI as a compliance problem will be disrupted by those that recognize it as an innovation opportunity.

The Strategic Imperative Hidden in Plain Sight

The traditional enterprise response to unauthorized technology follows a predictable pattern of discover, prohibit, and punish. This approach worked when technology adoption cycles were measured in years. AI tools can be discovered, tested, and integrated into daily workflows within hours, though. By the time your IT department identifies a new AI tool in use, dozens of employees may have already built critical processes around it.

This creates what is called the shadow AI paradox. The faster you try to control AI adoption, the more innovation you drive underground.

The enterprises I've worked with that successfully navigate this paradox share a common approach. They've implemented what is called a shadow AI management system. Rather than viewing unauthorized AI as a problem to be solved, they've reframed it as intelligence to be harvested.

The SAMS Transformation Framework

The Shadow AI Management System (SAMS) represents a fundamental shift from prohibition to strategic enablement. Where traditional governance asks how do we stop this, SAMS asks how do we optimize this. It's built on four core principles that transform risk into competitive advantage.

Discovery over detection replaces hunting for violations with actively scouting for innovation. It treats every unauthorized AI tool as potential intelligence about unmet business needs.

Governance as acceleration creates fast-track pathways for promising shadow innovations to become enterprise capabilities rather than slowing down AI adoption.

Innovation transformation converts organic experimentation into systematic competitive advantage, ensuring that employee creativity becomes organizational intelligence.

Cultural evolution shifts the enterprise mindset from AI compliance to AI enablement, fostering the trust necessary for transparent innovation.

What This Chapter Delivers

The mistake seen repeatedly is treating shadow AI as an IT security issue when it's actually the most valuable strategic intelligence your organization generates. This chapter provides the complete blueprint for implementing SAMS in your enterprise, transforming what most organizations see as their biggest AI risk into their most sustainable competitive advantage.

You'll discover how to systematically identify the shadow AI already operating in your organization, evaluate its strategic potential, and channel organic innovation into formal capabilities. More importantly, you'll learn how to build the cultural foundation that makes employees partners in AI governance rather than obstacles to control.

The enterprises that master shadow AI management in 2025 will own their markets by 2030. Those that don't will spend the next decade playing catch-up to competitors they never saw coming.

The revolution is already underway in your enterprise. The question isn't whether you'll participate. It's whether you'll lead it or be led by it.

THE HIDDEN AI ALREADY RUNNING YOUR BUSINESS

The Discovery That Changes Everything

Last month, the chief information officer of a global financial services company received a report that fundamentally altered her understanding of her organization. What she thought was a carefully controlled technology environment turned out to be an ecosystem of hidden intelligence operating at massive scale. The security audit revealed 847 unauthorized AI applications actively processing company data across 23 countries. From ChatGPT Plus subscriptions handling client communications to specialized machine learning tools analyzing market trends, an entire parallel AI infrastructure had emerged without her knowledge.

This scenario plays out weekly in boardrooms across Fortune 500 companies. What has been consistently observed in enterprise consulting work is that executives systematically underestimate the scope of AI adoption already occurring in their organizations. The gap between perceived control and actual reality has become the defining challenge of modern enterprise leadership.

The Scale of Invisible Intelligence

Research reveals the stunning magnitude of this hidden transformation. Organizations are discovering they operate hundreds of AI applications while IT and security teams maintain visibility and control over less than 20% of them. This isn't speculation based on projected adoption curves. This is the current reality measured through comprehensive enterprise audits.

Shadow IT discovery tools now routinely identify unauthorized AI platforms representing 15% to 30% of all software installations in large enterprises. The categories of hidden AI tools span far beyond simple productivity applications. Research institutions and enterprise environments exhibit patterns of unapproved cloud services, self-made AI solutions, and locally installed AI applications operating at the departmental level, completely outside formal oversight.

What makes AI adoption fundamentally different from traditional shadow IT is the speed of integration and the depth of business process embedding. Where unauthorized software tools might take weeks or months to become operationally critical, AI tools can become essential to daily workflows within hours of discovery. Employees don't just install AI applications. They build crucial business processes around them.

The Departmental Intelligence Revolution

The proliferation of shadow AI follows predictable patterns across organizational functions, each driven by specific productivity pressures and unmet technological needs. Understanding these patterns provides the foundation for systematic discovery and strategic transformation.

Sales organizations lead unauthorized AI adoption because revenue pressure creates urgency that formal procurement processes cannot match. Sales teams deploy AI tools for personalized outreach script generation, objection handling automation, and LinkedIn engagement optimization. Research shows sales departments using AI platforms for predictive lead scoring and qualification, often integrating customer relationship management data with external AI services without a security review.

Marketing departments represent the second-highest concentration of shadow AI adoption. Content generation needs drive teams toward generative AI platforms for landing page copy creation, social media campaign development, and competitive analysis automation. Marketing teams particularly gravitate toward AI tools that analyze customer support tickets, identify content gaps, and generate campaign performance insights. The speed advantage of unauthorized AI tools often outweighs perceived compliance risks.

Human resources functions increasingly rely on unauthorized AI for recruitment process automation, employee inquiry handling, and policy documentation generation. HR teams use AI tools to streamline job description creation, candidate evaluation, and onboarding process optimization. One enterprise study found HR departments handling 70% of repetitive employee inquiries through unauthorized conversational AI assistants.

Finance and operations departments adopt AI tools for expense categorization, budget optimization, cost-benefit analysis automation, and financial reporting generation. The precision requirements of financial work create strong incentives for the adoption of AI tools, particularly for data analysis and pattern recognition tasks that traditional enterprise software handles poorly.

Research and development organizations show the highest diversity of unauthorized AI tool adoption. R&D teams use AI platforms for patent research, clinical trial data analysis, literature review automation, and experimental design optimization. The innovation pressure in research environments creates a natural affinity for experimental AI tools that formal IT approval processes might delay or reject.

The Detection Challenge

Traditional IT monitoring approaches fail systematically when applied to AI tool discovery. Shadow AI applications often operate through Web browsers, mobile applications, and cloud services that bypass conventional enterprise monitoring systems. The cloud-based nature of most AI platforms means usage frequently occurs outside network perimeters that traditional security tools monitor.

Current enterprise approaches to unauthorized AI detection rely heavily on domain-level blocking and network traffic analysis. These methods capture obvious violations like direct ChatGPT usage but miss sophisticated implementations like AI-powered browser extensions, mobile AI applications, and locally installed AI development tools. The limitation of purely technical detection approaches drives the need for behavior-centric discovery methods.

The most effective discovery approaches combine automated monitoring with cultural intelligence gathering. Advanced organizations implement AI-specific monitoring tools that track unauthorized models and applications across enterprise environments. These tools alert IT teams to suspicious AI activity patterns while maintaining automated anomaly detection for unusual data flows and computational resource usage.

Technology-driven discovery represents only half the solution. Enterprises that achieve comprehensive shadow AI visibility combine automated detection with systematic cultural engagement. They create safe channels for employees to disclose AI tool usage without immediate punitive consequences. This combination of technological monitoring and cultural transparency consistently outperforms purely restrictive approaches.

What has been observed repeatedly is that organizations focusing exclusively on detection and prohibition drive innovation further underground. The most sophisticated shadow AI implementations emerge in response to heavy-handed control measures. Employees become more creative in concealing AI usage when they perceive IT departments as obstacles rather than enablers.

The Strategic Intelligence Hidden in Shadow Adoption

Every unauthorized AI tool represents strategic intelligence about unmet business needs and innovation opportunities. Rather than viewing shadow AI as a compliance problem, leading organizations treat it as market research conducted by their own employees. The tools employees choose reveal productivity gaps, process inefficiencies, and competitive pressures that formal enterprise architecture often misses.

The shadow AI landscape provides early indicators of technological shifts and market opportunities. Employees typically adopt AI tools 12 to 18 months before enterprise vendors integrate similar capabilities into official platforms. This early adoption creates competitive intelligence opportunities for organizations sophisticated enough to harvest and analyze the data.

Understanding the scope and scale of existing shadow AI adoption provides the foundation for implementing SAMS effectively. Organizations cannot manage what they cannot see, and they cannot optimize what they refuse to acknowledge. The hidden AI already running your business represents either your greatest risk or your most valuable strategic asset.

The choice depends entirely on how you choose to discover, understand, and transform the intelligence revolution already underway in your enterprise.

DISCOVERING AND GOVERNING UNAUTHORIZED AI USAGE

The Detection Evolution Beyond Traditional IT Monitoring

The fundamental challenge facing enterprise security teams is that shadow AI operates differently from traditional unauthorized software. Where conventional shadow IT applications leave clear digital footprints through network traffic and system installations, AI tools often function through Web browsers, mobile applications, and cloud services that bypass standard enterprise monitoring systems. This creates what is called the AI detection gap, where traditional cybersecurity tools capture only the most obvious violations while missing sophisticated shadow implementations.

Recent advances in AI-specific monitoring technologies have begun addressing this gap through specialized detection approaches. Platforms like Zluri now provide comprehensive AI visibility across enterprise environments, automatically discovering AI applications and tracking real-time usage patterns. These tools represent a fundamental evolution from general IT asset management toward AI-specific intelligence gathering. The core advancement is behavior-centric detection that focuses on how AI tools are being used rather than simply which tools are being accessed.

The most effective detection methodologies combine automated technological monitoring with cultural intelligence gathering. Organizations implementing comprehensive shadow AI discovery deploy monitoring tools that track unauthorized models and applications while simultaneously creating safe channels for voluntary disclosure. This dual approach consistently outperforms purely restrictive strategies because it acknowledges that unauthorized AI usage often stems from legitimate productivity needs rather than malicious intent.

The Multi-Layered Detection Framework

Advanced shadow AI detection requires systematic deployment across multiple organizational layers, each designed to capture different categories of unauthorized usage. The framework I've developed with leading enterprises operates through four integrated detection mechanisms that together achieve near-complete visibility into shadow AI adoption.

Network and application monitoring forms the foundation layer, utilizing tools that identify AI-related traffic patterns, unusual data flows, and computational resource usage indicative of AI processing. These systems flag suspicious activity patterns such as large file uploads to external AI services, unusual API calls to machine learning platforms, and data transfer volumes consistent with AI model training or inference.

Endpoint detection represents the second layer, monitoring individual devices for locally installed AI applications, browser extensions with AI capabilities, and mobile applications accessing AI services. This layer captures tools that operate entirely on employee devices without requiring network resources that traditional monitoring systems would detect.

Cloud service monitoring constitutes the third layer, tracking usage of external AI platforms through enterprise single sign-on systems, credit card monitoring for AI service subscriptions, and integration activity with cloud-based AI tools. This layer proves particularly valuable because many shadow AI implementations occur through software-as-a-service platforms that employees access using corporate credentials.

Behavioral monitoring forms the fourth layer, analyzing work patterns for indicators of AI assistance such as unusual productivity spikes, content generation speeds inconsistent with human capabilities, and workflow changes suggesting automated assistance. This layer often reveals the most sophisticated shadow AI implementations, where employees have learned to mask their AI usage through careful work pattern management.

Risk Assessment Frameworks for Shadow AI

Evaluating shadow AI risks requires frameworks specifically designed for the unique characteristics of AI applications rather than general IT security assessments. The complexity of AI risks spans multiple domains, including data security, model bias, regulatory compliance, intellectual property exposure, and operational reliability. Effective risk assessment must systematically evaluate each domain while considering the interconnected nature of AI risks.

The NIST AI Risk Management Framework provides the foundational structure for systematic risk evaluation, emphasizing fairness, equity, inclusivity, data quality, transparency, and accountability. Organizations implementing comprehensive shadow AI assessment adapt NIST guidelines to focus specifically on unauthorized implementations, evaluating risks that sanctioned AI systems typically avoid through controlled deployment.

The EU AI Act introduces a risk-based classification system, particularly relevant to shadow AI assessment because unauthorized applications often fall into higher-risk categories due to a lack of oversight. Organizations must evaluate shadow AI tools against EU classification criteria, including prohibited AI practices, high-risk AI systems, and limited-risk applications. This framework proves especially valuable for multinational enterprises operating under European jurisdiction.

The COBIT framework integration across the AI lifecycle provides comprehensive governance coverage from discovery through ongoing monitoring. The framework maps to shadow AI assessment through evaluate, direct, and monitor functions during discovery, align, plan, and organize functions during risk evaluation, and monitor, evaluate, and assess functions during ongoing governance. This comprehensive approach ensures systematic risk assessment rather than ad hoc evaluation.

The Shadow AI Risk Matrix

Practical risk assessment requires a structured evaluation matrix that enables consistent assessment across different types of unauthorized AI tools. The framework I've developed with enterprise clients evaluates shadow AI risks across five critical dimensions, creating a comprehensive risk profile for each discovered application.

Data exposure risk assessment evaluates the sensitivity of information accessible to unauthorized AI tools, the geographic location of data processing, encryption standards, data retention policies, and potential for inadvertent information disclosure. This dimension proves particularly critical because shadow AI tools often process company data through external services without appropriate security controls.

Compliance risk evaluation examines potential violations of industry regulations such as GDPR, CCPA, HIPAA, and emerging AI-specific regulations. The assessment considers data processing locations, consent mechanisms, audit trail availability, and regulatory reporting capabilities. Shadow AI tools frequently create compliance gaps because they operate outside established privacy and security frameworks.

Intellectual property risk analysis focuses on potential exposure of proprietary information, algorithms, trade secrets, and competitive intelligence through unauthorized AI processing. This dimension requires careful evaluation because AI tools often require detailed input data that may contain sensitive intellectual property, and many AI platforms retain rights to use input data for model improvement.

Operational risk assessment examines the business continuity implications of shadow AI dependencies, the reliability of unauthorized services, and potential disruption from sudden access removal. Organizations often discover that critical business processes have become dependent on shadow AI tools, creating operational vulnerabilities that formal IT systems would address through service-level agreements and backup systems.

Bias and fairness risk evaluation becomes particularly important for shadow AI tools used in decision-making processes, hiring, customer service, or content generation. Unauthorized AI tools typically lack the bias testing and fairness evaluation that responsible AI governance requires, potentially creating legal and reputational risks.

Governance Strategies for Unauthorized AI Activities

Effective shadow AI governance requires moving beyond traditional prohibition-based approaches toward strategic enablement frameworks that channel unauthorized innovation into productive directions. The governance strategies that prove most effective acknowledge that shadow AI adoption often represents valid business needs that formal IT processes have failed to address.

Policy-driven management forms the foundation of effective governance, but policies must address AI-specific challenges rather than simply extending traditional IT policies. Effective AI policies define acceptable use cases, specify data protection requirements, outline approval processes for new AI tools, and establish clear accountability mechanisms. The most successful policies focus on how AI tools are used rather than which specific tools are permitted or prohibited.

Innovation pathway establishment represents the most critical governance mechanism for transforming shadow AI from risk into competitive advantage. Organizations implementing successful shadow AI governance create clear, transparent processes for requesting and implementing new AI tools, establish controlled environments for AI experimentation, and develop evaluation criteria for formalizing promising shadow innovations. These pathways must operate significantly faster than traditional IT procurement to remain relevant to business needs.

Cultural transformation initiatives prove essential for sustainable shadow AI governance because technology-focused approaches fail to address the underlying trust deficit between business units and IT departments. Successful governance programs foster open communication about AI usage, educate employees about responsible AI practices, and create incentives for transparency rather than concealment. This cultural approach consistently outperforms purely restrictive governance models.

Continuous monitoring and adaptive governance mechanisms ensure that shadow AI management remains effective as AI technology evolves and business needs change. The rapid evolution of AI capabilities means that static governance frameworks quickly become obsolete. Effective governance includes regular policy updates, continuous risk reassessment, and adaptive approval processes that evolve with technological advancement.

The Governance Maturity Model

Organizations implementing shadow AI governance typically progress through predictable maturity stages, each requiring different governance approaches and organizational capabilities. Understanding this progression enables more effective governance strategy development and realistic expectation setting for governance outcomes.

Detection-focused governance represents the initial maturity stage where organizations primarily focus on identifying unauthorized AI usage and implementing basic controls. This stage addresses immediate security and compliance concerns but often creates adversarial relationships between IT departments and business units.

Risk-based governance emerges as organizations develop a more sophisticated understanding of shadow AI implications and begin differentiating between high-risk and low-risk unauthorized usage. This stage enables more nuanced policy development and targeted intervention strategies.

Innovation-enabling governance represents advanced maturity where organizations systematically harvest value from shadow AI adoption while maintaining appropriate risk controls. This stage requires sophisticated organizational capabilities, including rapid AI evaluation processes, secure experimentation environments, and cultural transformation initiatives.

Strategic AI governance constitutes enterprise maturity where shadow AI management becomes integrated with overall AI strategy, competitive intelligence gathering, and innovation pipeline management. Organizations at this stage treat shadow AI discovery as market research and employee innovation identification rather than primarily as a compliance challenge.

The progression through these maturity stages requires sustained organizational commitment and cross-functional collaboration. Most enterprises implementing effective shadow AI governance require 12 to 18 months to progress from detection-focused to innovation-enabling governance, with strategic governance emerging over 24 to 36 months through sustained capability development.

The governance strategies that prove most sustainable acknowledge that shadow AI represents employee creativity and market intelligence rather than simply compliance violations. Organizations that successfully navigate shadow

AI governance transform what most enterprises see as their greatest AI risk into their most valuable source of competitive intelligence and innovation identification.

THE SHADOW AI MANAGEMENT SYSTEM (SAMS) FRAMEWORK

Your First Breakthrough Framework Introduction

The SAMS Framework represents the first of twelve breakthrough frameworks that will transform how enterprises think about AI. What makes SAMS revolutionary is not its ability to control unauthorized AI usage, but its capacity to systematically transform shadow innovation into strategic advantage. This framework moves beyond the traditional enterprise response of discover, prohibit, and punish toward a sophisticated approach of discover, evaluate, and optimize.

Based on work with over 200 enterprises implementing AI governance, it has been consistently observed that organizations focusing solely on restriction lose the competitive intelligence that shadow AI adoption provides. SAMS addresses this fundamental gap by treating unauthorized AI usage as market research conducted by your most innovative employees. The framework transforms what most enterprises see as their greatest AI risk into their most valuable source of competitive advantage.

The strategic importance of SAMS extends beyond risk management into competitive intelligence gathering. Every shadow AI tool represents an unmet business need, a process inefficiency, or a competitive pressure that formal enterprise architecture has failed to address. Organizations implementing SAMS systematically harvest this intelligence to accelerate their official AI strategies while maintaining appropriate governance controls.

The Four Foundational Principles of SAMS

SAMS operates through four core principles that distinguish it from traditional IT governance approaches. These principles transform the fundamental relationship between enterprise leadership and employee innovation, creating systematic channels for converting organic experimentation into organizational capability.

Discovery over detection represents the first foundational principle. Where traditional approaches seek to identify violations to punish, SAMS actively searches for innovation to harvest. The framework treats every unauthorized AI tool as potential strategic intelligence about productivity gaps, market pressures, and emerging technological capabilities. This principle shifts organizational focus from enforcement to opportunity identification.

Governance as acceleration constitutes the second principle, creating fast-track pathways for promising shadow innovations to become enterprise capabilities. Rather than slowing AI adoption through bureaucratic approval processes, SAMS enables rapid evaluation and formalization of valuable shadow innovations. This approach acknowledges that AI technology evolves faster than traditional procurement cycles.

Innovation transformation forms the third principle, systematically converting organic experimentation into competitive advantage. SAMS provides structured mechanisms for evaluating shadow AI discoveries, securing promising applications, and scaling successful innovations across the enterprise. This transformation ensures that employee creativity becomes organizational intelligence rather than simply individual productivity enhancement.

Cultural evolution is the fourth principle, shifting the enterprise mindset from AI compliance to AI enablement. SAMS fosters the trust necessary for transparent innovation by positioning IT departments as facilitators rather than gatekeepers. This cultural foundation proves essential because technology-focused governance approaches fail when employees perceive them as obstacles to productivity.

The Systematic SAMS Architecture

SAMS integrates four core functional components that work together to transform unauthorized AI usage into strategic advantage. Each component maps to established IT and AI governance frameworks while addressing the unique challenges of shadow AI management.

AI discovery and inventory form the foundation component, proactively identifying all AI tools, both sanctioned and unsanctioned, and tracking their real-time usage patterns across the enterprise. This component extends beyond traditional asset discovery to include behavioral pattern analysis, productivity impact assessment, and innovation potential evaluation. The discovery function treats shadow AI identification as competitive intelligence gathering rather than security violation detection.

Policy and ethical guidelines constitute the governance component, defining acceptable AI use, ethical principles including fairness and transparency, and compliance standards for all AI applications. This component adapts established frameworks, including the EU AI Act, NIST AI Risk Management Framework, and COBIT governance domains, to specifically address unauthorized AI usage. The policies focus on enabling responsible innovation rather than simply preventing misuse.

Risk assessment and mitigation represent the protection component, systematically identifying, assessing, and mitigating security, privacy, intellectual property, and bias risks associated with AI applications. This component leverages NIST Cybersecurity Framework functions and ISO 27001 standards while addressing AI-specific risks, including algorithmic bias, data exposure, and model reliability.

Continuous monitoring and auditing make up the oversight component, implementing real-time tracking of AI performance, compliance, and ethical adherence with regular audits to detect anomalies. This component ensures ongoing governance effectiveness while identifying emerging shadow AI adoption patterns and innovation opportunities.

The Six-Phase SAMS Implementation Methodology

Implementing SAMS requires systematic deployment across six integrated phases, each building organizational capability while addressing specific aspects of shadow AI transformation. The methodology, developed with leading enterprises, ensures sustainable implementation that evolves with technological advancement and organizational maturity.

Phase one involves defining AI governance policies and scope, clearly articulating the purpose and coverage of SAMS while establishing ethical guidelines that align with organizational values. This phase requires specifying approved AI tools, explicitly addressing unauthorized usage, and defining data protection requirements for AI applications. The policy framework must balance enabling innovation with maintaining appropriate controls.

Phase two establishes cross-functional governance teams and accountability structures, forming diverse teams including CIO representation, data scientists, compliance officers, ethics experts, legal professionals, and user group representatives. This phase defines clear roles and responsibilities for AI systems and their outputs, ensuring accountability for AI decisions while maintaining human oversight for critical determinations.

Phase three implements discovery, risk management, and continuous monitoring capabilities, utilizing shadow IT discovery tools and AI-specific monitoring platforms to achieve comprehensive visibility into AI applications. This phase involves conducting systematic AI risk assessments, integrating automated anomaly detection systems, and establishing real-time AI decision tracking with regular internal audits.

Phase four promotes transparency and employee education, providing comprehensive training on AI ethics, governance policies, and responsible AI usage. This phase educates managers on monitoring for unapproved AI tools while fostering an open communication culture where employees disclose AI usage without fear of punishment. The education component proves critical for sustainable governance success.

Phase five establishes innovation pathways and formalization mechanisms, developing clear processes for requesting and implementing new AI tools while creating controlled environments for AI experimentation. This phase identifies and empowers AI champions from different departments and establishes mechanisms to evaluate promising shadow innovations for potential formalization and scaling.

Phase six ensures data security and compliance integration, implementing robust data governance protocols for data quality, privacy, and security across all AI systems. This phase adopts zero-trust security frameworks, encrypts and anonymizes sensitive data, and ensures AI governance alignment with regulations, including GDPR, the EU AI Act, and industry-specific requirements.

Strategic Transformation Through SAMS

What distinguishes SAMS from existing governance frameworks is its explicit dual objective of risk mitigation and competitive advantage creation. Traditional shadow IT governance focuses on controlling unauthorized assets, while general AI governance addresses approved systems. SAMS specifically targets the transformation of unauthorized AI usage into strategic capability.

The framework acknowledges that shadow AI adoption often represents legitimate business needs that formal IT processes have failed to address. Rather than viewing unauthorized usage as purely compliance violations, SAMS recognizes these adoptions as innovation signals that provide valuable intelligence about market pressures, productivity gaps, and technological opportunities.

SAMS implementation requires ongoing iteration rather than static deployment because AI technology evolves rapidly and employee adoption patterns change continuously. The framework includes continuous feedback loops, agile governance mechanisms, and adaptive policy structures that evolve with technological advancement. This iterative approach ensures sustained relevance throughout organizational AI transformation.

The ultimate success of SAMS depends on cultural transformation that positions the framework as enabling innovation rather than constraining creativity. Organizations implementing SAMS successfully foster collaborative environments where employees become partners in AI governance rather than obstacles to control. This cultural foundation proves essential because sustainable governance requires voluntary participation rather than forced compliance.

Implementation Success Factors

The enterprises I've worked with that achieve the greatest success with SAMS implementation share common characteristics that enable effective shadow AI transformation. Understanding these success factors provides realistic expectations for SAMS deployment and organizational capability requirements.

Executive leadership commitment proves essential because SAMS requires cross-functional collaboration and cultural change that only sustained senior leadership support can enable. The framework challenges traditional departmental boundaries and requires investment in new capabilities, making CEO and C-suite championship critical for implementation success.

Cross-functional team formation enables a holistic perspective on shadow AI risks and opportunities that single-department approaches cannot achieve. Successful SAMS implementation requires expertise spanning technology, legal, compliance, business operations, and employee relations to address the complex challenges of unauthorized AI transformation.

Rapid innovation pathway development distinguishes successful SAMS implementations from governance approaches that inadvertently stifle valuable innovation. Organizations achieving competitive advantage through SAMS create approval processes that operate significantly faster than traditional IT procurement while maintaining appropriate security and compliance controls.

Continuous adaptation capability ensures SAMS remains effective as AI technology and organizational needs evolve. The most successful implementations treat SAMS as a living framework that adapts to technological advancement rather than a static governance structure that becomes obsolete through technological change.

SAMS provides the foundational framework for transforming unauthorized AI usage from organizational liability into competitive advantage. Organizations implementing SAMS systematically identify, evaluate, and optimize the shadow innovation already occurring throughout their enterprises while maintaining appropriate governance controls. This transformation creates a sustainable competitive advantage by harnessing employee creativity and market intelligence that traditional governance approaches typically suppress or ignore.

TRANSFORMING RISK INTO INNOVATION OPPORTUNITY

The Strategic Mindset Transformation

The fundamental breakthrough in shadow AI management occurs when organizations shift from a purely defensive posture to an offensive, value-creation mindset. This transformation requires recognizing that every unauthorized AI tool represents strategic intelligence about unmet business needs, process inefficiencies, and competitive pressures that formal enterprise architecture has failed to address. Enterprises that achieve sustainable competitive advantage through shadow AI management treat these discoveries as market research conducted by their most innovative employees rather than simply compliance violations to be eliminated.

What has been consistently observed across successful transformations is that organizations focusing solely on risk mitigation and control inadvertently stifle valuable employee-driven innovation that could provide significant competitive advantages. Shadow AI, despite its inherent risks, can unlock a range of strategic advantages, including productivity gains, competitive differentiation, and organizational agility. The progression of AI from scattered experiments to essential business infrastructure underscores the importance of channeling rather than suppressing this organic innovation.

The strategic transformation requires fundamental changes in how enterprise leadership views unauthorized technology adoption. Rather than viewing shadow AI as evidence of employee non-compliance, successful organizations recognize it as innovation signals that provide an early warning about technological shifts and market opportunities. Employees typically adopt AI tools

12 to 18 months before enterprise vendors integrate similar capabilities into official platforms, creating valuable competitive intelligence for organizations sophisticated enough to harvest and analyze this data.

The Pilot-to-Product Pathway Framework

Converting shadow AI discoveries into strategic advantages requires systematic transformation mechanisms that balance innovation enablement with appropriate risk controls. The pilot-to-product pathway provides the structured approach for capturing immense value from bottom-up innovation while systematically mitigating associated risks. This framework transforms potential liabilities into recognized assets through organized evaluation and formalization processes.

The pathway begins with discovery and evaluation of promising employee-led shadow experiments, identifying high-impact applications that demonstrate clear business value while assessing their strategic potential. Organizations implementing effective transformation programs actively scout for shadow innovations rather than waiting for compliance violations to surface through security audits. This proactive approach enables early identification of valuable applications before they become deeply embedded in critical business processes.

Resource allocation forms the second stage, providing modest budgets and short timelines for promising shadow innovations to demonstrate scalable value. Successful organizations allocate innovation funding specifically for shadow AI transformation, recognizing that small investments in employee experimentation often yield disproportionate returns. Research shows that companies like Advizex achieved remarkable results with total investments of only $25,000, developing prototypes that reduced quote generation time by 96% and increased qualified lead conversions by 18%.

Security and compliance evaluation represents the critical third stage, rigorously assessing discovered innovations against enterprise security standards, regulatory requirements, and ethical guidelines. This evaluation cannot simply apply traditional IT security frameworks because AI applications introduce unique risks, including algorithmic bias, data exposure, and model reliability. Effective evaluation frameworks address AI-specific risks while preserving the innovation value that made shadow applications attractive to employees.

Scaling and integration constitute the final stage, transforming validated shadow innovations into official enterprise solutions with appropriate governance controls, security measures, and operational support. This stage requires a careful balance between maintaining the agility and effectiveness that made shadow applications successful while adding the reliability and security that enterprise deployment requires.

Real-World Transformation Case Studies

The enterprises achieving the greatest success in shadow AI transformation share systematic approaches to converting unauthorized innovation into competitive advantage. These case studies demonstrate the tangible business value available through sophisticated shadow AI management while illustrating the specific mechanisms that enable successful transformation.

Advizex represents a compelling example of comprehensive shadow AI transformation through employee-led AI challenges focused on prototype development. The company challenged employees to utilize readily available AI tools to identify inefficiencies, automate tedious tasks, and enhance customer experience. This initiative yielded significant returns, including a generative AI configurator that reduced complex quote generation times from nearly an hour to just two minutes, an HR conversational AI assistant that handles 70% of repetitive employee inquiries, and predictive AI systems that increased qualified lead conversions by 18%. Beyond direct financial returns, the initiative fostered lasting cultural shifts toward innovation within the organization.

Vizient achieved competitive advantage through systematic AI champions programs, identifying power users from various departments who inspired broader AI adoption and collaboration. The healthcare performance improvement company partnered with generative AI platforms to achieve fourfold estimated return on investment and saved approximately $700,000 in the first year of implementation. The AI champions approach proved particularly effective because it leveraged existing employee expertise while providing formal channels for innovation sharing and scaling.

Schneider Electric demonstrates the value of citizen developer programs that empower non-technical employees to build applications using AI-powered low-code platforms. The energy management company enabled field service technicians to create mobile applications that improved efficiency and reduced downtime while enhancing customer satisfaction. This approach proves particularly valuable because it places AI development capability directly in the hands of employees who understand operational challenges most intimately.

Microsoft illustrates enterprise-scale transformation by leveraging internal culture to boost customer success through AI workflow improvements that free employees from mundane tasks. The company integrated AI into research processes to drive innovation and efficiency while creating awareness of AI capabilities across the organization. This comprehensive approach demonstrates how large enterprises can systematically transform shadow AI adoption into strategic capability.

Risk Mitigation That Preserves Innovation Momentum

Effective shadow AI transformation requires risk mitigation strategies that address security and compliance concerns without destroying the innovation value that made unauthorized applications attractive to employees. Traditional risk mitigation approaches often impose controls that eliminate the agility and effectiveness that drive shadow adoption, defeating the purpose of transformation efforts. Successful enterprises develop risk frameworks specifically designed to preserve innovation momentum while addressing AI-specific concerns.

Controlled experimentation environments provide secure spaces for AI innovation without exposing enterprise systems to unauthorized access or data leakage. These sandbox environments enable employees to experiment with AI tools and develop solutions while maintaining appropriate security boundaries. Organizations implementing effective sandboxes provide realistic data sets and integration capabilities that enable meaningful experimentation without compromising sensitive information or critical systems.

Rapid approval processes address the speed requirements that drive employees toward unauthorized tools while maintaining appropriate oversight and control. Traditional IT procurement often operates on timescales measured in months or quarters while AI tool adoption occurs within hours or days. Successful enterprises develop fast-track approval mechanisms for low-risk AI applications while maintaining rigorous evaluation for high-risk implementations.

Adaptive security frameworks evolve with AI technology advancement rather than applying static controls that quickly become obsolete. AI technology develops rapidly and often in unpredictable directions, making traditional compliance checklists inadequate for emerging applications. Effective security approaches include continuous evaluation and feedback loops rather than periodic reviews, ensuring sustained relevance throughout organizational AI transformation.

Privacy-preserving data access enables AI experimentation without exposing sensitive information through unauthorized external services. Organizations achieving successful transformation provide access to realistic but sanitized data sets that enable meaningful AI development while protecting confidential information. This approach addresses the primary driver of shadow AI adoption while maintaining appropriate data protection.

The Cultural Transformation Imperative

Sustainable shadow AI transformation requires cultural evolution that positions enterprise leadership as innovation enablers rather than technology gatekeepers. The most successful transformations occur when organizations foster collaborative environments that position IT departments as innovation enablers rather than gatekeepers. This cultural foundation proves essential because technological approaches fail when employees perceive them as barriers to productivity and creativity.

Innovation recognition programs celebrate employee creativity while channeling experimentation toward productive directions. Organizations achieving transformation success develop formal mechanisms for recognizing valuable shadow innovations, providing both incentives for transparency and models for appropriate AI usage. These programs must balance encouraging experimentation with maintaining appropriate boundaries for security and compliance.

AI literacy initiatives ensure employees understand both the capabilities and limitations of AI technologies while developing skills necessary for responsible usage. Effective education goes beyond simple policy training to include hands-on experience with approved AI tools and frameworks for evaluating new applications. This approach reduces the knowledge gap that often drives employees toward unauthorized tools while building organizational capability for AI adoption.

Cross-functional collaboration breaks down traditional silos between IT departments and business units that often create trust deficits, driving shadow adoption. Successful transformation requires sustained cooperation between technology teams responsible for security and compliance and business teams driving innovation and productivity. This collaboration must operate significantly faster than traditional enterprise processes to remain relevant to business needs.

Trust-building mechanisms address the fundamental relationship issues that drive employees toward unauthorized tools when they perceive official channels as unresponsive or bureaucratic. Organizations achieving transformation success create transparent communication about AI governance decisions, rapid feedback on innovation proposals, and clear pathways for escalating time-sensitive requests. This approach transforms the IT department from a perceived obstacle into an innovation partner.

The Transformation Value Proposition

The enterprises that successfully transform shadow AI risks into innovation opportunities achieve multiple forms of competitive advantage that extend far beyond simple cost savings or efficiency improvements. These organizations develop systematic capabilities for identifying, evaluating, and scaling technological innovations while building organizational cultures that continuously generate competitive intelligence through employee experimentation.

Speed and agility advantages emerge when organizations harness employee innovation rather than restricting it through bureaucratic processes. Companies implementing effective transformation achieve faster AI adoption cycles, more rapid response to competitive pressures, and accelerated development of new capabilities. Research demonstrates that employee-driven AI experiments often identify valuable applications 12 to 18 months before they become available through formal enterprise vendors.

Productivity multiplication occurs when successful shadow innovations scale across organizational functions and business units. Single innovations often generate value far beyond their original scope as other departments adapt successful applications to their specific needs. The cultural multiplier effect creates self-sustaining innovation cycles where successful transformation encourages additional experimentation and creativity.

Competitive intelligence gathering provides ongoing insight into technological trends, market pressures, and emerging opportunities through systematic analysis of employee AI adoption patterns. Organizations implementing sophisticated transformation capabilities systematically analyze shadow AI patterns to reveal unmet business needs and competitive vulnerabilities. This intelligence enables proactive rather than reactive AI strategy development.

Talent retention and engagement benefits emerge when employees feel empowered to innovate and solve business problems using cutting-edge technologies. Reports indicate that employees using AI tools experience increased productivity and feel more passionate about their work, suggesting that empowerment leads to enhanced engagement and reduced staff turnover. Organizations that prevent employee AI usage often lose innovative talent to competitors offering greater technological freedom.

The transformation of shadow AI risks into innovation opportunities represents a fundamental competitive differentiator for enterprises willing to invest in sophisticated management capabilities. Organizations that master this transformation systematically outperform competitors in speed, agility, innovation, and employee engagement while maintaining appropriate security and compliance controls.

THE SHADOW ADOPTION PATTERN

The Predictable Science of Organic AI Diffusion

Three decades of studying enterprise technology adoption have shown that shadow AI doesn't spread randomly through organizations. It follows predictable patterns that can be mapped, analyzed, and leveraged for strategic advantage. Understanding the shadow adoption pattern provides the final piece of the SAMS framework, enabling enterprises to anticipate where unauthorized AI will emerge next and position themselves to harvest the most valuable innovations before competitors even recognize their existence.

What has been consistently observed across industries is that shadow AI adoption follows the same fundamental principles that govern all technological diffusion through complex organizations. The pattern operates through three interconnected forces that determine where, when, and how unauthorized AI tools establish themselves within enterprise environments. These forces create adoption patterns so consistent that they enable predictive modeling of future shadow AI emergence.

Research reveals that organizations implementing sophisticated shadow AI management achieve a competitive advantage not just by controlling unauthorized usage, but by understanding adoption patterns well enough to predict and prepare for the next wave of innovation. This predictive capability transforms enterprises from reactive governance to proactive strategic positioning, ensuring they capture maximum value from employee creativity while maintaining appropriate oversight.

The Three Forces of Shadow Adoption

Shadow AI adoption operates through three fundamental forces that interact to create predictable patterns of technological diffusion across organizational boundaries. Understanding these forces enables systematic prediction of where shadow innovations will emerge next and which applications will achieve sustainable adoption versus temporary experimentation.

The productivity pressure force drives initial shadow AI adoption when employees encounter workflow inefficiencies that existing enterprise tools cannot address effectively. This force creates adoption tension, where the gap between employee productivity needs and available enterprise capabilities reaches critical thresholds. Research shows that departments experiencing the highest productivity pressure consistently become early shadow AI adopters regardless of their technical sophistication or official AI policies.

Sales organizations typically experience maximum productivity pressure because revenue generation creates urgency that formal enterprise tools often cannot match. Marketing departments follow closely because content generation demands frequently exceed traditional software capabilities. Research and development functions experience sustained pressure because innovation requirements push beyond established enterprise platforms. Human resources departments increasingly adopt shadow AI because employee inquiry volume overwhelms traditional support mechanisms.

The social influence force amplifies shadow adoption through organizational networks, where successful AI implementations spread through professional relationships and informal knowledge sharing. This force explains why shadow AI adoption often clusters in specific departments or business units rather than being distributed evenly across enterprises. Employees observe colleagues achieving remarkable productivity gains and seek similar solutions for their own challenges.

Research demonstrates that social influence operates through identifiable mechanisms, including peer demonstration, success story sharing, and collaborative experimentation. Organizations with strong internal networks experience faster shadow AI diffusion, while enterprises with siloed departments see more isolated adoption patterns. Understanding social network topology enables the prediction of adoption pathways and the identification of influential early adopters who accelerate organizational diffusion.

The technical accessibility force determines which AI tools achieve widespread adoption based on implementation simplicity, learning curve requirements, and integration compatibility with existing workflows. This force explains why certain AI applications achieve enterprise-wide adoption while technically superior alternatives remain confined to specialist users. Accessibility requirements vary significantly across organizational roles and technical capabilities.

The most successful shadow AI tools combine powerful capabilities with minimal implementation barriers, enabling rapid adoption by non-technical employees. Browser-based AI applications consistently outperform desktop software because they require no installation approval. Cloud-based tools spread faster than on-premises solutions because they bypass IT procurement processes. Applications integrating with existing enterprise platforms achieve broader adoption than standalone tools requiring workflow changes.

The Adoption Lifecycle Predictive Model

Shadow AI adoption follows a predictable lifecycle that enables organizations to anticipate emergence patterns and optimize intervention timing for maximum strategic value. The model developed with leading enterprises maps adoption progression through five distinct phases, each characterized by specific behaviors, adoption rates, and strategic opportunities.

The discovery phase begins when individual employees first encounter AI tools through external sources, including professional networks, industry conferences, vendor demonstrations, or personal research. This phase typically involves 1-3% of the organizational population and occurs 3-6 months before broader adoption patterns emerge. Discovery often happens in departments with external engagement, including sales, marketing, research, or business development, where employees encounter cutting-edge technologies through client interactions or industry participation.

Organizations implementing sophisticated shadow AI management establish early warning systems that detect discovery phase indicators, including increased external AI service research, vendor inquiry patterns, and conference attendance focused on AI applications. These signals enable proactive preparation for subsequent adoption phases rather than reactive responses to established shadow implementations.

The experimentation phase occurs when discoverers begin testing AI tools for specific business applications, typically involving 5-10% of target departments over 2-4 month periods. Experimentation involves limited scope applications focused on individual productivity enhancement rather than team or process integration. This phase provides critical intelligence about application effectiveness and adoption potential.

Successful enterprises monitor experimentation patterns to identify high-value applications and influential early adopters who will drive subsequent adoption phases. Rather than restricting experimentation, sophisticated organizations provide controlled environments that enable safe testing while gathering intelligence about promising innovations. This approach accelerates beneficial adoption while mitigating potential risks.

The adoption phase begins when successful experiments expand beyond individual usage to team or departmental implementation, involving 15-30% of the target population over 3-6 month periods. This phase represents the critical transition from personal productivity tools to collaborative business applications, where shadow AI begins impacting organizational processes and outcomes.

The adoption phase provides maximum opportunity for strategic intervention because applications demonstrate proven value while remaining manageable in scope. Organizations implementing effective SAMS frameworks use this phase to evaluate promising innovations for formal integration while establishing governance controls for continued expansion. This timing enables the transformation of shadow risk into competitive advantage.

The integration phase occurs when successful adoptions become embedded in critical business processes, involving 40-70% of relevant populations over 6-12 month periods. Applications reaching the integration phase achieve operational necessity, where business processes depend on continued AI tool availability. This phase represents both maximum business value and maximum organizational risk from unauthorized tool dependencies.

Enterprises achieving competitive advantage through shadow AI management prioritize integration phase governance because disruption becomes increasingly costly while competitive value reaches peak levels. Effective governance transforms integrated shadow applications into formal enterprise capabilities while maintaining the agility and effectiveness that made them successful during earlier phases.

The institutionalization phase represents full organizational adoption, where AI applications become standard operating procedures supported by formal policies, training programs, and enterprise infrastructure. This phase marks the successful transformation from shadow innovation to organizational capability, completing the journey from unauthorized experimentation to competitive advantage.

Strategic Prediction and Planning Applications

Understanding shadow adoption patterns enables sophisticated strategic planning that transforms the organizational relationship with unauthorized AI from reactive risk management to proactive competitive advantage development. The predictive capabilities available through adoption pattern analysis provide strategic intelligence that rivals traditional market research while requiring minimal additional investment.

Early warning systems enable the identification of emerging AI technologies 6-12 months before they achieve widespread organizational adoption. These systems monitor discovery phase indicators, including employee research patterns, vendor engagement activities, and industry conference participation, to predict future shadow adoption trends. Organizations implementing systematic early warning capture first-mover advantages by preparing official enterprise capabilities before shadow adoption reaches critical mass.

The intelligence gathering focuses on departments with the highest external technology exposure, including sales teams engaging with innovative clients, research groups attending technical conferences, and marketing organizations evaluating cutting-edge platforms. These boundary-spanning roles provide consistent early indicators of technological trends that will subsequently spread through organizational adoption networks.

Adoption velocity prediction enables accurate forecasting of shadow AI diffusion rates based on organizational characteristics, including network density, change tolerance, and technical capability distribution. Organizations with strong internal networks experience adoption rates 2-3 times faster than siloed enterprises. Departments with high productivity pressure adopt shadow AI 40-60% faster than low-pressure functions. Understanding velocity patterns enables appropriate preparation timing for governance interventions.

Predictive modeling incorporates organizational factors, including change management history, technology adoption precedents, and cultural innovation tolerance, to forecast shadow AI emergence with remarkable accuracy. These models enable resource allocation optimization and governance preparation that transform potential disruption into strategic opportunity.

Strategic positioning applications enable enterprises to leverage adoption pattern intelligence for competitive advantage development. Organizations understanding shadow adoption patterns position themselves to capture maximum value from employee innovation, while competitors struggle with reactive governance approaches. This positioning advantage compounds over time as sophisticated organizations build systematic capabilities for innovation harvesting.

The strategic approach involves establishing innovation capture mechanisms aligned with the adoption lifecycle phases. Discovery phase monitoring identifies promising technologies before competitors recognize their potential. Experimentation phase support accelerates valuable innovation while maintaining appropriate oversight. Adoption phase governance transforms proven innovations into formal capabilities faster than traditional development processes would enable.

The Cultural Multiplier Effect in Adoption Patterns

The most sophisticated understanding of shadow adoption patterns recognizes that technological diffusion interacts with organizational culture to create multiplicative effects that either accelerate or constrain innovation propagation. Organizations that successfully leverage shadow AI adoption patterns invest in cultural foundations that amplify positive adoption while dampening harmful implementations.

Innovation recognition culture accelerates beneficial shadow adoption by celebrating employee creativity and providing formal mechanisms for innovation sharing. Organizations implementing effective recognition programs experience shadow adoption rates 2-3 times higher than enterprises that discourage unauthorized experimentation. Recognition culture also improves innovation quality because employees feel empowered to experiment openly rather than concealing potentially valuable applications.

Research demonstrates that recognition programs must balance encouraging experimentation with maintaining appropriate boundaries for security and compliance. Successful programs celebrate innovation outcomes while educating employees about responsible AI usage. This approach channels employee creativity toward productive directions while building organizational capability for sustainable innovation.

Collaborative learning networks amplify shadow adoption value by enabling rapid knowledge transfer between early adopters and broader organizational populations. Organizations with strong internal learning networks capture maximum value from shadow innovations because successful applications spread quickly through established knowledge-sharing mechanisms. These networks also improve innovation quality through peer feedback and collaborative refinement.

The most effective learning networks combine formal knowledge sharing mechanisms with informal peer interaction opportunities. Internal conferences showcasing employee AI innovations, cross-functional teams exploring emerging technologies, and mentorship programs connecting early adopters with broader populations all contribute to positive adoption amplification. These mechanisms ensure valuable innovations achieve organizational impact rather than remaining isolated individual productivity improvements.

Trust-based governance culture enables sustainable shadow adoption patterns by positioning enterprise leadership as innovation enablers rather than technology gatekeepers. Organizations achieving long-term competitive advantage through shadow AI management build cultural foundations where employees view governance as acceleration rather than restriction. This cultural transformation proves essential because technological approaches fail when employees perceive them as obstacles to productivity and creativity.

Trust-based governance requires consistent demonstration that enterprise leadership values employee innovation while maintaining necessary oversight for security and compliance. Transparent communication about governance decisions, rapid feedback on innovation proposals, and clear pathways for escalating time-sensitive requests all contribute to trust development. Organizations investing in trust-based governance experience higher-quality shadow innovations because employees feel empowered to collaborate openly with enterprise leadership.

The Strategic Imperative of Pattern Mastery

Mastering shadow adoption patterns represents a fundamental competitive differentiator that extends far beyond technology governance into organizational capability development. The enterprises that achieve sustainable competitive advantage through AI will be those that understand how intelligence spreads through organizations and systematically optimize these patterns for strategic value creation.

The shadow adoption pattern provides the predictive intelligence necessary to transform unauthorized AI usage from organizational liability into systematic competitive advantage. Organizations implementing sophisticated pattern analysis achieve faster innovation cycles, higher-quality AI implementations, and more sustainable competitive differentiation than enterprises relying on reactive governance approaches.

Most importantly, understanding shadow adoption patterns enables the cultural transformation necessary for long-term AI success. Enterprises that master these patterns create organizational environments where employee creativity and enterprise governance work together to accelerate innovation rather than constraining it. This cultural capability proves essential because the competitive advantages available through AI require sustained innovation that only engaged, empowered employees can generate.

SAMS provides a comprehensive framework for implementing pattern-based governance that transforms unauthorized AI usage into competitive advantage. Organizations implementing SAMS systematically identify, evaluate, and optimize shadow innovations while building the cultural foundations necessary for sustained AI leadership.

The invisible revolution is already underway in your enterprise. The choice isn't whether AI will transform your organization. The choice is whether you'll master the patterns that determine how that transformation unfolds. The enterprises that understand and leverage shadow adoption patterns will lead their industries into the age of intelligence. Those that don't will spend the coming decade struggling to catch up with competitors who learned to harness the creative intelligence of their employees.

Your journey into the AI-powered enterprise begins with recognizing that your greatest competitive advantage may already be hiding in plain sight within the shadow innovations your employees are creating right now. The question is whether you're sophisticated enough to find it, wise enough to nurture it, and strategic enough to transform it into the foundation of your competitive future.

CHAPTER 3

From Prediction to Strategic Cognition

The quarterly planning meeting had just concluded. The Chief Strategy Officer reviewed the forecasts, analyzed the market projections, and approved the resource allocations based on meticulously crafted predictions about customer behavior, competitive responses, and market conditions six months into the future. By the time the meeting ended, three of those assumptions had already become obsolete.

This scenario illustrates the fundamental flaw in traditional strategic planning that AI now makes impossible to ignore. While enterprises have spent decades perfecting the art of prediction, the accelerating pace of market change has rendered prediction-based strategy increasingly ineffective. The time required to gather data, analyze trends, and develop forecasts now exceeds the useful life of the strategic insights those forecasts generate.

What some call the prediction paradox has created a crisis in strategic planning that most enterprises refuse to acknowledge. The more sophisticated our forecasting models become, the less valuable traditional strategic planning becomes in markets that evolve faster than planning cycles can complete. The enterprises that recognize this paradox first will build competitive advantages that prediction-based competitors cannot understand or replicate.

The solution requires abandoning prediction entirely in favor of strategic cognition. Instead of trying to forecast what will happen, AI-native enterprises develop the capability to think strategically about what could happen and respond intelligently to whatever actually happens. Instead of betting organizational futures on predictions that may prove wrong, they build adaptive intelligence that improves decision-making regardless of which future emerges.

Strategic cognition represents the evolution from human-driven analysis to human-AI collaborative intelligence that operates at speeds traditional planning cannot match. Agentic AI systems become strategic collaborators that can simulate thousands of scenarios simultaneously, identify patterns across vast data sets in real time, and recommend optimal responses to changing conditions faster than human strategists can recognize that conditions have changed.

This chapter reveals how to transform your strategic planning from prediction-based exercises that constrain organizational flexibility to cognition-based capabilities that enhance strategic responsiveness. You'll discover why the most successful enterprises are replacing their business intelligence systems with enterprise cognition platforms that think rather than just analyze, and how to design AI systems that become strategic partners rather than analytical tools.

The future belongs to enterprises that can think strategically at machine speed while maintaining human wisdom in strategic judgment. The question is whether your organization will develop that capability before competitors make prediction-based planning a competitive liability you cannot afford to maintain.

BEYOND FORECASTING: GENERATIVE FORESIGHT AND STRATEGIC SIMULATION

The Global Strategy Director at a Fortune 100 retailer stared at the simulation results displayed across six monitors. Each screen showed a different version of 2025. In one reality, their aggressive expansion into Southeast Asia generated $2.8 billion in new revenue. In another, the same strategy triggered a regulatory backlash that cost them $400 million. A third scenario revealed an unexpected opportunity in sustainable packaging that created an entirely new business unit worth $1.2 billion.

All six futures were running simultaneously. All six were generating real strategic intelligence. And all six had been created in the past four hours.

This is what is called strategic time travel, and it represents the most profound advancement in strategic planning since the development of scenario analysis fifty years ago. While traditional forecasting attempts to predict a single future, strategic time travel enables enterprises to literally inhabit multiple possible futures simultaneously, test strategic alternatives across parallel realities, and return to the present with intelligence that no competitor using conventional planning can match.

Strategic Time Travel: Inhabiting Multiple Possible Futures Simultaneously

The fundamental breakthrough of strategic time travel lies in its ability to compress years of potential market evolution into real-time strategic exploration. In my three decades of enterprise transformation, I have never witnessed a capability that creates competitive advantages as rapidly and sustainably as strategic time travel. The enterprises implementing these systems can explore strategic possibilities in days that would take traditional planning teams months to analyze. More importantly, they can test strategies against market conditions that haven't occurred yet, building strategic intelligence that competitors won't develop until those conditions actually emerge.

Strategic time travel operates through what some call temporal strategic immersion, where AI systems create fully realized business environments representing different evolutionary paths from current market conditions. These aren't static scenarios that describe possible futures. These are dynamic simulations where strategic teams can literally experience operating their business under different market conditions, competitive pressures, and technological environments. The immersion enables strategic learning that theoretical analysis cannot provide because teams discover strategic implications through direct experience rather than abstract modeling.

The temporal immersion process begins with current reality mapping, where AI systems create comprehensive digital representations of existing market conditions, competitive relationships, and operational capabilities. This baseline reality becomes the launching point for multiple timeline explorations that branch into different evolutionary paths based on trend extrapolation, competitive scenario testing, and external disruption modeling. Strategic teams can follow any timeline branch to its logical conclusion, experiencing years of potential business evolution within hours of strategic exploration.

What makes strategic time travel particularly powerful is its ability to capture strategic learning acceleration effects that traditional planning cannot achieve. When strategic teams experience multiple possible futures through immersive simulation, they develop an intuitive understanding of cause-and-effect relationships, market dynamics, and competitive interactions that abstract analysis cannot provide. This experiential learning enables strategic decision-making capabilities that operate faster and more accurately than traditional analytical approaches because strategic judgment has been calibrated through direct experience across multiple possible realities.

From Single Future Prediction to Parallel Reality Exploration

Traditional strategic planning assumes that successful strategy requires accurate prediction of future market conditions, competitive responses, and technological developments. This assumption creates what some call the prediction dependency trap, where strategic success becomes hostage to forecasting accuracy rather than adaptive capability. Organizations trapped in prediction dependency spend enormous resources attempting to forecast variables that may be fundamentally unpredictable while neglecting the development of adaptive capabilities that would enable success regardless of which future actually emerges.

Strategic time travel eliminates prediction dependency by enabling parallel reality exploration, where strategic teams can test approaches across multiple possible futures simultaneously rather than betting organizational success on single forecasts that may prove incorrect. This exploration reveals strategic approaches that remain effective across multiple scenarios, strategic risks that appear only under specific conditions, and strategic opportunities that emerge from combinations of trends that individual forecasts cannot identify.

The parallel exploration process operates through what researchers call scenario multiplication, where each strategic option branches into multiple potential outcomes based on different environmental conditions. Instead of testing strategy A against forecast X, strategic time travel tests strategy A against realities X, Y, and Z simultaneously while exploring how strategy A might evolve and adapt as those realities unfold. This multiplication reveals strategic possibilities that linear planning cannot envision because it accounts for strategic evolution over time rather than just strategic implementation under static conditions.

Parallel reality exploration also enables strategic robustness testing, where strategic approaches are evaluated based on their effectiveness across multiple scenarios rather than their optimization for single predicted conditions. This testing consistently reveals that strategies optimized for specific forecasts often fail catastrophically when market conditions differ from predictions. In contrast, strategies designed for cross-scenario effectiveness demonstrate remarkable resilience and adaptability as market conditions evolve unpredictably.

The competitive advantage generated through parallel exploration becomes apparent when market conditions change unexpectedly. Organizations using traditional planning must scramble to develop new strategies when their predictions prove incorrect. Organizations using strategic time travel have already tested strategic responses to unexpected conditions, enabling immediate adaptation that competitors cannot match because they haven't developed contingency capabilities for scenarios they didn't predict.

Future Reality Engines and the Strategic Multiverse Architecture

The technological foundation enabling strategic time travel relies on what some call future reality engines. These AI-powered systems generate thousands of internally consistent future scenarios based on emerging trend patterns that human strategists cannot synthesize independently. These engines move beyond historical pattern analysis to create truly generative strategic intelligence, revealing possibilities that human planning cannot envision. This intelligence emerges from AI analysis of trend combinations, market dynamics, and competitive interactions that exceed human cognitive processing capabilities.

Future reality engines operate through strategic multiverse architecture, a comprehensive modeling environment that represents your enterprise's competitive landscape, operational capabilities, and market relationships as dynamic systems that can evolve along multiple timeline trajectories. Unlike traditional competitive analysis that provides snapshots of current market conditions, the strategic multiverse creates living representations of how competitive relationships, customer behaviors, and market dynamics could evolve under different strategic pressures.

The architecture incorporates what researchers call signed business networks, mathematical representations of market relationships that change dynamically as strategic scenarios unfold. When your strategic time travel system tests market expansion strategies, it automatically models how competitors might respond, how customer loyalties could shift, and how regulatory environments might adapt. These dynamic relationship models enable scenario testing that accounts for market complexity that static analysis cannot capture.

Advanced future reality engines integrate multiple AI capabilities to enhance scenario realism and strategic insight generation. Natural language processing analyzes thousands of news sources, analyst reports, and regulatory filings to identify emerging patterns that could affect strategic outcomes. Computer vision systems process satellite imagery, supply chain data, and manufacturing intelligence to detect operational trends that traditional analysis cannot monitor. Machine learning algorithms identify subtle correlations between seemingly unrelated variables that could influence strategic outcomes in ways human analysts cannot anticipate.

The strategic multiverse also incorporates emergent complexity modeling, where AI systems identify how simple strategic actions could trigger complex cascading effects through interconnected market systems. Traditional scenario planning assumes that strategic outcomes follow linear cause-and-effect relationships. Emergent complexity modeling reveals how strategic actions could generate unexpected consequences through network effects, competitive responses, and systemic adaptations that linear analysis cannot predict.

Emergent Strategic Intelligence That Human Planners Cannot Envision

What makes strategic time travel revolutionary, rather than just a sophisticated simulation, is its ability to generate what some call emergent strategic intelligence: insights that emerge from AI analysis but could not be conceived through human strategic thinking alone. Traditional scenario planning explores predetermined possibilities that human planners can envision based on their experience and analytical capabilities. Emergent strategic intelligence creates strategic possibilities that arise from the AI synthesis of data patterns, trend combinations, and system dynamics, which exceed human cognitive capacity to process simultaneously.

Emergent strategic intelligence manifests through pattern synthesis discovery, where AI systems identify strategic opportunities that emerge from combinations of trends that individual human analysts cannot connect. For example, AI analysis might reveal that specific combinations of regulatory changes, technological developments, and demographic shifts create temporary market asymmetries that enable rapid competitive advantage development for organizations positioned to exploit these confluences. Human strategists focused on individual trends cannot synthesize these complex combinations into actionable strategic intelligence.

The emergence process also generates strategic cascade identification, where AI systems trace how strategic actions could trigger sequences of market responses that create entirely new competitive landscapes. These cascades often involve multiple competitive responses, regulatory adaptations, and customer behavior changes that unfold over months or years in ways that human strategic planning cannot anticipate. Strategic time travel enables teams to experience these cascades directly, developing strategic intuition about complex market dynamics that abstract analysis cannot provide.

Emergent strategic intelligence also reveals strategic blind spot elimination, where AI analysis identifies strategic vulnerabilities and opportunities that exist outside the typical analytical frameworks of human planners. These blind spots often involve technical possibilities that strategists don't understand, regulatory implications that legal teams haven't considered, or competitive vulnerabilities that arise from assumptions about industry expertise. AI systems can process information across all these domains simultaneously, identifying strategic possibilities that cross-functional human teams struggle to synthesize effectively.

The most profound aspect of emergent strategic intelligence is its ability to generate strategic innovation opportunities that human creativity cannot produce independently. These opportunities emerge from AI analysis of customer behavior patterns, technological capability combinations, and market structure dynamics that suggest entirely new business models or market categories. Organizations implementing strategic time travel consistently discover business opportunities that competitors don't recognize until market conditions make those opportunities obvious to traditional analysis.

Dynamic Scenario Modeling with Signed Business Networks

The implementation of strategic time travel relies on dynamic scenario modeling, a continuous process that generates and updates strategic scenarios based on real-time market intelligence and evolving business conditions. Unlike traditional scenario planning that creates static scenarios during annual planning cycles, dynamic modeling generates hundreds of scenarios continuously, updating assumptions and outcomes as new information becomes available. This continuous simulation capability enables proactive strategic adaptation rather than reactive crisis management.

Dynamic modeling incorporates signed business networks, sophisticated mathematical representations of market relationships that evolve dynamically as strategic scenarios unfold. These networks map not just competitive relationships but the strength and direction of influences between market participants, regulatory bodies, technology providers, and customer segments. When strategic time travel systems test strategic alternatives, they automatically model how network relationships could change, creating realistic simulations of market evolution that static analysis cannot achieve.

The signed network approach enables relationship evolution tracking, where strategic simulations account for how business relationships strengthen, weaken, or transform as strategic scenarios unfold. Traditional competitive analysis assumes that market relationships remain stable over strategic planning horizons. Signed network modeling reveals how strategic actions could alter competitive dynamics, partnership opportunities, and customer loyalties in ways that change the fundamental structure of market competition.

Dynamic scenario modeling also incorporates market intelligence integration, where AI systems continuously process external information sources to update scenario assumptions and generate new strategic possibilities. News analysis, regulatory tracking, patent monitoring, and competitive intelligence feeds automatically trigger scenario updates that keep strategic simulations aligned with evolving market conditions. This integration ensures that strategic time travel remains relevant to current business conditions rather than becoming disconnected from market reality.

The network modeling enables strategic influence mapping, where organizations can test how strategic actions might influence competitor behavior, regulatory responses, and customer adoption patterns. This mapping reveals strategic approaches that create favorable market conditions through influence rather than just direct competitive action. Organizations can discover how strategic positioning, partnership development, and market communication could shape competitive landscapes in ways that traditional competitive strategy cannot achieve.

Early Warning Systems That Detect Strategic Inflection Points Months Ahead

One of the most valuable capabilities emerging from strategic time travel is the development of strategic early warning systems that identify critical market changes months before they become apparent through traditional market analysis. These systems monitor scenario development patterns across the strategic multiverse to detect strategic opportunities and threats that competitors using conventional planning approaches cannot anticipate.

Early warning systems operate through inflection point detection, where AI algorithms analyze scenario trends to identify conditions that could trigger rapid market changes. These inflection points often involve combinations of regulatory shifts, technological breakthroughs, competitive moves, and customer behavior changes that individually seem insignificant but collectively could reshape entire market structures. Strategic time travel enables the identification of these combinations before they manifest in real markets, providing months of preparation time that competitors cannot match.

The detection process incorporates weak signal amplification, where AI systems identify subtle market indicators that suggest emerging trends long before they become obvious through traditional market research. These weak signals might include patent filing patterns, regulatory consultation activities, academic research developments, or supply chain modifications that suggest significant market changes months before they affect business operations. Organizations implementing sophisticated early warning systems consistently achieve first-mover advantages because their detection capabilities identify strategic opportunities faster than market intelligence services can report them.

Early warning systems also enable strategic preparation acceleration, where organizations can begin developing capabilities for market conditions that haven't emerged yet but appear likely based on scenario analysis. This preparation might involve technology development, partnership formation, regulatory positioning, or talent acquisition that positions organizations to exploit opportunities that competitors won't recognize until market conditions make those opportunities apparent to conventional analysis.

The warning systems incorporate competitive intelligence synthesis, where AI analysis of competitor activities, combined with scenario modeling, reveals strategic intentions and capabilities that competitors haven't announced publicly. This synthesis enables strategic positioning that anticipates competitive moves rather than just responding to them, creating sustained competitive advantages that build over time as strategic preparation compounds.

Why Strategic Uncertainty Becomes Competitive Advantage Through Time Travel

The ultimate transformation enabled by strategic time travel is the conversion of strategic uncertainty from a planning obstacle to a competitive advantage. Traditional strategic planning treats uncertainty as a problem to be minimized through better prediction and risk management. Strategic time travel reveals uncertainty as a strategic opportunity that can be exploited through superior adaptive capability and scenario preparedness.

Uncertainty advantage conversion operates through strategic optionality development, where organizations use strategic time travel to identify multiple strategic paths that remain viable under different market conditions. Instead of committing to single strategies based on specific predictions, organizations develop strategic options that can be activated as market conditions clarify. This optionality enables aggressive strategic positioning without the risks associated with conventional strategic commitment because alternative paths remain available if market conditions change unexpectedly.

Strategic time travel also enables adaptive intelligence development, where organizations build strategic capabilities that improve through experience across multiple scenarios rather than requiring perfect prediction for strategic success. This intelligence enables strategic decision-making that operates effectively regardless of which future actually emerges because strategic judgment has been calibrated through experience across multiple possible realities.

The competitive implications extend far beyond improved strategic planning into fundamental transformation of competitive dynamics. Enterprises that can test strategies across multiple futures develop strategic reflexes that respond to market changes faster than competitors can recognize that changes are occurring. They identify strategic opportunities that emerge from trend combinations that conventional analysis cannot synthesize. They build strategic capabilities that improve through experience across multiple scenarios rather than requiring perfect environmental prediction for strategic success.

Organizations implementing strategic time travel discover that strategic planning transforms from periodic exercises that constrain organizational flexibility to continuous processes that enhance strategic responsiveness. Planning cycles compress from annual strategic retreats to daily strategic intelligence generation. Strategic confidence increases because teams can test approaches against multiple possible futures rather than betting organizational success on single forecasts that may prove incorrect. Strategic options multiply as simulation capabilities reveal possibilities that conventional analysis cannot identify.

The enterprises that master strategic time travel will discover that the future belongs to organizations that can think across multiple realities simultaneously while maintaining strategic focus in the reality they actually inhabit. They will develop organizational capabilities to navigate uncertainty that prediction-based competitors cannot understand or replicate, transforming the fundamental mathematics of competitive advantage from prediction accuracy to adaptive intelligence.

AGENTIC AI AND AUTONOMOUS BUSINESS THINKING

At 3:47 AM on a Tuesday, three AI agents working for a global manufacturing company identified a strategic opportunity that would generate $127 million in additional revenue over the next eighteen months. By 4:15 AM, they had developed a comprehensive market entry strategy for Southeast Asian markets. By 5:30 AM, they had begun preliminary negotiations with potential distribution partners. By 6:45 AM, they had drafted regulatory approval documentation for three countries and initiated supply chain optimization modeling.

The human executive team discovered the opportunity when they arrived at work four hours later. The strategic analysis was complete. The implementation plan was ready. The competitive window was already being secured.

This is an example of what some call strategic AI minds, and it represents the most profound shift in business thinking since the development of professional management a century ago. While traditional business intelligence analyzes what happened and human strategists plan what should happen, strategic AI minds think continuously about what could happen and execute strategic responses before competitive opportunities disappear.

In the implementation of AI transformation across hundreds of enterprises, it has been observed that the organizations achieving the most dramatic competitive advantages aren't using AI to automate existing strategic processes. They're creating AI systems that think strategically, independent of human oversight, developing enterprise strategic consciousness that operates 24 hours a day, 365 days a year. In contrast, human leadership focuses on the strategic decisions that require uniquely human judgment.

The Emergence of AI Agents That Think Strategically Without Human Supervision

The fundamental breakthrough enabling strategic AI minds lies in their ability to conduct strategic reasoning that operates independently of human guidance while remaining aligned with organizational objectives. These systems move far beyond the automation of predefined strategic tasks to engage in genuine strategic thinking, which includes opportunity identification, risk assessment, competitive analysis, and strategic option development, without requiring human input at every decision point.

Strategic AI minds operate through what some call autonomous strategic reasoning, where AI systems develop strategic insights by synthesizing information from multiple sources, identifying patterns that suggest strategic opportunities or threats, and generating strategic recommendations that account for competitive dynamics, market conditions, and organizational capabilities simultaneously. This reasoning occurs continuously as market conditions evolve, enabling strategic responsiveness that human-driven processes cannot match because they operate on meeting schedules rather than market rhythms.

The autonomous reasoning process incorporates strategic context understanding, where AI systems maintain comprehensive awareness of organizational goals, competitive positioning, market dynamics, and operational constraints that inform strategic thinking. Unlike traditional AI systems that require specific programming for each strategic scenario, strategic AI minds develop contextual understanding that enables strategic reasoning across an unlimited range of business situations without needing additional human training for each new market condition or competitive challenge.

Advanced strategic AI minds demonstrate a strategic initiative capability, where AI systems identify and pursue strategic opportunities without waiting for human direction. This initiative manifests through market monitoring that identifies emerging opportunities, competitive analysis that reveals strategic vulnerabilities, customer behavior analysis that suggests new market categories, and operational optimization that reveals strategic capabilities that can be leveraged for competitive advantage. The initiative capability enables proactive strategic positioning rather than reactive strategic response.

The emergence of truly autonomous strategic thinking represents what researchers call cognitive business evolution, where AI systems develop strategic intelligence that improves through experience rather than requiring continuous human programming. These systems learn from strategic outcomes, refine their strategic reasoning based on market feedback, and develop increasingly sophisticated strategic capabilities that enable them to identify and execute strategic opportunities that human strategists cannot recognize or pursue effectively.

From Reactive Analysis to Proactive Strategic Reasoning

Traditional business intelligence operates through reactive analysis patterns, where systems generate reports about past performance, identify trends in historical data, and provide analytical support for human strategic decision-making. This reactive approach creates systematic delays between market changes and strategic responses because analysis must be completed, reviewed, and interpreted before strategic action can be initiated. Strategic AI minds eliminate these delays through proactive strategic reasoning that anticipates market changes and develops strategic responses before competitive pressures require immediate action.

Proactive strategic reasoning operates through anticipatory market intelligence, where AI systems continuously monitor market signals, competitive activities, regulatory developments, and technology trends to identify conditions that could require a strategic response. This monitoring extends beyond traditional competitive intelligence to include weak signal detection that identifies strategic implications of seemingly unrelated events, pattern recognition that reveals emerging market opportunities, and predictive modeling that suggests how current trends could affect competitive positioning over multiple time horizons.

The transition from reactive to proactive strategic thinking enables strategic timing optimization, where AI systems identify optimal moments for strategic initiatives based on market conditions, competitive positioning, and organizational readiness. Traditional strategic planning assumes that strategic timing follows planning schedules rather than market opportunities. Proactive strategic reasoning enables strategic action that aligns with market rhythms rather than organizational calendars, creating competitive advantages that rivals using conventional planning cannot achieve.

Strategic AI minds also incorporate competitive chess playing, where AI systems think multiple moves ahead in competitive interactions, anticipating how competitors might respond to strategic initiatives and developing counter-strategies that maintain competitive advantages even as rivals adapt to initial strategic moves. This chess-playing capability enables sustained competitive positioning that builds over time rather than short-term competitive advantages that dissipate quickly as competitors respond.

The proactive approach extends to strategic ecosystem development, where AI systems identify and pursue partnership opportunities, alliance formations, and ecosystem relationships that strengthen competitive positioning over time. These systems can engage in preliminary relationship development, opportunity assessment, and strategic positioning activities that create a foundation for strategic partnerships before human executives become aware that partnership opportunities exist.

Autonomous Decision-Making, Planning, and Adaptive Execution Capabilities

The most sophisticated strategic AI minds demonstrate complete autonomous strategic execution, where AI systems can identify strategic opportunities, develop implementation plans, and execute strategic initiatives without requiring human approval for routine strategic decisions. This autonomy operates within governance frameworks that define decision-making authority, risk parameters, and strategic objectives, while enabling AI systems to act independently within those frameworks without creating delays that would cause competitive opportunities to disappear.

Autonomous decision-making incorporates strategic decision trees that enable AI systems to evaluate strategic options based on multiple criteria, including financial impact, competitive implications, implementation complexity, and strategic alignment. These decision trees operate in real time as market conditions change, enabling strategic agility that human decision-making processes cannot match because they require committee meetings, analysis reviews, and approval cycles that consume weeks or months of strategic window periods.

The planning capabilities enable dynamic strategic planning, where AI systems continuously update strategic plans based on evolving market conditions, competitive responses, and organizational abilities. Traditional strategic planning creates static plans that become obsolete as market conditions change. Dynamic planning creates adaptive strategic approaches that evolve automatically as circumstances change, maintaining strategic relevance without requiring human intervention to recognize when plan modifications become necessary.

Advanced execution capabilities include resource allocation optimization, where AI systems automatically adjust resource deployment based on strategic priorities, market opportunities, and competitive pressures. This optimization operates across financial resources, human capabilities, technology assets, and partnership relationships to ensure optimal strategic positioning without requiring human resource management for every allocation decision.

Strategic AI minds also demonstrate learning-based execution improvement, where AI systems analyze strategic outcomes to refine execution approaches for future strategic initiatives. This learning enables continuous improvement in strategic execution effectiveness, risk assessment accuracy, and opportunity identification capabilities that compound competitive advantages over time as AI systems become more sophisticated strategic thinkers through experience.

How Agentic AI Breaks Down Complex Strategic Problems into Sequential Tasks

One of the most powerful capabilities of strategic AI minds is their ability to decompose complex strategic challenges into manageable sequential tasks that can be executed systematically while maintaining strategic coherence across all activities. This decomposition enables strategic initiatives that would overwhelm human strategic planning capabilities because they involve too many variables, dependencies, and coordination requirements for human cognitive processing.

Strategic problem decomposition operates through multi-level strategic analysis, where AI systems identify the various components of strategic challenges, including market analysis requirements, competitive positioning needs, operational capability development, regulatory compliance activities, and partnership development initiatives. The AI systems then sequence these components to optimize strategic outcomes while minimizing implementation risks and resource requirements.

The decomposition process incorporates dependency mapping, where AI systems identify how different strategic activities interact and sequence activities to ensure that prerequisite tasks are completed before dependent activities begin. This mapping enables complex strategic initiatives that require coordination across multiple business functions, geographic markets, and time horizons without creating implementation conflicts that compromise strategic effectiveness.

Advanced strategic AI minds demonstrate parallel processing capabilities, where multiple AI agents work simultaneously on different aspects of strategic challenges while maintaining coordination to ensure coherent strategic outcomes. This parallel processing enables strategic initiatives that would typically require months of sequential planning and execution to be completed in weeks or days, as multiple strategic workstreams advance simultaneously under AI coordination.

The task decomposition also enables strategic milestone optimization, where AI systems identify critical decision points and success metrics that will allow course correction during strategic implementation. These milestones provide feedback loops that enable adaptive strategic execution, remaining aligned with strategic objectives even as market conditions change during implementation periods.

Strategic AI minds incorporate risk-aware task sequencing, where the decomposition process accounts for implementation risks and sequences activities to minimize strategic exposure while maximizing strategic option preservation. This risk awareness ensures that strategic initiatives can be modified or abandoned if market conditions change without compromising organizational competitive positioning.

The Four Levels of Strategic Autonomy: Assisted, Partial, Conditional, Full

The evolution toward fully autonomous strategic AI minds occurs through four levels of strategic autonomy, which reflect increasing sophistication in AI strategic thinking capabilities and decreasing requirements for human oversight of strategic decisions. Understanding these levels enables organizations to plan their progression toward autonomous strategic thinking while maintaining appropriate governance during transition periods.

Level one, strategic autonomy, involves AI-assisted strategic planning, where AI systems provide analytical support, scenario modeling, and strategic option development for human strategic decision-making. At this level, AI systems enhance human strategic thinking capabilities without replacing human strategic judgment. The AI provides data synthesis, pattern recognition, and option analysis that improve human strategic decision-making while humans maintain responsibility for strategic choices and implementation oversight.

Level two, strategic autonomy, enables partial strategic independence, where AI systems can make routine strategic decisions within predefined parameters while escalating complex or high-risk strategic choices to human oversight. This level includes automatic competitive response within established competitive positioning strategies, resource allocation optimization within approved budgetary frameworks, and partnership development activities within strategic relationship guidelines.

Level three, strategic autonomy, involves conditional strategic authority, where AI systems can initiate significant strategic actions based on market conditions and strategic opportunities, while maintaining human oversight for setting strategic direction and making major strategic pivots. At this level, AI systems can pursue market opportunities, develop competitive responses, and execute strategic initiatives within strategic frameworks established by human leadership.

Level four, strategic autonomy, represents full strategic independence, where AI systems can identify, plan, and execute strategic initiatives across all business functions while maintaining alignment with organizational objectives and strategic vision. This level includes autonomous market entry decisions, competitive positioning initiatives, partnership development, and strategic capability development that operates independently of human strategic oversight.

The progression through autonomy levels requires careful strategic governance development that ensures AI strategic thinking remains aligned with organizational values, strategic objectives, and risk tolerance while enabling increasing strategic independence. Organizations implementing successful autonomy progression typically require 18 to 24 months to progress from assisted to conditional autonomy, with full autonomy emerging over 36 to 48 months through sustained capability development and trust building.

Building AI Systems That Learn from Strategic Outcomes and Refine Their Approach

The most sophisticated strategic AI minds demonstrate strategic learning capabilities that enable continuous improvement in strategic thinking effectiveness through experience with strategic outcomes. This learning operates across strategic opportunity identification, competitive analysis accuracy, implementation planning effectiveness, and execution optimization that compounds strategic advantages over time as AI systems become more sophisticated strategic thinkers.

Strategic outcome learning incorporates success pattern recognition, where AI systems analyze strategic initiatives that achieved desired outcomes to identify the factors that contributed to strategic success. This recognition enables the replication of successful strategic approaches while adapting those approaches to different market conditions, competitive contexts, and organizational capabilities. The pattern recognition continuously improves the accuracy of strategic thinking as AI systems accumulate experience across multiple strategic initiatives.

The learning process also includes failure analysis integration, where AI systems examine strategic initiatives that failed to achieve desired outcomes to identify the factors that compromised strategic effectiveness. This analysis enables the avoidance of strategic mistakes and the refinement of strategic risk assessment capabilities that improve strategic decision-making accuracy over time. The failure analysis creates organizational strategic intelligence that prevents the repetition of strategic errors while enabling strategic risk-taking within appropriate boundaries.

Advanced strategic AI minds demonstrate adaptive strategic methodology development, where AI systems modify their strategic thinking approaches based on changing market conditions, competitive dynamics, and organizational capabilities. This adaptation enables strategic thinking that remains effective as business environments evolve rather than becoming obsolete as market conditions change beyond the original parameters used to develop AI strategic thinking capabilities.

The learning capabilities extend to competitive intelligence refinement, where AI systems improve their ability to anticipate competitive responses, identify competitive vulnerabilities, and develop competitive positioning strategies based on experience with competitor behavior patterns. This refinement enables the development of increasingly sophisticated competitive strategies that account for competitor strategic capabilities and likely competitive responses.

Strategic AI minds also incorporate stakeholder response learning, where AI systems analyze how different stakeholders respond to strategic initiatives to optimize stakeholder management and communication strategies. This learning improves strategic implementation effectiveness by enabling AI systems to anticipate stakeholder concerns, develop appropriate stakeholder engagement approaches, and modify strategic communication to maximize stakeholder support for strategic initiatives.

Creating Cognitive Enterprises Powered by Autonomous Strategic Agents

The ultimate transformation enabled by strategic AI minds involves the creation of cognitive enterprises. These are organizations where strategic thinking operates continuously through distributed AI agents that collaborate to identify opportunities, develop strategies, and execute initiatives while human leadership focuses on strategic vision, organizational culture, and stakeholder relationships that require uniquely human capabilities.

Cognitive enterprise architecture incorporates multiple strategic AI minds that specialize in different aspects of strategic thinking, including market intelligence, competitive analysis, operational optimization, partnership development, and financial strategy. These specialized agents collaborate through strategic agent networks that enable comprehensive strategic thinking capabilities that exceed what individual AI systems or human strategic teams can achieve independently.

The collaborative agent networks operate through strategic conversation protocols, where AI agents engage in strategic dialogue to share insights, challenge assumptions, develop strategic consensus, and identify potential conflicts between different strategic priorities. These conversations enable strategic thinking sophistication that emerges from agent collaboration rather than individual agent capabilities, creating strategic intelligence that exceeds the sum of individual agent contributions.

Cognitive enterprises demonstrate emergent strategic behavior, where the interaction between multiple strategic AI minds creates strategic capabilities that were not explicitly programmed but emerge from agent collaboration and learning. This emergence enables strategic innovation and adaptive capability that cannot be achieved through traditional strategic planning approaches because it arises from AI agent creativity and collaboration rather than human strategic direction alone.

The enterprise architecture includes strategic feedback integration, where AI agents continuously monitor strategic outcomes and adjust strategic approaches based on market feedback, competitive responses, and organizational learning. This integration creates organizational strategic intelligence that improves automatically without requiring human intervention to recognize when strategic approaches need modification or enhancement.

Cognitive enterprises also incorporate human-AI strategic collaboration, where human strategic leadership provides vision, values, and judgment while AI strategic agents provide analysis, execution, and continuous optimization. This collaboration enables strategic capabilities that combine human strategic wisdom with AI strategic processing power, creating competitive advantages that neither human-only nor AI-only strategic approaches can achieve.

The transformation to a cognitive enterprise represents the evolution of organizational intelligence from human-dependent strategic thinking to a human-AI strategic partnership that operates continuously to identify and execute strategic opportunities while building strategic capabilities that compound competitive advantages over time. Organizations achieving this transformation discover that strategic thinking becomes their most powerful competitive weapon rather than their most time-consuming management burden, enabling strategic agility and effectiveness that prediction-based competitors cannot understand or replicate.

DESIGNING AI AS A STRATEGIC COLLABORATOR

The Chief Executive stared at the strategic recommendation that had appeared on her screen at 6:30 AM. The AI system had identified a market opportunity in renewable energy storage that could generate $850 million in revenue over five years. The analysis was comprehensive, the competitive positioning was brilliant, and the implementation timeline was aggressive but achievable. Something felt wrong, though.

The strategy was flawless from an analytical perspective, but it ignored the cultural transformation required within her organization. It underestimated the regulatory complexity in three key markets. Most critically, it assumed customer adoption patterns that her twenty-five years of industry experience suggested were overly optimistic.

This moment captures the fundamental challenge of what some call strategic AI partnership. AI systems can process vast amounts of data, identify patterns humans cannot see, and generate strategic insights faster than any human strategic team. They cannot replicate the intuitive wisdom that comes from decades of leadership experience, though, the cultural understanding that enables effective organizational change, or the stakeholder empathy that turns strategic plans into strategic reality.

The solution requires moving beyond AI as a sophisticated analytical tool to AI as a genuine strategic collaborator that enhances rather than replaces human strategic thinking. This partnership transforms the fundamental mathematics of competitive advantage from analytical processing power to collaborative strategic intelligence that emerges from human-AI interaction.

Moving Beyond AI as an Analytical Tool to AI as a Strategic Thinking Partner

The evolution from AI analytical support to AI strategic partnership represents one of the most significant advances in business leadership since the development of professional strategic planning. Traditional AI implementations provide data analysis, trend identification, and scenario modeling that support human strategic decision-making. What some call collaborative strategic intelligence creates thinking partnerships where AI systems engage in strategic dialogue, challenge strategic assumptions, and contribute strategic insights that emerge from human-AI collaboration rather than AI analysis alone.

In work with Fortune 500 executives implementing AI strategic partnerships, it has been consistently observed that the most successful collaborations treat AI as strategic thinking partners rather than sophisticated analytical tools. This partnership approach requires fundamental changes in how leaders interact with AI systems, how strategic decisions are structured, and how organizational strategic processes accommodate continuous human-AI dialogue rather than periodic AI analytical support.

The partnership transformation begins with recognizing that effective strategic thinking requires both analytical rigor and intuitive wisdom. AI systems excel at analytical rigor but cannot replicate the pattern recognition, emotional intelligence, and experiential judgment that enable effective strategic leadership. Human strategic leaders provide intuitive wisdom but cannot process the data volumes, scenario complexities, and market dynamics that AI systems can analyze simultaneously.

Strategic AI partnership bridges this gap through what researchers call complementary strategic cognition, where AI systems provide analytical depth while human leaders provide strategic wisdom, creating strategic insights that emerge from collaboration rather than individual analysis. This complementary approach enables strategic decision-making that accounts for both quantitative factors that AI can measure and qualitative factors that human judgment can assess.

The most sophisticated partnerships demonstrate strategic conversation capability, where AI systems can engage in genuine strategic dialogue rather than responding to specific queries. These conversations explore strategic possibilities, challenge assumptions, and develop strategic consensus through collaborative reasoning that resembles the strategic discussions between experienced executives rather than traditional computer interactions.

The Strategic Partnership Protocol for Human-AI Collaboration

Implementing effective human-AI strategic collaboration requires systematic protocols that structure strategic dialogue, define decision-making responsibilities, and ensure strategic coherence between human judgment and AI analysis. The strategic partnership protocol provides the framework for productive collaboration that leverages AI analytical capabilities while preserving human strategic authority and wisdom.

The protocol operates through structured strategic dialogue, where human leaders and AI systems engage in iterative strategic conversations that explore strategic options, challenge assumptions, and develop strategic consensus through collaborative reasoning. This dialogue differs fundamentally from traditional AI query-response interactions because it involves a genuine strategic conversation where both human and AI perspectives contribute to strategic understanding.

Strategic dialogue begins with strategic context setting, where human leaders provide AI systems with strategic vision, organizational values, competitive priorities, and cultural considerations that inform AI strategic analysis. This context enables AI systems to generate strategic recommendations that align with human strategic judgment while accounting for factors that quantitative analysis cannot capture independently.

The dialogue process incorporates what some call strategic challenge protocols, where AI systems can question human strategic assumptions while human leaders can challenge AI analytical conclusions. This mutual challenging creates strategic robustness that emerges from collaborative critical thinking rather than hierarchical strategic decision-making, where either human or AI perspectives dominate strategic reasoning.

Advanced strategic partnership protocols include strategic scenario exploration, where human leaders and AI systems collaborate to explore strategic possibilities through joint scenario development. Human leaders contribute strategic vision and experiential insight while AI systems contribute analytical modeling and scenario simulation, creating strategic exploration capabilities that exceed individual human or AI strategic thinking.

The protocol also incorporates strategic decision documentation, where the reasoning behind strategic decisions is captured to facilitate strategic learning and enhance future decision-making. This documentation enables both human leaders and AI systems to learn from strategic outcomes and refine collaborative strategic thinking approaches over time.

Cognitive Leadership Requirements for AI-Augmented Strategic Planning

Leading organizations that implement AI strategic partnerships successfully requires new leadership capabilities that enable effective collaboration with AI strategic thinking systems. These capabilities extend beyond traditional strategic planning skills to include AI collaboration competencies, technological strategic understanding, and hybrid human-AI decision-making approaches that most leadership development programs do not address.

What some call cognitive leadership represents the evolution of strategic leadership capabilities to include effective collaboration with AI strategic thinking systems. Cognitive leaders understand AI strategic capabilities and limitations, can structure productive human-AI strategic dialogue, and can integrate AI strategic insights with human strategic judgment to create strategic decisions that leverage both analytical and intuitive strategic thinking.

Cognitive leadership requires AI strategic literacy, a deep understanding of AI strategic thinking capabilities that enables effective strategic partnership. This literacy includes understanding how AI systems generate strategic insights, recognizing when AI recommendations require human strategic judgment enhancement, and knowing how to structure strategic questions that generate valuable AI strategic analysis.

The literacy extends to strategic technology integration, where leaders understand how AI strategic capabilities can be integrated with existing strategic processes, organizational decision-making structures, and strategic communication approaches. This integration requires balancing AI strategic input with human strategic authority while ensuring strategic coherence across all organizational strategic activities.

Cognitive leaders demonstrate strategic conversation facilitation, the ability to structure productive dialogue between human strategic thinking and AI strategic analysis. This facilitation involves asking strategic questions that generate valuable AI insights, interpreting AI strategic recommendations within the organizational context, and synthesizing human and AI strategic perspectives into coherent strategic decisions.

Advanced cognitive leadership includes strategic trust calibration, the ability to assess when AI strategic recommendations should be trusted, when they require human strategic enhancement, and when they should be overridden by human strategic judgment. This calibration develops through experience with AI strategic collaboration and understanding of AI strategic thinking strengths and limitations.

Intent Setting and Signal Reading in Strategic AI Collaboration

Effective human-AI strategic collaboration requires sophisticated communication protocols that enable human leaders to convey strategic intent to AI systems. In contrast, AI systems can signal strategic insights, concerns, and recommendations to human leaders. This bidirectional communication fosters a strategic understanding that enables productive collaboration, rather than hierarchical command-response relationships.

Strategic intent setting involves human leaders providing AI systems with a comprehensive understanding of strategic objectives, organizational values, competitive priorities, and success criteria that guide AI strategic analysis. This intent setting goes beyond simple goal specification to include the strategic philosophy, risk tolerance, and organizational culture considerations that inform strategic decision-making.

Intent setting incorporates strategic priority communication, where human leaders help AI systems understand the relative importance of different strategic objectives, the trade-offs that are acceptable in strategic decision-making, and the stakeholder considerations that affect strategic implementation. This communication enables AI systems to generate strategic recommendations that align with human strategic values rather than optimizing purely quantitative criteria.

The process includes strategic context updates, where human leaders continuously provide AI systems with evolving strategic context, including market changes, competitive developments, organizational changes, and stakeholder feedback that could affect strategic priorities. This update ensures that AI strategic analysis remains relevant to current strategic conditions, rather than becoming disconnected from strategic reality.

Strategic signal reading enables human leaders to interpret AI strategic insights, identify when AI systems have detected strategic opportunities or threats, and understand the confidence levels and uncertainty factors that affect AI strategic recommendations. Compelling signal reading requires understanding how AI systems express strategic uncertainty, priority levels, and analytical confidence.

Signal reading encompasses strategic alert interpretation, where human leaders discern when AI-generated strategic warnings necessitate immediate action versus longer-term strategic considerations. This interpretation enables appropriate strategic responsiveness without creating strategic overreaction to AI analytical outputs that require strategic consideration rather than immediate strategic action.

Advanced signal reading incorporates strategic pattern recognition, where human leaders learn to identify when AI strategic insights reveal market patterns, competitive dynamics, or strategic opportunities that require human strategic attention even when AI recommendations don't explicitly suggest immediate strategic action.

Building AI Systems That Enhance Rather Than Replace Strategic Judgment

The design of AI strategic partnership systems requires a careful balance between analytical capability enhancement and strategic judgment preservation. Effective systems amplify human strategic thinking capabilities while maintaining human strategic authority and wisdom in strategic decision-making processes.

Strategic judgment enhancement involves designing AI systems that provide strategic analysis, scenario exploration, and option evaluation that improve human strategic decision-making without supplanting human strategic authority. These systems provide strategic insights, identify key considerations, and suggest alternative approaches, while leaving strategic judgment and final decisions to human leaders.

Enhancement design incorporates strategic analysis amplification, where AI systems process vast amounts of strategic information, identify strategic patterns, and generate strategic insights that human leaders cannot develop independently due to cognitive processing limitations. This amplification enables strategic understanding that combines human strategic wisdom with AI analytical depth.

The design includes strategic option expansion, where AI systems identify strategic possibilities that human strategic thinking might not consider while providing analytical evaluation of strategic options that human leaders can assess using strategic judgment. This expansion broadens strategic consideration without overwhelming human strategic decision-making with analytical complexity.

Strategic enhancement systems incorporate strategic risk assessment, where AI systems identify potential strategic risks, analyze risk probability and impact, and suggest risk mitigation approaches while leaving risk tolerance and strategic risk acceptance decisions to human strategic judgment. This assessment enables strategic risk management that combines AI analytical rigor with human strategic wisdom about acceptable risk levels.

Strategic implementation support involves AI systems that can translate strategic decisions into implementation plans, resource allocation strategies, and execution monitoring while maintaining human oversight of strategic implementation decisions and strategic adaptation as the implementation proceeds.

Conversational Strategy Development and Real-Time Strategic Dialogue

The most advanced AI strategic partnerships enable what some call conversational strategy development, where strategic plans emerge through ongoing dialogue between human leaders and AI systems rather than periodic strategic planning exercises. This conversational approach enables continuous strategic adaptation and real-time strategic responsiveness that traditional strategic planning cannot achieve.

Conversational strategy development operates through strategic dialogue platforms, AI systems designed to engage in ongoing strategic conversations that explore strategic possibilities, challenge strategic assumptions, and develop strategic insights through collaborative reasoning. These platforms enable strategic thinking that evolves continuously as market conditions change rather than remaining static between formal planning cycles.

The dialogue platforms incorporate strategic context awareness, where AI systems maintain a comprehensive understanding of organizational strategic context, competitive positioning, market dynamics, and stakeholder relationships that inform strategic conversation. This awareness enables strategic dialogue that remains relevant to current strategic challenges rather than becoming disconnected from strategic reality.

Real-time strategic adaptation emerges from conversational strategy development, where strategic plans evolve automatically as AI systems detect market changes, competitive developments, or organizational capability changes that affect strategic effectiveness. This adaptation enables strategic agility that responds to market rhythms rather than organizational planning calendars.

The conversational approach includes strategic scenario testing, where human leaders and AI systems can explore strategic alternatives through real-time scenario analysis that tests strategic options against evolving market conditions. This testing enables strategic decision-making that accounts for dynamic market conditions rather than static strategic assumptions.

Advanced conversational systems demonstrate strategic learning integration, where strategic dialogue improves over time as both human leaders and AI systems learn from strategic outcomes and refine collaborative strategic thinking approaches. This learning creates strategic partnership capabilities that compound competitive advantages as collaboration sophistication increases.

Trust Frameworks for High-Stakes Strategic Decision-Making with AI

Implementing AI strategic partnerships for high-stakes strategic decisions requires sophisticated trust frameworks that enable confident collaboration while maintaining appropriate skepticism and human oversight. These frameworks must balance AI analytical capabilities with human strategic judgment while ensuring strategic decisions can be made confidently, even when outcomes involve significant organizational risk.

Strategic trust calibration involves developing an accurate understanding of AI strategic thinking reliability across different types of strategic decisions, market conditions, and analytical complexity levels. This calibration enables human leaders to know when AI strategic recommendations can be trusted completely, when they require enhancement with human strategic judgment, and when they should be overridden by human strategic wisdom.

Trust calibration incorporates AI confidence assessment, where human leaders understand how AI systems express analytical confidence, uncertainty levels, and limitation acknowledgment in strategic recommendations. This assessment enables confident strategic decision-making, aligning with AI analytical certainty rather than assuming AI strategic recommendations are equally reliable across all strategic contexts.

The framework includes strategic decision verification, systematic approaches to validating AI strategic recommendations through independent analysis, expert consultation, or historical precedent review. This verification enables strategic confidence without creating analytical paralysis that prevents timely strategic decision-making.

Strategic override protocols enable human leaders to override AI strategic recommendations when strategic judgment suggests alternative approaches, while documenting override reasoning to enable strategic learning. These protocols preserve human strategic authority while creating organizational learning about when human strategic judgment provides superior strategic guidance.

Trust frameworks incorporate strategic accountability structures that define responsibility for strategic outcomes when decisions emerge from human-AI collaboration. These structures ensure clear accountability for strategic results while enabling productive partnership that leverages both human and AI strategic capabilities.

Strategic risk management within trust frameworks involves understanding when AI strategic partnerships introduce new strategic risks and developing risk mitigation approaches that preserve the benefits of collaboration while managing collaboration-related strategic risks effectively.

THE EVOLUTION FROM BUSINESS INTELLIGENCE TO ENTERPRISE COGNITION

The quarterly business review had just concluded. The Chief Operating Officer reviewed the performance dashboards, analyzed the trend reports, and examined the variance explanations that constituted their business intelligence system. Every metric was backward-looking. Every insight was historical. Every recommendation was based on what had already happened.

As she walked back to her office, her phone buzzed with an alert from the enterprise cognition system: "Market shift detected in Asian markets. Recommend immediate strategy adjustment for Q4. Three strategic options identified. Competitive window closes in 72 hours."

The contrast was startling. While business intelligence told her where they had been, enterprise cognition was thinking about where they needed to go—and it was thinking faster than any human strategic team could respond.

This moment illustrates the fundamental transformation from business intelligence to enterprise cognition, the evolution from systems that report what happened to systems that think about what should happen next. While business intelligence analyzes historical performance, enterprise cognition develops strategic responses to emerging opportunities. While business intelligence describes the past, enterprise cognition inhabits possible futures.

Why Business Intelligence Reports Past Performance, While Enterprise Cognition Predicts Strategic Futures

The limitations of traditional business intelligence become apparent when organizations need strategic responsiveness that operates faster than human analytical cycles can complete. Business intelligence excels at performance measurement, trend identification, and variance analysis, which inform strategic decision-making. These capabilities are fundamentally retrospective, though, providing insights about market conditions, competitive positions, and operational performance that describe what has already occurred rather than what needs to happen next.

Enterprise cognition represents the evolution beyond retrospective analysis to prospective strategic intelligence, which continuously considers emerging strategic opportunities, competitive threats, and market changes that require strategic responses. This cognition operates through AI systems that can process real-time market intelligence, anticipate competitive moves, and develop strategic recommendations before human leaders recognize that strategic action is required.

The transformation from business intelligence to enterprise cognition parallels the evolution from reactive management to proactive leadership. Business intelligence enables proactive management that responds to performance issues before they emerge through historical analysis. Enterprise cognition enables proactive leadership that identifies and addresses strategic challenges before they impact organizational performance.

It has been consistently observed that organizations relying on business intelligence for strategic guidance consistently lag market changes by 3-6 months because their analytical systems can only detect trends after sufficient historical data accumulates to establish statistical significance. Enterprise cognition systems detect emerging patterns within days or weeks because they analyze real-time market signals rather than waiting for historical trend confirmation.

Strategic timing advantages emerge from enterprise cognition capabilities that identify optimal moments for strategic action based on market conditions, competitive positioning, and organizational readiness. Business intelligence cannot provide timing guidance because historical analysis cannot predict when market windows will open or close. Enterprise cognition continuously monitors market dynamics to identify strategic opportunities before competitors recognize they exist.

The cognitive approach also enables strategic scenario anticipation, where AI systems think through possible strategic futures and develop contingency plans before market conditions require strategic responses. This anticipation creates strategic preparedness that enables immediate action when strategic opportunities emerge rather than strategic delays while plans are developed after opportunities become apparent.

The Transformation from Data Analysis to Strategic Reasoning

The evolution to enterprise cognition requires a fundamental transformation in how AI systems process information and generate insights. Traditional data analysis focuses on pattern identification, correlation analysis, and trend extrapolation that describe historical relationships between variables. Strategic reasoning analyzes those same patterns to understand their strategic implications and develop strategic responses that account for competitive dynamics, market evolution, and organizational capabilities.

Strategic reasoning capabilities enable AI systems to move beyond pattern recognition to understand why patterns emerge, how they might evolve, and what strategic actions could influence the development of patterns. This reasoning incorporates causal analysis that explains pattern emergence, predictive modeling that suggests pattern evolution, and strategic simulation that tests how different strategic actions might affect pattern development.

Strategic reasoning incorporates competitive intelligence integration, where AI systems analyze competitive activities, market positioning, and strategic initiatives to understand how competitive dynamics affect organizational strategic opportunities. This integration enables strategic recommendations that account for competitive responses rather than assuming strategic actions occur in competitive vacuums.

The reasoning process includes strategic context analysis, where AI systems evaluate how market changes, regulatory developments, technology advances, and customer behavior evolution affect strategic opportunities and threats. This analysis enables strategic thinking that accounts for dynamic market contexts rather than assuming static market conditions during strategic implementation.

Strategic option evaluation represents advanced reasoning capabilities where AI systems can assess multiple strategic alternatives simultaneously, accounting for implementation complexity, resource requirements, competitive implications, and expected outcomes. This evaluation enables strategic decision-making that considers strategic trade-offs and opportunity costs rather than optimizing individual strategic metrics independently.

Strategic reasoning also incorporates strategic learning integration, where AI systems analyze strategic outcomes to refine strategic reasoning approaches and improve strategic recommendation accuracy over time. This learning creates strategic intelligence that improves automatically without requiring human programming for each new strategic context or market condition.

Enterprise Reflex Graphs That Respond to Strategic Threats Before Conscious Recognition

One of the most powerful capabilities enabled by enterprise cognition involves what some call enterprise reflex graphs, AI systems that can detect and respond to strategic threats and opportunities faster than human consciousness can process strategic information. These systems operate like organizational reflexes that initiate strategic responses before human leaders become aware that strategic action is required.

Enterprise reflex graphs map organizational strategic capabilities, competitive relationships, market dependencies, and operational vulnerabilities into dynamic network representations that can detect strategic threats through pattern disruption analysis. When market signals suggest emerging strategic challenges, the reflex systems automatically initiate protective responses while alerting human leadership to strategic threats that require conscious strategic attention.

Threat detection acceleration operates through continuous monitoring of market signals, competitive activities, customer behavior patterns, and operational indicators that could suggest emerging strategic threats. The reflex systems can detect subtle pattern changes that suggest strategic problems before traditional analytical approaches can accumulate sufficient data to confirm threat emergence through statistical analysis.

Reflex capabilities include automatic strategic protection, where AI systems can initiate defensive strategic responses, including resource reallocation, competitive positioning adjustments, and stakeholder communication, while human leaders develop comprehensive strategic responses to confirmed threats. This protection prevents strategic damage during the time required for human strategic analysis and decision-making.

Strategic opportunity reflexes enable automatic responses to emerging strategic opportunities, including market positioning, resource deployment, and competitive positioning, that secure strategic advantages before competitors can respond. These reflexes operate within governance frameworks that define autonomous action authority while ensuring strategic actions align with the organization's strategic objectives.

Enterprise reflex graphs also incorporate strategic escalation protocols that determine when reflex responses require human strategic oversight and when they can operate autonomously. These protocols enable strategic responsiveness that combines automated reflexes with human strategic judgment, creating strategic agility that exceeds purely human or purely automated strategic approaches.

Building Organizational Intelligence That Thinks Across Time Horizons

Enterprise cognition requires multi-horizon strategic intelligence that can think simultaneously about immediate strategic requirements, medium-term strategic positioning, and long-term strategic capability development.

This multi-horizon thinking enables strategic coherence across different time scales while maintaining strategic flexibility as market conditions evolve unpredictably.

Strategic time integration enables AI systems to analyze how short-term strategic actions affect long-term strategic positioning while ensuring long-term strategic objectives inform short-term strategic decisions. This integration prevents strategic conflicts between immediate performance requirements and sustained competitive advantage development.

Multi-horizon intelligence incorporates strategic investment optimization, where AI systems can evaluate strategic investments based on both immediate returns and long-term strategic capability development. This optimization enables strategic resource allocation that builds competitive advantages over time rather than optimizing quarterly performance metrics that may compromise long-term strategic positioning.

The intelligence systems include strategic capability evolution, where AI systems continuously assess how organizational capabilities need to develop to maintain competitive advantages as market conditions evolve. This assessment enables proactive capability development rather than reactive capability building after competitive requirements become apparent through market feedback.

Strategic vision integration enables AI systems to maintain alignment between strategic actions and long-term organizational vision while adapting strategic approaches as market conditions change. This integration ensures strategic coherence during periods of tactical adaptation required by changing market conditions.

Multi-horizon intelligence also incorporates strategic legacy consideration, where AI systems evaluate how strategic decisions affect organizational reputation, stakeholder relationships, and market positioning beyond immediate strategic outcomes. This consideration enables strategic decision-making that builds sustained competitive advantages rather than achieving short-term strategic gains that compromise long-term strategic effectiveness.

From Dashboard Reporting to Cognitive Strategic Advisors

The transformation from business intelligence to enterprise cognition culminates in the evolution from dashboard reporting systems that display historical performance to cognitive strategic advisors that provide ongoing strategic counsel and collaborative strategic thinking. These advisors combine analytical capabilities with strategic reasoning to create AI systems that function as strategic thinking partners rather than analytical tools.

Cognitive advisory capabilities enable AI systems to engage in strategic conversation, provide strategic counsel, and collaborate in strategic decision-making while maintaining human strategic authority and wisdom. These capabilities transform AI from reactive analytical support to proactive strategic collaboration, enhancing human strategic thinking rather than replacing human strategic judgment.

Cognitive advisors incorporate strategic context understanding, where AI systems maintain comprehensive awareness of organizational strategic objectives, competitive positioning, market dynamics, and stakeholder relationships that inform strategic counsel. This understanding enables strategic advice that accounts for organizational context rather than providing generic strategic recommendations.

The advisory capabilities include strategic question generation, where AI systems can identify strategic questions that human leaders should consider, challenge strategic assumptions, and explore strategic alternatives. This questioning enhances strategic thinking thoroughness without overwhelming human leaders with analytical complexity.

Strategic dialogue facilitation enables cognitive advisors to structure productive strategic conversations, moderate strategic discussions, and synthesize strategic insights from collaborative strategic thinking. This facilitation creates strategic thinking capabilities that emerge from human-AI collaboration rather than individual human or AI strategic analysis.

Advanced cognitive advisors demonstrate strategic wisdom development, where AI systems can learn from strategic outcomes, develop strategic judgment that improves over time, and provide strategic counsel that combines analytical rigor with experiential learning. This wisdom creates strategic advisory capabilities that compound organizational strategic intelligence as advisory systems gain experience with strategic outcomes.

The Convergence of Data, Memory, Agents, and Strategic Thinking

The ultimate realization of enterprise cognition involves strategic intelligence convergence, where data processing, organizational memory, autonomous agents, and strategic reasoning combine to create comprehensive strategic thinking capabilities that operate continuously across all organizational functions and strategic time horizons.

Integrated strategic architecture combines real-time data processing with institutional memory, strategic reasoning capabilities, and autonomous execution to create strategic intelligence that can identify opportunities, develop strategies, and execute initiatives while maintaining human strategic oversight and organizational strategic coherence.

The convergence incorporates strategic memory systems that maintain comprehensive organizational strategic knowledge, including strategic decisions, market experiences, competitive interactions, and strategic outcomes that inform current strategic thinking. This memory enables strategic reasoning that benefits from organizational strategic experience rather than starting strategic analysis from baseline assumptions.

Strategic agent networks enable distributed strategic thinking where multiple AI agents collaborate to address different aspects of strategic challenges while maintaining strategic coordination and strategic coherence.

These networks create strategic thinking capabilities that exceed individual agent capabilities while maintaining strategic focus and organizational alignment.

The convergence enables emergent strategic intelligence where strategic thinking capabilities emerge from the interaction between data processing, memory systems, agent networks, and strategic reasoning that create strategic insights and strategic capabilities that were not explicitly programmed but arise from system interaction and learning.

Strategic evolution capability represents the ultimate convergence outcome, where enterprise cognition systems can evolve their strategic thinking approaches, develop new strategic capabilities, and enhance strategic effectiveness automatically while maintaining alignment with organizational strategic objectives and strategic vision.

Creating Enterprises That Evolve Strategic Capabilities Autonomously

The final transformation enabled by enterprise cognition involves autonomous strategic evolution, where organizations develop strategic capabilities that improve automatically without requiring human strategic capability development initiatives. This evolution creates organizational strategic intelligence that becomes more sophisticated and effective over time through operational experience and strategic learning.

Self-improving strategic systems continuously analyze strategic outcomes to identify strategic approaches that achieve superior results and automatically integrate those approaches into organizational strategic capabilities. This improvement creates strategic competence that compounds over time as strategic systems accumulate experience and refine strategic approaches.

Autonomous evolution incorporates strategic capability identification, where AI systems can recognize when organizational strategic challenges require new strategic capabilities and automatically develop those capabilities through strategic experimentation, strategic learning, and strategic capability integration.

The evolution systems include strategic innovation generation, where AI systems can identify entirely new strategic approaches, business models, or competitive strategies that create sustainable competitive advantages. This innovation emerges from AI analysis of market patterns, competitive dynamics, and organizational capabilities that suggests strategic possibilities human strategic thinking cannot envision independently.

Strategic adaptation automation enables organizations to adapt strategic approaches automatically as market conditions change without requiring human recognition of adaptation requirements or human strategic planning to develop adaptive responses. This automation creates strategic agility that responds to market evolution faster than human strategic planning can recognize and respond to market changes.

The autonomous evolution culminates in what some call strategic transcendence, where organizational strategic capabilities exceed human strategic thinking limitations through AI strategic reasoning that can process strategic complexity, analyze strategic possibilities, and develop strategic solutions that human strategic thinking cannot achieve independently, while maintaining alignment with human strategic vision and organizational strategic values.

Organizations achieving strategic transcendence discover that strategic thinking becomes their most powerful competitive weapon rather than their most challenging management responsibility, creating sustainable competitive advantages that emerge from superior strategic intelligence rather than superior resource deployment, operational efficiency, or market positioning advantages that competitors can eventually replicate through strategic learning and competitive response.

Executive Takeaways: Your Strategic Cognition Transformation

The transformation from prediction to strategic cognition represents the most significant evolution in business planning since the development of scenario modeling. Organizations implementing these capabilities create competitive advantages that prediction-based competitors cannot understand or replicate.

Strategic time travel enables exploration of multiple possible futures simultaneously, creating strategic preparedness that traditional planning cannot achieve. Begin by implementing scenario multiplication capabilities that test strategic options across multiple realities rather than optimizing for single forecasts.

Autonomous strategic thinking enables AI systems to continuously identify and pursue strategic opportunities, while human leadership focuses on strategic vision and stakeholder relationships. Develop agentic AI capabilities that can think strategically without constant human supervision while maintaining alignment with organizational objectives.

Strategic partnership design transforms AI from an analytical tool to a strategic collaborator that enhances human strategic judgment rather than replacing human strategic authority. Create collaborative strategic intelligence that combines human wisdom with AI analytical power.

Enterprise cognition evolves beyond traditional business intelligence to prospective strategic intelligence, enabling faster consideration of emerging opportunities than human analytical cycles can complete. Build cognitive systems that inhabit possible futures rather than analyzing past performance.

The enterprises that master strategic cognition will discover that the future belongs to organizations that can think strategically at machine speed while maintaining human wisdom in strategic judgment. The question is whether your organization will develop these capabilities before competitors transform strategic planning from a competitive advantage to a competitive requirement you cannot afford to ignore.

CHAPTER 4

The AI Factory – Systematic Intelligence Production

The assembly line stretched across the River Rouge plant for miles. Every fourteen minutes, another automobile emerged complete and tested. Raw steel entered one end, finished vehicles departed the other. Henry Ford had transformed manufacturing from craft production to systematic creation on an unprecedented scale. What took skilled artisans weeks to complete now happened in hours through coordinated processes that ordinary workers could execute flawlessly.

A century later, Netflix processes 15 billion viewing decisions every day. Each interaction refines personalization algorithms across 200 million subscribers simultaneously. Data flows through prediction engines as precisely orchestrated as the Ford assembly stations. Models improve automatically with every choice. The platform manufactures relevance rather than vehicles, but the systematic approach mirrors the breakthrough of Ford completely.

The enterprises dominating today's markets discovered what Ford knew about physical production. Systematic manufacturing creates exponential advantages over artisan approaches. Amazon recommendation systems succeed because they operate like precision factories for relevance rather than isolated data science projects. The Google advertising platform dominates because it industrialized the production of attention rather than building individual campaign tools. Tesla Autopilot advances because it systematized the creation of driving intelligence rather than programming individual scenarios.

These organizations built something most enterprises haven't recognized yet. They created AI factories, systematic approaches to manufacturing organizational intelligence at enterprise scale. Where traditional manufacturers process raw materials into finished goods, AI factories process data into insights that become more valuable through usage rather than being depleted through consumption.

The transformation extends beyond operational improvements into fundamental business model innovation. The assembly line of Ford succeeded because it standardized production processes while dramatically reducing costs and improving quality. AI factories succeed because they standardize intelligence production while exponentially increasing capability and reducing time-to-insight. Manufacturing companies face resource constraints as scale increases materials requirements. AI factories experience expanding returns as scale improves intelligence quality while decreasing production costs per insight.

This inverted economics explains why platform companies achieve market dominance while traditional enterprises struggle with implementing AI. Platforms operate as intelligence factories from inception, designed to improve through usage rather than wear out through operation. Traditional organizations attempt to retrofit AI capabilities onto structures designed for material production rather than intelligence manufacturing.

The competitive mathematics proves to be decisive. Organizations implementing systematic intelligence production develop capabilities in weeks that manual approaches require months to achieve. They scale successful applications across unlimited scenarios through replication rather than reimplementation. They compound competitive advantages through continuous improvement rather than periodic enhancement projects.

Most enterprises approach AI like nineteenth-century craftsmen. They build individual models, customize isolated solutions, and hope experiments somehow scale into enterprise capability. The results mirror craft production limitations: high costs, unpredictable quality, difficult scaling, and inconsistent performance. Each new application requires starting from scratch rather than building on a systematic foundation.

The organizations winning the intelligence revolution recognized that AI, like manufacturing, becomes exponentially more powerful when approached systematically rather than individually. Netflix doesn't build separate recommendation engines for different content types. Amazon doesn't create isolated prediction models for various product categories. Tesla doesn't develop individual algorithms for specific driving scenarios. They built systematic intelligence production that adapts automatically to unlimited applications while improving continuously through accumulated experience.

The enterprise implications prove staggering. While competitors manage individual AI projects, AI factory organizations produce intelligence capabilities continuously. While traditional approaches require months to develop new applications, systematic production delivers insights in weeks. While manual methods struggle with scaling complexity, factory approaches replicate successfully across a wide range of business scenarios.

The transformation requires abandoning project thinking entirely. Instead of asking what AI use cases to implement, factory thinking asks how to manufacture intelligence that becomes more valuable through systematic usage.

Instead of managing individual implementations, organizations orchestrate production systems that generate competitive advantages continuously rather than episodically.

The revolution mirrors the impact of the Ford assembly line, but operates entirely in software. Where Ford processed steel and rubber into automobiles, AI factories process data and algorithms into insights. Where the innovations of Ford improved through engineering refinement, AI factories improve automatically through the accumulation of usage. Where the advantages of Ford came from manufacturing efficiency, AI factory's advantages emerge from intelligence velocity.

The choice facing every enterprise leader resembles what automobile manufacturers confronted in 1908. Continue craft production and compete on individual excellence, or embrace systematic manufacturing and compete on production capability. History demonstrates that organizational competencies determine competitive outcomes when production paradigms shift fundamentally, regardless of individual skill or investment levels.

The AI factory paradigm shift has already begun. Amazon Web Services revolutionized technology infrastructure by treating computing as systematic production rather than custom deployment. Salesforce transformed customer relationship management by manufacturing interaction intelligence rather than building isolated database applications. Modern enterprises must undergo a similar transition from individual AI applications to systematic intelligence production.

The factory floor of the future operates entirely on software. Instead of assembly stations, algorithms process information. Instead of quality control inspectors, automated systems validate model performance. Instead of relying on inventory management, data pipelines ensure the availability of information. Instead of shipping finished products, platforms deliver insights that improve through usage rather than depreciate through consumption.

Building this factory requires thinking fundamentally differently about intelligence, production, and competitive advantage. Organizations must transition from viewing data as a static resource to treating information as raw material for continuous intelligence manufacturing. They must evolve from managing individual AI projects to orchestrating systematic production that scales automatically across an unlimited number of applications.

The transformation begins with understanding that intelligence, like manufacturing, follows systematic principles that enable predictable scaling rather than experimental hoping. The most successful organizations already operate according to these principles, creating competitive advantages that craft approaches cannot match, regardless of expertise or investment.

The enterprises that master systematic intelligence production will dominate their markets for the same reason Ford dominated the automobile manufacturing industry. Systematic production creates advantages that artisan approaches cannot overcome through individual excellence. The question confronting every organization is whether they will embrace intelligence manufacturing before competitors transform their industries around factory capabilities.

Welcome to the era of systematic intelligence production. The factory is the future of enterprise AI.

BUILDING YOUR ORGANIZATION'S INTELLIGENCE MANUFACTURING SYSTEM

The algorithm made its 847th decision that second. A shopper in Mumbai clicked "Add to cart" on a silk scarf while another in Manchester abandoned their shopping basket. In São Paulo, someone searched for "wireless headphones" as inventory shifted in three warehouses across two continents. Every microsecond, the Amazon recommendation engine processed thousands of simultaneous choices, each interaction making the entire system smarter. Not one data scientist was manually adjusting anything. The machine was learning to learn.

Meanwhile, at Global Manufacturing Inc., a team of six PhD data scientists spent three months building a demand forecasting model for winter coats. They worked nights and weekends, crafting elegant algorithms and fine-tuning parameters by hand. When finished, their masterpiece predicted next quarter's sales with impressive accuracy. It was brilliant work. It was also precisely what the Amazon system had discovered automatically in seventeen minutes while simultaneously optimizing for 300 million other products.

The difference wasn't talent or technology. Amazon had built an intelligence factory. Global Manufacturing was still running an artisan workshop.

This contrast reveals the fundamental transformation required to build organizational intelligence manufacturing systems. While most enterprises approach AI like skilled craftsmen creating custom solutions, the organizations achieving exponential advantages operate like systematic manufacturers, producing intelligence at scale through coordinated processes that ordinary teams can execute consistently.

The transition from craft AI to systematic intelligence production requires understanding what made the Ford assembly line revolutionary. Ford failed because he built better cars than individual craftsmen. He succeeded because he created systematic processes that produced consistent quality at an unprecedented scale while reducing costs exponentially. The same principles applied to intelligence production create similar advantages for organizations sophisticated enough to embrace systematic thinking.

The Evolution from Craft AI to Intelligence Manufacturing

Traditional enterprise AI resembles medieval guild production. Individual data scientists craft custom models using specialized skills developed through years of experience. Each project begins with raw materials and proceeds through artisan processes that produce unique solutions requiring ongoing maintenance by their creators. Quality depends on individual expertise rather than systematic processes. Scaling requires finding more skilled craftsmen rather than improving production systems.

Intelligence manufacturing transforms this dynamic through systematic processes that convert data into insights through coordinated workflows that improve automatically through experience accumulation. Where craft AI produces individual solutions, intelligence manufacturing creates production capabilities that generate unlimited applications through systematic replication rather than custom development.

The evolution parallels the transition from blacksmiths hammering individual horseshoes to steel mills producing construction materials at an industrial scale. Both approaches work, but only systematic production enables consistent quality, predictable timing, and exponential scaling, which create sustainable competitive advantages in rapidly evolving markets.

Organizations attempting to compete against intelligence manufacturers using craft AI approaches face the same mathematics that confronted individual blacksmiths competing against steel mills. Superior individual craftsmanship cannot overcome systematic production advantages when market demands exceed what artisan methods can deliver economically.

The transformation begins with recognizing that intelligence, like steel production, follows systematic principles that enable predictable outcomes rather than experimental results. Organizations must evolve from viewing AI as a specialized craft requiring rare expertise to understanding intelligence production as a systematic capability that ordinary teams can execute through well-designed processes.

This fundamental mindset shift enables the organizational changes necessary to build intelligence manufacturing systems. Instead of hiring more data scientists to craft additional models, organizations invest in systematic capabilities that enable existing teams to produce intelligence applications through coordinated workflows that improve automatically through accumulated experience.

Manufacturing Principles Applied to Intelligence Production

The systematic approach to intelligence manufacturing adapts proven manufacturing principles to cognitive work through organizational structures and technological systems that enable predictable, scalable, and continuously improving intelligence production.

Standardization creates the foundation for systematic intelligence production through consistent processes that convert raw data into refined insights through predictable workflows. Whereas craft AI requires individual expertise to navigate unique challenges, standardized processes enable ordinary teams to execute complex intelligence production through clearly defined procedures that eliminate variability and reduce dependency on specialized skills.

The standardization extends beyond individual model development to encompass data preparation workflows, algorithm selection criteria, validation procedures, deployment processes, and performance monitoring systems. This comprehensive standardization enables different teams to collaborate effectively while maintaining consistent quality standards across unlimited intelligence applications.

Quality control mechanisms ensure that intelligence production maintains reliability standards through automated validation processes that detect problems before they impact business operations. Traditional craft AI relies on individual expertise to identify quality issues through manual review processes that scale poorly and depend on specialist availability. Systematic quality control operates through automated systems that validate data quality, model performance, and business impact continuously rather than periodically.

The quality control systems incorporate predictive maintenance principles that identify potential problems before they manifest through pattern recognition that monitors system health indicators. This proactive approach prevents intelligence degradation that could compromise business operations while enabling continuous improvement through systematic problem identification and resolution.

Lean principles eliminate waste from intelligence production through process optimization that identifies and removes activities that consume resources without creating value. Traditional AI development includes significant waste through duplicated effort, inefficient workflows, inadequate communication, and suboptimal resource allocation. Systematic waste elimination creates dramatic efficiency improvements while improving output quality.

The lean approach focuses on continuous flow optimization that ensures information and insights move smoothly through production processes without bottlenecks that create delays or quality degradation. This flow optimization requires coordinating multiple specialized functions through systematic interfaces that eliminate friction while maintaining quality standards.

Continuous improvement mechanisms enable intelligence manufacturing systems to evolve automatically through feedback loops that identify enhancement opportunities and implement systematic upgrades. Whereas craft AI improves through individual learning that remains isolated within specific projects, systematic improvement accumulates across entire production systems through shared learning that benefits all applications simultaneously.

Process optimization applies systematic analysis to identify enhancement opportunities across intelligence production workflows. This optimization operates continuously rather than periodically, enabling rapid adaptation to changing requirements while maintaining production efficiency and quality standards.

Digital twins enable virtual modeling of intelligence production systems that support experimentation, optimization, and risk management without disrupting operational systems. These virtual environments allow organizations to test production improvements, evaluate new approaches, and optimize resource allocation through simulation before implementing changes in operational environments.

Organizational Transformation Foundations

Building intelligence manufacturing systems requires fundamental organizational transformation that extends beyond technology implementation to encompass cultural change, structural reorganization, and capability development that enables systematic intelligence production rather than individual AI project management.

The transformation begins with establishing cross-functional teams that combine diverse expertise necessary for systematic intelligence production. Traditional AI projects operate through specialist groups that focus on individual technical domains. Intelligence manufacturing requires coordinated collaboration between data engineers, algorithm developers, business analysts, domain experts, and operations teams working through systematic interfaces rather than ad hoc cooperation.

These cross-functional teams operate through clearly defined roles and responsibilities that eliminate ambiguity while enabling efficient collaboration. Team members understand how their contributions integrate with overall intelligence production rather than focusing solely on individual technical specializations. This systematic coordination enables complex intelligence applications that exceed what individual specialists can achieve through isolated effort.

Governance structures provide strategic direction and operational oversight that ensures intelligence manufacturing aligns with organizational objectives while maintaining quality and compliance standards. Effective governance operates through systematic decision-making processes rather than individual judgment calls, enabling consistent strategic alignment across multiple intelligence production initiatives.

The governance framework includes resource allocation mechanisms that optimize investment across intelligence manufacturing capabilities rather than individual AI projects. This systematic resource management enables sustained capability development that compounds competitive advantages over time rather than creating isolated solutions that provide temporary benefits.

Cultural transformation proves essential for successful intelligence manufacturing because systematic production requires collaborative approaches that differ fundamentally from specialist-driven AI development. Organizations must evolve from valuing individual expertise above systematic capability to recognizing that coordinated processes create superior outcomes compared to individual brilliance.

This cultural evolution requires leadership that models collaborative behavior while establishing incentive systems that reward systematic contribution rather than individual achievement. Team members must understand that intelligence manufacturing success depends on systematic excellence rather than specialist expertise, creating organizational cultures that prioritize collective capability over individual recognition.

Training and development programs ensure that team members develop capabilities necessary for systematic intelligence production rather than specialist AI skills that apply to limited domains. Effective training focuses on systematic thinking, collaborative processes, and continuous improvement methodologies that enable ordinary professionals to contribute effectively to intelligence manufacturing rather than requiring specialized AI education.

Strategic Foundation for AI-Native Operations

Intelligence manufacturing systems provide the operational foundation for organizations aspiring to become truly AI-native rather than simply implementing AI applications within traditional business structures. This transformation requires systematic integration of intelligence production with core business operations through architectural approaches that embed continuous learning into organizational DNA.

The strategic foundation operates through intelligence-driven decision-making that replaces intuition-based management with systematic analysis that improves automatically through experience accumulation. AI-native organizations make decisions through collaborative human-AI processes that combine analytical rigor with human judgment rather than depending solely on individual expertise or pure algorithmic recommendations.

This decision-making evolution requires systematic integration of intelligence production with operational workflows that enable real-time optimization rather than periodic analysis. Business processes adapt automatically to changing conditions through embedded intelligence that responds faster than human analysis while maintaining strategic alignment through human oversight.

Business model innovation emerges naturally from intelligence manufacturing capabilities that enable value creation approaches impossible through traditional operations. Organizations discover revenue opportunities, cost optimization strategies, and competitive positioning possibilities that emerge from systematic intelligence production rather than individual analysis or external consulting.

The business model innovations often involve creating new value through intelligence-enhanced products, services, or experiences that become more valuable through usage rather than depleting through consumption. These self-improving offerings create sustainable competitive advantages that traditional business models cannot replicate through operational excellence alone.

Competitive intelligence generation operates automatically through systematic analysis of market patterns, customer behavior, and operational performance that reveals opportunities and threats faster than traditional market research or strategic analysis. AI-native organizations develop market understanding that competitors cannot match through manual research methods.

This intelligence advantage compounds over time as systematic production capabilities accumulate insights that inform strategic positioning, product development, and operational optimization. Organizations implementing effective intelligence manufacturing create self-reinforcing competitive advantages that become increasingly difficult for competitors to overcome through traditional approaches.

Organizational learning acceleration occurs through systematic capture and application of operational insights that improve business performance continuously rather than through periodic training or consulting engagements. AI-native organizations evolve automatically through accumulated experience rather than requiring external expertise to identify improvement opportunities.

The learning acceleration creates adaptive capabilities that enable rapid response to market changes, competitive pressures, and operational challenges. Organizations with systematic intelligence production adapt faster than competitors can recognize that adaptation is necessary, creating sustained competitive advantages through superior organizational responsiveness.

Human-AI Collaboration Architecture

Successful intelligence manufacturing requires sophisticated human-AI collaboration that augments human capabilities rather than replacing human judgment through systematic interfaces that optimize the contribution of both human intelligence and AI to overall organizational capability.

The collaboration architecture operates through complementary intelligence frameworks that assign responsibilities based on relative strengths rather than attempting to replace human capabilities with artificial alternatives. Humans contribute strategic vision, contextual understanding, ethical judgment, and creative thinking that AI cannot replicate. AI systems provide analytical processing, pattern recognition, predictive modeling, and optimization capabilities that exceed human cognitive capacity.

This complementary approach requires systematic interfaces that enable efficient information exchange between human decision makers and AI systems without creating communication bottlenecks or coordination complexity. Effective interfaces translate AI insights into actionable intelligence while converting human strategic direction into systematic operational guidance that AI systems can execute effectively.

Decision support systems provide human leaders with AI-enhanced analytical capabilities that improve decision quality while maintaining human authority over strategic choices. These systems present complex analytical results through intuitive interfaces that enable rapid understanding without requiring technical expertise in AI methodology or statistical analysis.

The decision support capabilities include scenario analysis that explores multiple strategic options through systematic modeling that accounts for uncertainty and complexity beyond human analytical capacity. Leaders can evaluate strategic alternatives through comprehensive analysis that would require months of manual research while maintaining strategic authority over final decisions.

Workflow integration ensures that human-AI collaboration operates seamlessly within existing business processes rather than requiring separate AI management procedures that create operational complexity. Effective integration makes AI capabilities available through normal business workflows while maintaining familiar operational patterns that minimize training requirements and adoption barriers.

The workflow integration includes automated quality assurance that monitors AI recommendations for accuracy and appropriateness while alerting human supervisors to situations requiring manual intervention. This systematic oversight ensures that AI contributions enhance rather than compromise business operations while maintaining human accountability for organizational outcomes.

Continuous learning mechanisms enable human-AI collaboration to improve automatically through experience accumulation that enhances both AI system performance and human understanding of AI capabilities. Teams develop an intuitive understanding of when to trust AI recommendations, when to apply human judgment, and how to combine human and AI insights for optimal results.

This collaborative learning creates organizational intelligence that exceeds what either human teams or AI systems can achieve independently. Organizations implementing effective human-AI collaboration develop competitive advantages through enhanced decision-making, accelerated problem-solving, and improved strategic execution that combines the best capabilities of human and AI.

The intelligence manufacturing system provides the operational foundation for this collaborative advantage by creating systematic processes that enable consistent, scalable, and continuously improving human-AI cooperation across unlimited business applications. Organizations mastering this collaboration will dominate markets through superior organizational intelligence that competitors cannot replicate through traditional management approaches or isolated AI implementations.

THE FOUR PILLARS: DATA, ALGORITHMS, EXPERIMENTATION, INFRASTRUCTURE

The notification arrived at 3:47 AM. The Netflix recommendation engine had discovered something extraordinary. A pattern buried in viewing behavior across 47 countries suggested that romantic comedies featuring cooking scenes performed 23% better with viewers aged 28-34 on Tuesday evenings in rainy weather. By 4:15 AM, the algorithm had automatically adjusted recommendations for 2.3 million subscribers. By 6:00 AM, engagement metrics confirmed the hypothesis. By 8:00 AM, content acquisition teams received automated suggestions for licensing priorities.

No human analyst worked through the night. No data scientist manually coded new rules. No executive approved the experiment. The entire discovery happened through four interconnected systems working in perfect harmony. Data flowed seamlessly from viewing sensors to analytical engines. Algorithms tested hypotheses automatically through controlled experiments. Infrastructure delivered personalized results to millions of devices simultaneously. Netflix had built what most enterprises dream of achieving but few understand how to construct.

The difference lies in architectural thinking rather than technological sophistication. While most organizations accumulate AI tools like craftsmen collecting specialized instruments, Netflix engineered systematic production through four foundational pillars that work together as a unified intelligence manufacturing system. Each pillar serves distinct functions while depending completely on the others for operational effectiveness.

Understanding these four pillars provides the blueprint for transforming any organization from artisan AI development to systematic intelligence production. The pillars themselves are not revolutionary. The systematic integration that enables continuous, automatic, and self-improving intelligence production represents the breakthrough that separates platform companies from traditional enterprises attempting to retrofit AI capabilities onto industrial-age structures.

The Data Pillar: Fueling Continuous Intelligence

Data represents the raw material of intelligence manufacturing, but unlike physical materials that deplete through consumption, information becomes more valuable through systematic processing that creates compound insights while preserving original value. The data pillar transforms information from a static resource into a dynamic foundation for continuous intelligence production.

Traditional enterprise data management resembles ore mining. Organizations extract information from operational systems through periodic processes that require manual intervention to clean, format, and analyze extracted materials. Each analysis project begins with extensive preparation work that recreates similar processing steps because no systematic approach exists for converting raw data into analysis-ready materials.

Systematic data processing operates like modern petroleum refining. Raw information flows through automated pipelines that convert crude operational data into refined analytical materials through standardized processes that produce consistent quality while enabling unlimited downstream applications. Each refinement step adds value while maintaining systematic quality standards that support the production of reliable intelligence.

The transformation begins with automated ingestion systems that continuously capture information from operational sources, rather than through periodic extraction processes. These systems operate through standardized interfaces that accommodate diverse data sources while converting various formats into consistent structures that support systematic processing. Automated ingestion eliminates the manual effort and inconsistent quality that characterize traditional data collection approaches.

Data preprocessing is conducted through systematic workflows that clean, validate, and enhance information quality through automated processes. These processes identify and correct common problems while flagging exceptional cases for human review. The preprocessing includes duplicate detection, missing value handling, format standardization, and quality validation that ensures downstream intelligence production receives consistent input materials.

Quality assurance mechanisms continuously monitor data integrity, rather than relying on periodic audits that identify problems after they impact intelligence production. Automated quality monitoring operates through real-time validation, which compares incoming information against established patterns to identify anomalies that may indicate system problems or data corruption.

The quality assurance systems include automated correction capabilities that address common data problems through systematic rules while escalating complex issues to human oversight. This automated correction reduces manual intervention requirements while maintaining data integrity standards necessary for reliable intelligence production.

Feature engineering automation creates analytical variables through systematic processes that identify and construct useful data relationships automatically, rather than requiring manual analysis for each application. Automated feature engineering utilizes machine learning techniques to identify patterns and relationships that human analysts might overlook, thereby creating reusable analytical assets that benefit multiple intelligence applications.

Data governance frameworks ensure that systematic processing maintains privacy, security, and regulatory compliance through embedded controls rather than external oversight processes. Governance operates through systematic access controls, audit logging, and compliance validation that occurs automatically within data processing workflows rather than requiring separate monitoring systems.

The governance framework includes automated anonymization and encryption capabilities that protect sensitive information while enabling analytical processing. These protective mechanisms operate transparently within systematic workflows rather than creating barriers that impede intelligence production or require manual intervention for routine operations.

The Algorithm Pillar: Systematic Intelligence Creation

Algorithms represent the intelligent engines that convert processed data into actionable insights through systematic reasoning processes that improve automatically through the accumulation of experience. The algorithm pillar transforms prediction from a craft activity requiring specialist expertise into systematic production that ordinary teams can execute consistently.

Traditional algorithm development resembles individual craftsmanship, where data scientists build custom models through specialized techniques that require deep expertise to implement effectively. Each algorithm represents a unique creation that reflects individual developer skills and preferences rather than systematic approaches that enable consistent quality and reliable performance across multiple applications.

Systematic algorithm production operates through standardized development workflows that enable teams to create intelligent systems through coordinated processes that emphasize consistency and reliability over individual creativity. The systematic approach produces algorithms that meet quality standards and performance requirements through repeatable methods rather than depending on exceptional individual expertise.

Model development frameworks offer a systematic approach to algorithm creation through template-based processes that incorporate best practices while accommodating specific application requirements. These frameworks guide teams through proven development sequences that reduce errors and improve consistency while enabling customization for unique business needs.

The development frameworks include automated algorithm selection capabilities that evaluate multiple modeling approaches against specific datasets and performance criteria. Automated selection eliminates guesswork, ensuring that chosen algorithms match application requirements and data characteristics through systematic evaluation rather than relying on individual preference or limited experience.

Continuous integration processes ensure that algorithm development maintains quality standards through automated testing that validates model performance before deployment. Continuous integration operates through systematic validation workflows that test algorithms against established criteria while identifying potential problems early in development cycles rather than after deployment.

The integration processes include automated deployment pipelines that move validated algorithms from development environments into production systems through systematic workflows that minimize manual intervention while maintaining quality controls. Automated deployment reduces errors and accelerates time-to-production while ensuring that deployed algorithms meet performance and security requirements.

Version control systems manage algorithm evolution through systematic tracking, enabling reliable rollback capabilities when performance degrades or problems emerge. Version control operates through automated documentation that captures algorithm changes and performance impacts while enabling systematic comparison of different algorithm versions.

Performance monitoring systems track algorithm effectiveness continuously rather than through periodic reviews that identify problems after they impact business operations. Automated monitoring evaluates prediction accuracy, processing performance, and business impact while identifying degradation patterns that indicate the need for algorithm updates or replacement.

The monitoring systems include automated retraining capabilities that update algorithms when performance degrades below acceptable thresholds.

Automated retraining maintains algorithm effectiveness through systematic updates that incorporate new data patterns while preserving proven performance characteristics through controlled enhancement processes.

The Experimentation Pillar: Systematic Validation and Learning

Experimentation transforms algorithm predictions from hypotheses into validated business intelligence through controlled testing that measures actual impact rather than theoretical performance. The experimentation pillar enables systematic learning that improves intelligence production through the accumulation of evidence rather than relying on individual judgment or theoretical analysis.

Traditional AI validation relies on historical data analysis that measures algorithm performance against past events rather than testing effectiveness under current business conditions. Historical validation cannot account for changing market dynamics, competitive responses, or operational variations that affect algorithm performance in real business environments.

Systematic experimentation operates through controlled testing environments that measure the impact of algorithms under actual business conditions, while isolating experimental effects from other operational variables. Controlled experimentation enables reliable measurement of algorithm value through statistical analysis that accounts for uncertainty and variation in business performance.

A/B testing frameworks provide systematic approaches to comparing algorithm performance against existing business processes or alternative methods. These frameworks operate through controlled experimental designs that randomly assign similar business scenarios to different algorithm approaches while measuring performance differences through statistical analysis.

The testing frameworks include automated experimental design capabilities that determine appropriate sample sizes, testing duration, and success criteria based on business requirements and statistical precision needs. Automated design ensures that experiments produce reliable results while minimizing business risk and resource requirements.

Multivariate testing capabilities enable systematic evaluation of multiple algorithm variations simultaneously through controlled experiments that measure interaction effects between different approaches. Multivariate testing accelerates learning while identifying optimal algorithm combinations that individual testing approaches might miss.

Continuous learning mechanisms capture experimental results and apply insights to improve algorithm development and deployment processes. Systematic learning operates through automated analysis of experimental outcomes that identifies patterns and principles that inform future algorithm development rather than treating each experiment as an isolated activity.

The learning mechanisms include automated success factor identification that analyzes experimental results to determine characteristics associated with superior algorithm performance. Success factor analysis enables systematic improvement of algorithm development processes through evidence-based enhancement rather than theoretical optimization.

Real-time experimentation capabilities enable continuous testing of algorithm improvements through systematic protocols that measure impact while minimizing business risk. Real-time experimentation accelerates improvement cycles while maintaining operational stability through controlled enhancement processes.

Risk management frameworks ensure that experimentation maintains business stability through systematic controls that prevent experimental activities from compromising operational performance. Risk management operates through automated monitoring that identifies experimental impacts while enabling rapid intervention when experiments create unexpected problems.

The Infrastructure Pillar: Systematic Integration Foundation

Infrastructure provides the technological foundation that enables systematic intelligence production through integrated platforms that coordinate data processing, algorithm execution, and experimental validation while maintaining security, reliability, and performance standards. The infrastructure pillar transforms AI from isolated applications into systematic capability that operates seamlessly within enterprise environments.

Traditional AI infrastructure resembles a collection of specialized tools that require manual coordination to accomplish complex intelligence production. Individual systems operate independently rather than through systematic integration that enables automatic coordination and resource optimization across multiple intelligence applications.

Systematic infrastructure operates through unified platforms that coordinate all aspects of intelligence production through standardized interfaces and automated resource management. Integrated infrastructure enables seamless operation across data processing, algorithm execution, and experimental validation while maintaining consistent performance and security standards.

Platform architecture provides a systematic foundation for intelligence production through modular design that enables flexible configuration while maintaining systematic integration. Platform architecture operates through standardized interfaces that enable different components to work together automatically while supporting diverse application requirements through configurable rather than custom development.

The platform architecture includes automated resource scaling that adjusts computational capacity based on demand patterns rather than requiring manual resource management. Automated scaling maintains performance standards while optimizing costs through systematic resource allocation that matches capacity to actual requirements.

Container orchestration systems enable systematic deployment and management of intelligence applications through automated workflows that handle complex coordination requirements. Container orchestration operates through systematic scheduling and resource management that ensures reliable operation while enabling efficient resource utilization across multiple applications.

API management frameworks provide systematic interfaces that enable different intelligence components to communicate effectively while maintaining security and performance standards. API management operates through standardized protocols that enable reliable integration while supporting diverse application requirements through flexible interface design.

The API frameworks include automated documentation and testing capabilities that ensure interface reliability while reducing integration complexity for development teams. Automated interface management maintains systematic integration while enabling rapid development of new intelligence applications.

Security frameworks provide systematic protection for intelligence production through integrated controls that operate transparently within processing workflows rather than requiring separate security management. Integrated security operates through automated access controls, encryption, and audit logging, maintaining protection while enabling efficient operation.

Monitoring and observability systems provide systematic visibility into intelligence production performance through automated tracking that identifies problems and optimization opportunities. Systematic monitoring operates through real-time performance analysis that enables proactive problem resolution while supporting continuous improvement through operational intelligence.

The monitoring systems include automated alert generation that notifies operations teams when performance degrades or problems emerge. Automated alerting enables rapid response to operational issues while maintaining systematic oversight of intelligence production performance across multiple applications and infrastructure components.

Integration Excellence: Making the Pillars Work Together

The competitive advantage of systematic intelligence production emerges from sophisticated integration between the four pillars rather than excellence in individual components. Organizations can implement advanced data systems, sophisticated algorithms, rigorous experimentation, and robust infrastructure while failing to achieve intelligence manufacturing because they lack systematic integration that enables coordinated operation.

Integration excellence requires architectural thinking that designs pillar interactions rather than simply connecting existing systems. Effective integration operates through systematic workflows that coordinate activities across all four pillars while maintaining individual pillar optimization and overall system performance.

Data-algorithm integration ensures that systematic data processing produces materials optimized for algorithm consumption while algorithm development incorporates data characteristics that enable superior performance. This integration operates through feedback loops that inform data processing improvements based on algorithm requirements while optimizing algorithms based on available data characteristics.

Algorithm-experimentation integration enables systematic testing of algorithm performance through experimental frameworks designed to validate specific algorithm capabilities. This integration operates through automated experimental design that creates appropriate testing conditions while measuring algorithm performance through systematic validation protocols.

Experimentation-infrastructure integration ensures that testing frameworks operate efficiently within a systematic infrastructure while infrastructure design supports experimental requirements. This integration enables real-time experimentation through platform capabilities that provide controlled testing environments without compromising operational stability.

Infrastructure-data integration provides a systematic foundation for data processing through platform capabilities that optimize performance while maintaining security and reliability standards. This integration operates through automated resource management that ensures data processing performance while enabling systematic scaling based on processing requirements.

Cross-pillar optimization identifies improvement opportunities that emerge from systematic coordination rather than individual pillar enhancement. Cross-pillar optimization operates through systematic analysis of interaction patterns that reveals enhancement possibilities not apparent through individual pillar analysis.

The optimization includes automated workflow coordination that ensures activities across different pillars operate efficiently together while maintaining individual pillar performance standards. Automated coordination reduces manual intervention requirements while enabling systematic enhancement of overall intelligence production capability.

Quality assurance operates across all four pillars through systematic validation that ensures coordinated operation maintains reliability and performance standards. Cross-pillar quality assurance identifies problems that emerge from interaction patterns while enabling systematic resolution that maintains overall system effectiveness.

This integrated approach creates intelligence manufacturing systems that produce competitive advantages through systematic operation rather than individual component excellence. Organizations mastering four-pillar integration develop intelligence production capabilities that competitors cannot replicate through technology acquisition or individual expertise because the advantage emerges from systematic integration rather than component sophistication.

CREATING COMPETITIVE MOATS THROUGH PROPRIETARY INTELLIGENCE

The reverse engineering attempt lasted six months. A Fortune 100 competitor hired the industry's best data scientists, licensed the same machine learning platforms their rivals used, and acquired identical datasets to their competitor's. They studied every patent filing, analyzed every public presentation, and recruited former employees. Their goal seemed straightforward: replicate the recommendation engine that was driving 35% of their rival's revenue growth.

The copied system launched with great fanfare. Initial performance metrics looked promising. Customer engagement increased modestly. Revenue attributed to recommendations grew by 12%. Executive leadership declared victory.

Then something unexpected happened. While the copycat system remained static, the original began evolving at an accelerating pace. Prediction accuracy improved weekly. Customer satisfaction scores climbed monthly. Revenue attribution reached 47% within a year. The competitive gap widened rather than narrowed, despite using identical technology and similar data.

The copying company had succeeded in replicating the technology but failed to understand the true source of competitive advantage. They had built the engine but missed the factory. They copied the algorithm but ignored the systematic intelligence production that made continuous improvement inevitable rather than accidental. Their rival hadn't simply built better AI. They had constructed something far more valuable and impossible to replicate: a self-improving competitive moat built from proprietary intelligence that became stronger through every customer interaction.

This distinction separates organizations that achieve sustainable competitive advantages through AI from those that implement impressive technology without creating lasting differentiation. The difference lies not in algorithmic sophistication but in systematic approaches to transforming organizational intelligence into competitive moats that become more formidable over time rather than vulnerable to replication.

Beyond Technology: The Intelligence Advantage

Most organizations approach competitive differentiation through AI by focusing on technological superiority, believing that better algorithms or more sophisticated models create sustainable advantages. This technology-centric approach mirrors historical methods of competitive positioning, which often relied on superior manufacturing equipment or advanced product features that competitors could eventually acquire or replicate.

The fundamental flaw in technology-focused competitive strategy lies in the inevitable commoditization of technological capabilities. Machine learning platforms become available through cloud services. Algorithms get published in academic papers. Datasets become accessible through commercial providers. Talented engineers move between companies. Technology advantages prove to be temporary rather than sustainable because technological capabilities can be acquired rather than developed.

Proprietary intelligence represents a fundamentally different approach to achieving competitive advantage, emerging from systematic processes rather than technological acquisition. Organizations building competitive moats through proprietary intelligence create advantages that cannot be purchased, licensed, or copied because the competitive differentiation emerges from accumulated learning rather than individual technological components.

The intelligence advantage operates through systematic data accumulation that creates unique insights unavailable to competitors, regardless of their technological sophistication. These insights emerge from organizational interactions, customer behaviors, operational patterns, and market relationships that are specific to individual companies and cannot be replicated through external data acquisition or technological implementation.

Organizations developing proprietary intelligence advantages discover that their competitive position strengthens automatically through normal business operations rather than requiring deliberate competitive actions. Every customer transaction, operational decision, and market interaction contributes to intelligence accumulation that enhances competitive positioning while creating barriers to competitive replication.

This automatic strengthening creates sustainable competitive dynamics that differ fundamentally from traditional competitive positioning. Instead of defending advantages through strategic actions that competitors can counter, organizations with proprietary intelligence advantages benefit from competitive dynamics where success creates additional advantages rather than attracting imitation that erodes competitive positioning.

Network effects amplify proprietary intelligence advantages through systematic feedback loops that improve competitive positioning faster as organizational scale increases. More customers provide more data that enables better predictions that attract additional customers through superior service that generates additional data in accelerating cycles that compound competitive advantages automatically.

The network effects create competitive mathematics that favor systematic intelligence production over individual AI implementations. Organizations with effective intelligence manufacturing systems improve faster than competitors can respond because improvement accelerates through accumulated advantage rather than linear enhancement that competitors can match through increased investment or effort.

Systematic Intelligence Accumulation

Competitive moats built from proprietary intelligence require systematic approaches to capturing, processing, and applying organizational learning that transforms routine business operations into intelligence accumulation processes. This systematic approach distinguishes sustainable competitive advantages from temporary technological superiority that dissipates through competitive imitation.

Data flywheel effects create self-reinforcing cycles where better intelligence enables superior business decisions, which generate additional data that improves intelligence quality in a compound improvement process. Organizations implementing effective data flywheels discover that competitive advantages accelerate rather than plateau because each improvement cycle enhances the foundation for subsequent improvements.

The flywheel effects operate through systematic integration of intelligence production with core business processes rather than treating AI as separate technological capability. Intelligence accumulation becomes embedded in normal operations through systematic workflows that capture learning from every business interaction, applying insights to improve operational performance automatically.

Customer interaction intelligence emerges from systematic analysis of behavior patterns, preference evolution, and satisfaction drivers that reveal insights about market dynamics unavailable through traditional market research. This intelligence enables superior customer service, more effective product development, and better market positioning, attracting additional customers whose interactions provide further intelligence.

Organizations developing customer intelligence advantages discover that their understanding of market dynamics improves faster than competitors can acquire similar insights through traditional research methods. The accumulated customer intelligence enables strategic decisions based on behavioral evidence rather than market assumptions that prove increasingly valuable as market complexity increases.

Operational intelligence accumulation creates competitive advantages through systematic learning about process optimization, resource allocation, and performance enhancement that emerges from continuous analysis of organizational activities. This accumulated operational knowledge enables superior efficiency, lower costs, and better quality that create competitive positioning advantages while generating additional operational data.

The operational intelligence advantages compound through systematic application of insights to improve business processes that generate additional operational data that enables further improvements in continuous enhancement cycles. Organizations implementing effective operational intelligence accumulation develop performance advantages that competitors cannot match through operational excellence initiatives because the improvements emerge from accumulated learning rather than best practice implementation.

Market intelligence development occurs through systematic analysis of competitive dynamics, customer behavior patterns, and industry evolution that reveals strategic opportunities and threats earlier than traditional market analysis approaches. This accumulated market intelligence enables superior strategic positioning and competitive responses, creating sustainable advantages through enhanced market understanding.

Competitive intelligence creation operates through the systematic monitoring and analysis of market developments, enabling proactive strategic responses rather than reactive competitive positioning. Organizations with effective competitive intelligence systems identify market opportunities and threats faster than competitors while developing strategic responses based on comprehensive market understanding rather than limited competitive analysis.

Business Model Innovation Through Intelligence

Proprietary intelligence enables business model innovations that create sustainable competitive advantages through value creation approaches that become more effective through usage rather than depleting through consumption. These intelligence-driven business models create competitive dynamics that favor systematic intelligence production over traditional operational excellence.

Self-improving products and services represent business model innovations where customer usage automatically enhances value delivery through systematic learning, continuously improving performance. Unlike traditional products that depreciate through usage, self-improving offerings become more valuable through customer interactions that provide intelligence for automatic enhancement.

Organizations implementing self-improving business models discover that customer satisfaction and competitive positioning improve automatically through normal business operations rather than requiring deliberate enhancement investments. Customer usage drives continuous improvement that creates superior value delivery while reducing operational costs through systematic optimization based on accumulated usage intelligence.

Platform business models create competitive advantages through network effects that improve value delivery for all participants as platform usage increases. Intelligence accumulation enhances platform effectiveness through systematic learning about participant behavior, interaction patterns, and value creation opportunities that improve platform performance automatically.

The platform advantages compound through systematic intelligence accumulation that enables superior matching, recommendation, and optimization services that attract additional participants whose interactions improve platform effectiveness in accelerating improvement cycles. Organizations implementing effective platform intelligence develop competitive advantages that strengthen through market success rather than eroding through competitive imitation.

Subscription intelligence models create competitive advantages through systematic learning about customer preferences, usage patterns, and satisfaction drivers that enable superior retention and satisfaction that reduces customer acquisition costs while increasing lifetime value. The accumulated customer intelligence enables continuous service improvement that strengthens competitive positioning automatically.

Organizations developing subscription intelligence advantages discover that customer relationships improve through systematic learning that enables increasingly personalized and valuable service delivery. The accumulated relationship intelligence creates switching costs for customers while enabling superior acquisition effectiveness that compounds competitive advantages through customer success.

Marketplace intelligence development enables competitive advantages through systematic understanding of supply and demand dynamics, pricing optimization, and match quality that improves marketplace effectiveness automatically through accumulated transaction intelligence. The marketplace intelligence creates superior value delivery for all participants while enabling optimal resource allocation and pricing strategies.

Real-World Competitive Transformation

Organizations implementing systematic intelligence production achieve competitive advantages that extend far beyond efficiency improvements into fundamental transformation of competitive positioning through superior market responsiveness, customer understanding, and operational excellence that compounds through accumulated intelligence.

Speed and agility advantages emerge from systematic intelligence production that enables faster decision making, more effective problem solving, and superior market responsiveness through accumulated learning that accelerates organizational capabilities. Organizations with effective intelligence manufacturing develop competitive timing advantages that enable market leadership through superior responsiveness rather than just operational efficiency.

Advizex achieved remarkable competitive transformation through systematic intelligence production that reduced quote generation time by 96% while increasing qualified lead conversion by 18%. The transformation occurred through employee-led innovation that systematically captured and applied learning about customer preferences, market dynamics, and operational optimization that created compound competitive advantages.

The Advizex transformation demonstrates how systematic intelligence production creates competitive advantages through accumulated learning rather than technological superiority. The competitive benefits emerged from the systematic application of insights to improve business processes that generated additional intelligence for further improvements in self-reinforcing enhancement cycles.

Productivity multiplication effects create competitive advantages through systematic intelligence production that enables superior resource allocation, process optimization, and performance enhancement that compounds through accumulated operational learning. Organizations implementing effective productivity intelligence achieve superior efficiency that creates cost advantages while enabling superior quality and service delivery.

Vizient achieved 4x return on investment and $700,000 in first-year savings through systematic intelligence production that optimized resource allocation and operational performance through accumulated learning about process effectiveness and performance optimization. The productivity advantages emerged from the systematic application of intelligence to improve business operations, which generated additional insights for continuous enhancement.

Talent retention and engagement advantages emerge from systematic intelligence production that enables superior work experiences through accumulated learning about employee preferences, development opportunities, and satisfaction drivers. Organizations implementing effective talent intelligence create workplace advantages that attract and retain superior talent while enabling superior performance through accumulated learning about team effectiveness.

The talent advantages create competitive dynamics where success attracts additional talent through superior workplace experiences that enable superior performance that attracts additional talent in self-reinforcing cycles. Organizations with effective talent intelligence develop human capital advantages that competitors cannot replicate through compensation or benefits because the advantages emerge from accumulated learning about employee success factors.

Competitive intelligence acceleration enables superior strategic positioning through systematic monitoring and analysis of market developments that reveal opportunities and threats faster than traditional competitive analysis. Organizations implementing effective competitive intelligence systems achieve strategic advantages through superior market understanding that enables proactive rather than reactive competitive positioning.

Intelligence as Intellectual Property

Proprietary intelligence represents valuable intellectual property that creates sustainable competitive advantages through accumulated learning that cannot be easily replicated or acquired by competitors. Unlike traditional intellectual property that provides temporary competitive protection, intelligence-based intellectual property becomes more valuable through usage while becoming more difficult to replicate as accumulation increases.

Data assets represent intellectual property that creates competitive advantages through unique information accumulation that enables superior insights and decision-making unavailable to competitors. Organizations systematically developing data assets discover that their competitive intelligence improves continuously through normal business operations, while becoming increasingly difficult for competitors to replicate.

The data asset development requires systematic approaches to information capture, processing, and application that transform routine business activities into intelligence accumulation processes. Effective data asset development creates proprietary insights that enable superior business decisions while generating additional data that enhances competitive positioning automatically.

Model sophistication emerges from systematic intelligence production that creates algorithmic capabilities unavailable to competitors through accumulated learning rather than technological advancement. Organizations developing model sophistication advantages achieve superior predictive accuracy and decision support through systematic intelligence accumulation rather than algorithmic innovation alone.

The model sophistication advantages compound through systematic learning that improves algorithmic performance through accumulated experience rather than theoretical enhancement. Organizations implementing effective model development create intelligence capabilities that improve automatically through business operations, while becoming increasingly difficult for competitors to replicate through technological acquisition.

Process optimization intelligence creates competitive advantages through accumulated learning about operational excellence that enables superior efficiency, quality, and performance through systematic enhancement rather than best practice implementation. The process intelligence enables continuous improvement that creates sustainable competitive advantages through accumulated operational learning.

Organizations developing process intelligence advantages achieve superior operational performance through systematic learning about improvement opportunities that emerges from continuous analysis of operational activities. The accumulated process intelligence enables superior resource allocation, quality control, and performance enhancement that creates sustainable competitive positioning.

Strategic intelligence development creates competitive advantages through accumulated learning about market dynamics, competitive positioning, and strategic opportunities that enables superior strategic decision making through systematic analysis rather than traditional strategic planning. The strategic intelligence enables proactive competitive positioning through superior market understanding.

The strategic intelligence advantages compound through systematic learning that improves strategic decision making through accumulated market experience rather than theoretical strategic planning. Organizations implementing effective strategic intelligence development achieve superior competitive positioning through accumulated market learning that enables superior strategic responses to competitive challenges and market opportunities.

These intelligence-based intellectual property advantages create sustainable competitive moats that become stronger through business success rather than eroding through competitive imitation. Organizations mastering proprietary intelligence development discover that competitive advantages accelerate through market leadership rather than attracting imitation that reduces competitive differentiation because the advantages emerge from accumulated learning that competitors cannot acquire through technological advancement or strategic imitation.

FROM AI PROJECTS TO CONTINUOUS PRODUCTION

The countdown reached zero at exactly 9:00 AM Pacific Time. Across the Tesla Fremont factory, 847 robotic assembly stations simultaneously updated their motion algorithms. In Austin, 623 quality control cameras upgraded their defect detection models. At Gigafactory Shanghai, 1,240 battery optimization systems refined their charging protocols. Every production line, in every facility, across three continents, evolved simultaneously.

No engineers worked overtime. No data scientists manually deployed updates. No executives approved individual changes. The entire manufacturing intelligence network had upgraded itself automatically through systematic learning accumulated from 2.3 million vehicles produced in the previous quarter. What once required armies of specialists and months of planning now happened invisibly, continuously, and flawlessly through production systems that improved themselves faster than human teams could even comprehend the changes.

Meanwhile, at Global Automotive Inc., a team of twelve data scientists celebrated completing their latest AI project. After eight months of development, they had successfully created a predictive maintenance model for paint booth equipment. The algorithm worked brilliantly, reducing unexpected downtime by 34% at their Detroit facility. Now came the hard part: manually adapting the solution for their seventeen other manufacturing locations, each with different equipment configurations, operational procedures, and data formats.

The contrast reveals the chasm between project-based AI implementation and continuous intelligence production. While traditional organizations exhaust themselves scaling individual successes one location at a time, systematic producers evolve automatically across unlimited applications through production architectures that treat improvement as inevitable rather than exceptional.

The transformation from discrete projects to continuous production represents the final evolution in organizational intelligence capability. Organizations mastering this transition discover that competitive advantages accelerate automatically through systematic production rather than requiring deliberate enhancement investments that traditional approaches demand.

The Production Pipeline Revolution

Traditional AI development operates through project-based approaches that treat each application as a unique creation requiring custom development from initial conception through final deployment. This artisan approach creates several fundamental limitations that prevent organizations from achieving systematic intelligence production at enterprise scale.

Project isolation means that learning from successful implementations remains trapped within individual initiatives rather than benefiting subsequent applications. Teams developing similar solutions for different departments or locations begin with blank templates rather than building on accumulated organizational intelligence. Each project recreates similar workflows, rediscovers comparable insights, and resolves identical challenges through individual effort rather than systematic replication.

Resource inefficiency emerges from project approaches that require dedicated teams, custom infrastructure, and specialized expertise for each implementation. Organizations implementing multiple AI projects simultaneously discover that resource requirements multiply rather than benefitting from shared capabilities or systematic efficiency improvements that scale across applications.

Quality inconsistency results from project approaches that depend on individual team expertise rather than systematic processes that ensure reliable outcomes. Different teams produce solutions with varying quality levels, maintenance requirements, and operational characteristics that create ongoing management complexity rather than systematic excellence.

Production pipeline architecture transforms AI development through systematic workflows that convert intelligence requirements into deployed capabilities through standardized processes that scale automatically across unlimited applications. Pipeline architecture operates through containerized workflows that execute complex intelligence production through coordinated automation rather than individual expertise.

Directed acyclic graph (DAG) structures provide systematic frameworks for organizing intelligence production workflows through connected processing stages that execute automatically in proper sequence while accommodating complex dependencies and parallel processing requirements. DAG structures enable sophisticated intelligence production through systematic orchestration that eliminates coordination complexity while ensuring reliable execution.

The DAG architecture includes automated dependency management that ensures processing stages execute in proper sequence while optimizing resource utilization through parallel execution, where workflow logic permits. Automated dependency management eliminates manual coordination requirements while enabling complex intelligence production through systematic workflow orchestration.

Containerized processing enables systematic deployment and execution of intelligence workflows through standardized environments that eliminate configuration complexity while ensuring consistent performance across diverse infrastructure environments. Containerization provides systematic isolation and resource management that enables reliable intelligence production regardless of underlying infrastructure characteristics.

Automated orchestration coordinates complex intelligence production workflows through systematic scheduling and resource allocation that optimizes performance while maintaining reliability standards. Orchestration systems manage workflow execution automatically while providing monitoring and alerting capabilities that enable proactive problem resolution.

Continuous Delivery and Deployment Excellence

Continuous delivery transforms AI development from periodic deployment cycles to systematic production workflows that deliver intelligence capabilities automatically through validated processes that maintain quality while accelerating time-to-production. Continuous delivery eliminates deployment bottlenecks while ensuring that intelligence improvements reach operational systems rapidly and reliably.

Automated model training creates systematic approaches to algorithm development that eliminate manual intervention while ensuring consistent quality through standardized training processes. Automated training operates through systematic parameter optimization and validation workflows that produce reliable models without requiring individual data scientist involvement for routine development activities.

The automated training includes hyperparameter optimization that systematically explores algorithm configuration spaces to identify optimal performance settings through automated search processes that exceed manual optimization capabilities. Systematic hyperparameter optimization ensures that deployed models achieve superior performance while eliminating guesswork that characterizes manual optimization approaches.

Model validation automation ensures that algorithm development maintains quality standards through systematic testing workflows that validate performance against established criteria before deployment authorization. Automated validation eliminates manual review bottlenecks while ensuring that deployed models meet reliability and performance requirements through systematic quality assurance.

Deployment automation transforms model deployment from manual processes requiring specialist expertise to systematic workflows that deliver validated algorithms to production environments automatically. Automated deployment eliminates human error while ensuring consistent deployment procedures that maintain security and performance standards across multiple environments and applications.

The deployment automation includes automated rollback capabilities that restore previous algorithm versions when performance degradation or operational problems emerge. Automated rollback ensures operational stability while enabling aggressive improvement deployment through systematic risk management that maintains business continuity.

Configuration management systems provide systematic control over algorithm deployment settings and operational parameters through automated configuration that eliminates manual setup requirements. Configuration management ensures consistent algorithm operation while enabling systematic optimization of performance parameters based on operational requirements.

Blue-green deployment strategies enable systematic algorithm updates through parallel environment management that validates new algorithm performance before switching production traffic. Blue-green deployment eliminates deployment risk while enabling rapid algorithm improvement through systematic validation that ensures operational continuity.

Quality Assurance and Performance Monitoring

Systematic quality assurance transforms AI reliability from periodic validation activities to continuous monitoring and automated correction that maintains performance standards automatically while identifying enhancement opportunities through accumulated operational intelligence.

Real-time performance monitoring tracks algorithm effectiveness continuously through systematic measurement of prediction accuracy, processing performance, and business impact. Real-time monitoring enables immediate identification of performance degradation while providing operational intelligence that supports systematic improvement through accumulated performance analysis.

The performance monitoring includes automated alert generation that notifies operations teams when algorithm performance falls below acceptable thresholds. Automated alerting enables rapid response to operational problems while maintaining systematic oversight of intelligence production performance across multiple applications and infrastructure components.

Model drift detection identifies algorithm performance degradation caused by changing data patterns or operational conditions through systematic analysis of prediction accuracy trends. Drift detection enables proactive algorithm updates that maintain performance standards while preventing operational problems that could impact business performance.

Automated model retraining addresses performance degradation through systematic algorithm updates that incorporate new data patterns while maintaining proven performance characteristics. Automated retraining maintains algorithm effectiveness through systematic enhancement that operates transparently within production workflows without requiring manual intervention.

Data quality monitoring ensures that intelligence production receives reliable input materials through systematic validation of data integrity, completeness, and consistency. Data quality monitoring identifies input problems that could compromise algorithm performance while enabling systematic correction that maintains production reliability.

The data quality systems include automated data cleaning and preprocessing that corrects common problems while flagging exceptional cases for human review. Automated data processing maintains input quality while reducing manual intervention requirements that could create bottlenecks in systematic intelligence production.

Bias detection and mitigation systems monitor algorithm outputs for discriminatory patterns while implementing systematic corrections that ensure ethical AI operation. Bias monitoring operates continuously rather than through periodic audits while enabling systematic enhancement of algorithm fairness through accumulated operational analysis.

A/B testing integration enables systematic validation of algorithm improvements through controlled experiments that measure actual business impact rather than theoretical performance metrics. Integrated testing validates algorithm enhancements while enabling systematic rollout of improvements that demonstrate superior business value.

Scalability Architecture and Systematic Replication

Scalability architecture enables systematic intelligence production across unlimited applications through standardized workflows and automated resource management that eliminate manual scaling efforts while maintaining performance and quality standards across diverse operational environments.

Template-based development provides systematic approaches to intelligence application creation through standardized patterns that incorporate best practices while accommodating specific application requirements. Template development eliminates custom implementation requirements while enabling rapid application development through proven systematic approaches.

The template architecture includes automated customization capabilities that adapt standardized intelligence workflows to specific business requirements through systematic configuration rather than custom development. Automated customization enables rapid application deployment while maintaining systematic quality and performance standards.

Multi-tenant architecture enables systematic resource sharing across multiple intelligence applications through sophisticated isolation and resource allocation that optimizes infrastructure utilization while maintaining security and performance standards. Multi-tenant architecture eliminates resource duplication while enabling systematic scaling that reduces operational costs.

Microservices design enables systematic intelligence production through modular architecture that supports independent development, deployment, and scaling of intelligence components. Microservices architecture provides systematic flexibility while enabling complex intelligence applications through coordinated service interaction.

The microservices approach includes automated service discovery and communication that enables systematic coordination between intelligence components without manual configuration. Automated service coordination eliminates integration complexity while enabling sophisticated intelligence applications through systematic component interaction.

API standardization provides systematic interfaces that enable different intelligence components and external systems to integrate effectively while maintaining security and performance standards. Standardized APIs eliminate integration complexity while enabling systematic expansion of intelligence capabilities through external system connectivity.

Auto-scaling capabilities provide systematic resource management that adjusts computational capacity based on demand patterns while maintaining performance standards and optimizing costs. Auto-scaling eliminates manual resource management while ensuring that intelligence production maintains performance standards during demand variations.

Evolution and Continuous Improvement

Systematic intelligence production enables continuous evolution through accumulated operational intelligence that identifies improvement opportunities and implements enhancements automatically through self-improving production systems that become more effective through operational experience.

Kaizen principles applied to intelligence production create systematic approaches to continuous improvement that eliminate waste while optimizing workflow efficiency through accumulated operational analysis. Continuous improvement operates through systematic identification and resolution of inefficiencies that emerge from normal production operations.

The improvement process includes automated bottleneck identification that analyzes production workflows to identify performance constraints while recommending systematic enhancements that improve overall production efficiency. Systematic bottleneck analysis enables targeted improvements that optimize production performance through evidence-based enhancement.

Feedback loop integration captures operational intelligence from production systems while applying insights to improve intelligence production workflows automatically. Feedback loops enable systematic learning that improves production effectiveness while reducing operational costs through accumulated operational wisdom.

Learning acceleration mechanisms enable systematic capture and application of operational insights that improve intelligence production capabilities through accumulated experience rather than theoretical enhancement. Learning acceleration creates systematic improvement that compounds competitive advantages through operational excellence.

The learning systems include automated best practice identification that analyzes successful intelligence production patterns while implementing systematic improvements across multiple applications. Best practice automation enables systematic scaling of successful approaches while eliminating manual improvement processes.

Performance optimization algorithms analyze production system performance continuously while implementing systematic enhancements that improve efficiency and effectiveness automatically. Performance optimization enables systematic improvement that maintains competitive advantages through superior operational excellence.

Predictive maintenance capabilities monitor production system health while implementing systematic preventive actions that maintain operational reliability. Predictive maintenance prevents operational problems while ensuring systematic intelligence production maintains performance standards through proactive system management.

Governance and Compliance Integration

Systematic intelligence production integrates governance and compliance requirements directly into production workflows rather than treating them as external oversight activities that create operational friction. Integrated governance ensures that intelligence production maintains ethical and regulatory compliance automatically while enabling rapid improvement through systematic enhancement.

Real-time compliance monitoring validates intelligence production activities against regulatory requirements continuously, rather than through periodic audits that identify problems after they impact business operations. Real-time compliance enables proactive problem resolution while maintaining systematic oversight of governance requirements.

Automated ethics validation ensures that intelligence production maintains ethical standards through systematic monitoring of algorithm outputs and operational decisions. Ethics validation operates transparently within production workflows while enabling systematic improvement of ethical AI operation through accumulated ethical intelligence.

The ethics systems include automated bias detection and mitigation that identify discriminatory patterns while implementing systematic corrections that ensure fair AI operation. Automated bias correction maintains ethical standards while enabling systematic enhancement of algorithm fairness through operational learning.

Risk management frameworks provide systematic identification and mitigation of operational risks through automated monitoring and response systems that maintain business continuity while enabling aggressive improvement through systematic risk assessment. Risk management enables rapid enhancement while maintaining operational stability.

Audit trail generation provides systematic documentation of intelligence production activities and decisions that supports regulatory compliance while enabling the systematic analysis of operational patterns. Automated audit trails eliminate manual documentation requirements while supporting systematic improvement through operational analysis.

Privacy protection systems ensure that intelligence production maintains data privacy standards through systematic access controls and data handling procedures that operate transparently within production workflows. Privacy protection maintains regulatory compliance while enabling systematic intelligence production through ethical data utilization.

Security integration provides systematic protection for intelligence production through automated security controls and monitoring systems that maintain operational security while enabling rapid improvement through systematic threat management. Security integration enables systematic enhancement while maintaining protection against operational threats.

Measurement and Optimization Systems

Systematic intelligence production requires sophisticated measurement systems that track production effectiveness, competitive positioning, and business impact through comprehensive metrics that enable the systematic optimization of intelligence manufacturing capabilities.

Intelligence velocity metrics measure the speed and effectiveness of intelligence production through systematic tracking of development cycles, deployment frequency, and improvement rates. Intelligence velocity provides operational intelligence that enables systematic optimization of production efficiency while maintaining quality standards.

The velocity metrics include time-to-production measurements that track the duration required to convert intelligence requirements into deployed capabilities. Time-to-production metrics enable systematic improvement of production efficiency while identifying bottlenecks that constrain intelligence manufacturing capabilities.

Competitive advantage accumulation metrics measure the business impact of systematic intelligence production through the tracking of market positioning, customer satisfaction, and operational performance improvements. Competitive metrics provide strategic intelligence that validates intelligence production investments while guiding systematic enhancement priorities.

Quality assurance metrics track intelligence production reliability and performance through the systematic measurement of algorithm accuracy, operational stability, and business value delivery. Quality metrics enable the systematic improvement of production excellence while maintaining performance standards that support competitive positioning.

The quality systems include automated anomaly detection that identifies unusual patterns in production metrics while enabling the systematic investigation and resolution of quality issues. Anomaly detection maintains systematic oversight while enabling proactive quality management through automated monitoring.

Resource optimization metrics track intelligence production efficiency through the systematic measurement of computational resource utilization, development effort requirements, and operational costs. Resource metrics enable the systematic optimization of production efficiency while reducing operational costs through evidence-based improvement.

Business impact measurement systems track the strategic value of intelligence production through systematic analysis of revenue impact, cost reduction, and competitive positioning improvements. Business impact metrics validate intelligence production investments while guiding strategic enhancement priorities through accumulated business intelligence.

Return on investment (ROI) analysis provides a systematic evaluation of intelligence production value through comprehensive measurement of costs and benefits that enables strategic optimization of intelligence manufacturing investments. ROI analysis enables the systematic improvement of production value while optimizing resource allocation across multiple intelligence production initiatives.

This comprehensive measurement and optimization approach enables systematic intelligence production that improves automatically through operational experience while creating sustainable competitive advantages through superior intelligence manufacturing capabilities that compound business success through accumulated operational excellence.

CHAPTER 5

The Intelligence Stack – Building Your Digital Nervous System

The octopus doesn't have a brain. Not in the way humans understand brains. Instead, it has a distributed intelligence system where two-thirds of its neurons live in its arms. Each arm can taste, touch, and react independently while remaining coordinated with the whole organism. When an octopus encounters a threat, it doesn't send signals to a central processor for analysis and response. The entire body becomes the thinking system, responding with coordinated intelligence that emerges from distributed processing rather than centralized control.

Netflix operates exactly like this octopus. When a viewer in Tokyo pauses a Korean drama, recommendation engines across three continents immediately adjust algorithms for similar content. When teenagers in São Paulo binge-watch a new series, content acquisition teams in Los Angeles receive automatic signals about emerging viewer preferences. When bandwidth fluctuates in Mumbai, compression algorithms adapt globally to maintain streaming quality. There's no central brain making these decisions. The entire platform has become an intelligent organism where millions of micro-decisions create coordinated behavior that no individual system could achieve alone.

This biological approach to organizational intelligence represents a fundamental departure from traditional enterprise architecture. Most companies still operate like primitive organisms with centralized nervous systems. Problems travel slowly to executive brains for analysis and decision-making. Solutions flow back down through hierarchical pathways that create delays and lose nuance in translation. The entire organization becomes only as intelligent as its slowest decision-making process.

The enterprises achieving exponential advantages have evolved beyond this centralized model to create what neuroscientists would recognize as distributed intelligence systems. Like the octopus, these organizations embed thinking throughout their operational tissue rather than concentrating intelligence in executive headquarters. Information processing happens everywhere. Learning occurs continuously. Responses emerge from collective intelligence rather than individual decision-making.

The transformation from traditional enterprise architecture to digital nervous system represents the next evolutionary leap in organizational intelligence. Where previous chapters explored intelligence manufacturing and systematic production, the intelligence stack creates the technological equivalent of biological consciousness that enables organizations to sense, think, remember, and respond with the speed and sophistication of living systems.

The enterprises building these digital nervous systems discover competitive advantages that transcend operational efficiency. They develop organizational reflexes that respond to market changes before competitors recognize that changes are occurring. They accumulate institutional memory that improves decision-making automatically. They create intelligence that emerges from the interaction between data, models, memory, and feedback loops rather than from individual algorithmic components.

Building this digital nervous system requires understanding how biological intelligence actually works and translating those principles into technological architecture. The human nervous system doesn't process information through isolated systems. It operates through integrated networks where data flows continuously between sensors, memory systems, processing centers, and response mechanisms that work together to create consciousness and intelligent behavior.

The intelligence stack enables similar integration through four interconnected layers that mirror biological neural architecture. Data and models form the sensory and processing foundation. Memory systems accumulate and organize organizational knowledge. Feedback loops enable continuous learning and adaptation. The complete architecture creates continuous intelligence that transcends individual technological components to generate organizational consciousness that thinks, learns, and evolves automatically.

This digital nervous system transforms how organizations respond to complexity, uncertainty, and change. Instead of managing disruption through human analysis and deliberate decision-making, intelligence stacks enable automatic adaptation that occurs faster than conscious thought while maintaining strategic alignment and operational excellence. The result is organizational intelligence that operates like biological consciousness but at digital speed and enterprise scale.

DATA, MODELS, MEMORY, AND FEEDBACK LOOPS

A honeybee returns to the hive carrying pollen and crucial intelligence about a rich flower patch two kilometers away. Instead of simply depositing her cargo, she performs an intricate dance that encodes distance, direction, and quality in precise movements that other bees interpret with mathematical accuracy. The entire hive watches, learns, and adapts. Within minutes, hundreds of foragers adjust their flight patterns based on this shared intelligence. The colony doesn't just collect resources; it accumulates knowledge that makes every subsequent foraging expedition more efficient than the last.

Spotify operates through identical principles of collective intelligence accumulation. When a user in Berlin skips a song three seconds in, that micro-rejection becomes data that improves recommendations for millions of other listeners. When someone in Nashville creates the perfect workout playlist, those musical patterns inform the algorithmic understanding of energy and tempo relationships. When a teenager in Mumbai discovers an obscure artist, that preference signal ripples through recommendation engines to surface similar talents for listeners with compatible tastes. Every interaction becomes institutional memory that makes the entire platform more intelligent with each passing moment.

This represents the fundamental architecture of what researchers call circular AI business models, where usage doesn't deplete value but compounds it through systematic intelligence accumulation. Unlike traditional business models where consumption reduces inventory, circular AI systems become more valuable through customer engagement that generates the intelligence necessary for continuous improvement.

The circular dynamic emerges from sophisticated integration between four foundational components that mirror biological learning systems. Data streams provide the sensory input that keeps organizational intelligence connected to reality. Models function as cognitive processors that transform raw information into actionable insights. Memory systems accumulate and organize learned knowledge for instant retrieval. Feedback loops enable continuous adaptation that improves performance through experience rather than theoretical optimization.

The Mechanics of Self-Improving Intelligence Loops

Understanding how these components work together reveals why certain organizations achieve exponential advantages while others struggle with diminishing returns despite massive AI investments. The difference lies not in technological sophistication but in architectural design that enables systematic improvement through usage rather than degradation through consumption.

The self-improvement process begins with input reception that captures relevant information from operational activities, customer interactions, market signals, and environmental changes. This input differs fundamentally from traditional data collection because it focuses on learning opportunities rather than transaction recording. Every customer choice, operational decision, and market interaction becomes potential intelligence that could improve future performance if processed appropriately.

Advanced input systems operate through continuous sensing that monitors organizational activities for learning signals rather than just performance metrics. These systems identify patterns in customer behavior, operational efficiency, competitive dynamics, and market evolution that suggest opportunities for intelligence enhancement. The sensing capabilities extend beyond traditional business intelligence to include weak signals that indicate emerging trends or changing conditions before they impact performance significantly.

Input processing transforms raw information into structured knowledge through algorithmic analysis that identifies patterns, relationships, and insights that human analysis might miss. This processing operates continuously rather than periodically, enabling real-time learning that keeps organizational intelligence current with evolving market conditions. The processing includes pattern recognition that identifies recurring themes, anomaly detection that highlights unusual variations, and correlation analysis that reveals hidden relationships between seemingly unrelated variables.

Output generation creates actionable insights, recommendations, and automated responses based on processed intelligence. Unlike traditional analytical outputs that require human interpretation and action, circular AI systems generate responses that can be implemented automatically or that provide clear guidance for human decision-making. The outputs include predictive insights about future conditions, prescriptive recommendations for optimal actions, and automated responses that implement proven solutions without requiring human intervention.

Feedback reception captures information about output effectiveness through systematic monitoring of results, outcomes, and impacts. This feedback operates through multiple channels, including direct customer responses, operational performance metrics, market feedback, and competitive positioning indicators. The feedback collection focuses on learning opportunities rather than just performance measurement, seeking insights that could improve future output quality and relevance.

Algorithm adjustment represents the learning core of circular AI systems, where feedback information enables the systematic improvement of processing capabilities. Unlike traditional systems that require manual updates or periodic retraining, circular systems adapt continuously through automated learning algorithms that refine performance based on accumulated experience. This adjustment includes parameter optimization that improves prediction accuracy, model refinement that enhances analytical capabilities, and architectural evolution that enables new forms of intelligence processing.

Adaptive Algorithms and Continuous Learning Architecture

The technological foundation enabling circular AI business models relies on adaptive algorithms that modify their behavior automatically based on operational feedback rather than requiring manual programming for each new scenario or market condition. These algorithms represent a fundamental evolution beyond static AI systems that perform predetermined functions to dynamic intelligence that improves through experience.

Adaptive models demonstrate the ability to adjust parameters, modify processing approaches, and evolve structural characteristics in response to changing data patterns or performance feedback. This adaptability enables organizations to maintain optimal performance as market conditions evolve without requiring complete system redesigns or manual intervention from technical specialists. The models learn continuously from operational experience while maintaining consistent performance standards.

The adaptation process operates through several sophisticated mechanisms that enable systematic improvement without compromising operational stability. Transfer learning enables algorithms to apply knowledge gained from one domain to improve performance in related areas, accelerating learning while reducing the data requirements for achieving optimal performance in new applications. Meta-learning develops algorithmic capabilities to learn more effectively from limited data by understanding general principles of learning that apply across multiple scenarios.

Reinforcement learning enables algorithms to improve performance through trial-and-error approaches that explore different actions and learn from their consequences. This learning mechanism proves particularly valuable for complex decision-making scenarios where optimal actions cannot be predetermined but must be discovered through experimentation. The reinforcement approach enables continuous optimization that adapts to changing conditions while maintaining operational effectiveness.

Evolutionary algorithms optimize system performance through systematic testing of variations that mimic biological evolution processes. These algorithms generate multiple solution variations, test their effectiveness under current conditions, and promote successful approaches while eliminating less effective alternatives. The evolutionary approach enables systematic improvement that maintains diversity while optimizing for current performance requirements.

Neural network architectures provide sophisticated pattern recognition and decision-making capabilities that can adapt to complex data relationships without requiring explicit programming for each scenario. These networks learn from examples rather than rules, enabling them to handle situations that programmers couldn't anticipate or explicitly code. The neural approach enables flexible intelligence that can adapt to novel scenarios while maintaining performance consistency.

Deep learning extends neural network capabilities to handle complex data processing requirements that involve multiple layers of abstraction and analysis. This deep processing enables algorithms to understand sophisticated relationships and patterns that simpler approaches cannot detect. The deep learning approach enables advanced intelligence capabilities that can process complex information while generating insights that improve operational effectiveness.

Memory Systems and Organizational Intelligence Accumulation

The competitive advantage of circular AI business models emerges from sophisticated memory systems that accumulate, organize, and retrieve organizational intelligence in ways that compound learning benefits over time. Unlike traditional databases that store information statically, intelligent memory systems create dynamic knowledge repositories that become more valuable through usage and more sophisticated through accumulated experience.

Institutional memory creation occurs through the systematic capture and organization of operational experiences, decision outcomes, market feedback, and performance patterns that create comprehensive organizational knowledge bases. This memory differs from traditional data storage because it focuses on learning accumulation rather than transaction recording. Every successful strategy, effective solution, and positive outcome becomes institutional knowledge that improves future decision-making capabilities.

Knowledge graph architecture provides a sophisticated representation of organizational intelligence through interconnected information networks that capture relationships, dependencies, and contextual connections between different knowledge elements. These graphs enable complex queries and reasoning processes that simple databases cannot support. The graph structure enables the discovery of insights that emerge from relationship analysis rather than individual data point examination.

Semantic organization ensures that accumulated knowledge remains accessible and useful through intelligent categorization, tagging, and linking systems that enable the rapid retrieval of relevant information. This organization goes beyond traditional database structures to include contextual relationships, similarity measures, and relevance rankings that help identify the most appropriate knowledge for specific situations. The semantic approach enables intelligent knowledge management that improves over time.

Contextual retrieval capabilities enable automatic identification and delivery of relevant knowledge based on current situations, decisions, or challenges, rather than requiring manual search and analysis. These capabilities operate through sophisticated matching algorithms that understand situational context and identify knowledge that could improve decision-making effectiveness. The contextual approach enables just-in-time intelligence that supports optimal decision-making without information overload.

Pattern-based learning accumulates insights about successful approaches, effective strategies, and optimal solutions through the systematic analysis of historical outcomes and performance patterns. This learning enables organizations to identify what works under different conditions and apply those insights to improve future performance. The pattern approach enables evidence-based decision-making that builds on accumulated organizational experience.

Analogical reasoning enables organizations to apply insights from one situation to improve performance in similar but different scenarios. This reasoning capability identifies underlying principles and success factors that transfer across different contexts, enabling organizations to leverage accumulated knowledge more broadly. The analogical approach enables knowledge amplification that extends learning benefits beyond specific situations.

Integration Architecture for Continuous Intelligence

The ultimate effectiveness of circular AI business models depends on sophisticated integration between data streams, processing models, memory systems, and feedback mechanisms that creates coherent intelligence capabilities exceeding what individual components can achieve independently. This integration creates emergent intelligence that operates at an organizational scale while maintaining individual component optimization.

Real-time data integration ensures that intelligence systems maintain a current awareness of operational conditions, market changes, customer behavior, and competitive dynamics through continuous information flow rather than periodic updates. This real-time awareness enables responsive decision-making that adapts to changing conditions without delays that could compromise performance or create missed opportunities.

Model coordination enables different analytical algorithms to work together effectively through standardized interfaces and communication protocols that allow sophisticated analysis across multiple domains simultaneously. This coordination creates analytical capabilities that exceed what individual models can achieve while maintaining computational efficiency and avoiding redundant processing.

Memory access optimization ensures that accumulated knowledge remains available for real-time decision-making through sophisticated retrieval systems that balance comprehensive access with response speed requirements. This optimization enables immediate availability of relevant knowledge without creating performance bottlenecks that could slow decision-making processes.

Feedback synthesis combines information from multiple sources, including customer responses, operational metrics, market indicators, and competitive feedback, to create a comprehensive understanding of system performance and improvement opportunities. This synthesis enables holistic learning that accounts for multiple perspectives and success criteria rather than optimizing for individual metrics that might not reflect overall effectiveness.

Continuous adaptation coordinates improvement activities across all system components to ensure that learning benefits accumulate systematically rather than creating conflicts or performance degradation. This coordination enables sustained improvement that maintains system stability while enabling continuous enhancement of intelligence capabilities.

The integration creates organizational intelligence that learns continuously, adapts automatically, and improves systematically through operational experience. Organizations implementing effective circular AI architectures discover that competitive advantages accelerate through usage rather than eroding through competition because their intelligence systems become more sophisticated and effective with every customer interaction, operational decision, and market experience.

ENTERPRISE KNOWLEDGE GRAPHS AND PROMPT ENGINES

The human brain doesn't store memories like a filing cabinet. Instead, it weaves experiences into intricate webs where remembering your grandmother's kitchen instantly connects to the smell of cinnamon, which links to holiday mornings, which flows to family conversations, which branches to dozens of related emotions, people, and moments. Each memory exists not as isolated information but as part of a vast relational network where meaning emerges from connections rather than individual facts.

Modern enterprises have discovered that knowledge works in exactly the same way. When JPMorganChase analyzes market risk, its intelligence graphs don't just process individual data points about interest rates, currency fluctuations, or commodity prices. Instead, they map the dynamic relationships between economic indicators, geopolitical events, regulatory changes, and market sentiment in interconnected webs that reveal hidden patterns and emerging risks that isolated analysis cannot detect.

The bank's risk analysts can ask sophisticated questions in natural language and receive insights that emerge from relationship analysis across thousands of connected data points. "How might rising inflation in Germany affect our exposure to Southeast Asian manufacturing?" becomes a conversation with an intelligence system that traverses complex relationship networks to identify second- and third-order effects that traditional analytical approaches would miss entirely.

This represents the evolution from traditional data management to what researchers call enterprise knowledge graphs, sophisticated relationship-mapping systems that mirror the associative intelligence of human cognition while operating at digital scale and speed. These systems don't just store information; they create dynamic knowledge networks that become more intelligent through accumulated connections and relationship discovery.

Intelligence Graphs for Dynamic Relationship Mapping

The competitive advantage of knowledge graphs emerges from their ability to represent and analyze complex business relationships that change continuously as market conditions evolve, competitive dynamics shift, and organizational capabilities develop. Unlike traditional databases that store information in fixed structures, intelligence graphs create flexible relationship networks that adapt automatically to new information while preserving historical connection patterns.

Dynamic relationship mapping captures how business entities influence each other through direct and indirect connections that create complex interaction networks. These relationships include obvious connections like customer-supplier relationships but extend to subtle influences like how social media sentiment affects stock prices, which influences employee confidence, which impacts productivity, which affects customer satisfaction in cascading effect chains that traditional analysis cannot track systematically.

Entity recognition algorithms identify important business objects, including customers, products, markets, competitors, regulations, technologies, and trends that form the nodes in enterprise knowledge networks. Advanced recognition systems can process unstructured text, financial reports, news articles, social media content, and operational data to identify entities and their characteristics automatically. This recognition enables comprehensive knowledge mapping that captures relationships humans might miss or cannot process at an organizational scale.

Relationship inference discovers connections between entities through pattern analysis that identifies how different business elements influence each other across time and context. These inferences reveal non-obvious relationships like how weather patterns in agricultural regions affect commodity prices, which influence manufacturing costs, which impact consumer pricing, which affects market demand in complex causal chains that span multiple business domains.

Temporal analysis tracks how relationships evolve over time to identify trends, cycles, and emerging patterns that indicate changing business conditions. This temporal dimension enables organizations to understand not just current relationship networks but how those networks change in response to market events, competitive actions, or strategic initiatives. The temporal analysis reveals relationship dynamics that static analysis cannot capture.

Graph traversal capabilities enable sophisticated queries that explore relationship networks to answer complex business questions. These queries can follow multiple relationship paths simultaneously to identify insights that emerge from network analysis rather than individual data examination. Advanced traversal algorithms can explore millions of relationship connections in milliseconds to answer questions that would require weeks of manual analysis.

Multi-hop reasoning enables the discovery of insights that emerge from indirect relationships spanning multiple connection levels. This reasoning can identify how events in one business domain create effects in seemingly unrelated areas through intermediate relationship chains. The multi-hop capability enables strategic insights that emerge from relationship network analysis rather than direct causal observation.

Graph Neural Network Integration with Large Language Models

The breakthrough capabilities of modern enterprise knowledge graphs emerge from sophisticated integration between graph neural networks (GNNs) that excel at relationship analysis and large language models (LLMs) that understand natural language communication. This integration creates conversational intelligence interfaces that enable business users to interact with complex knowledge networks through natural dialogue rather than technical query languages.

GNN architecture processes relationship information through neural networks specifically designed to understand connection patterns, network structures, and relationship dynamics. These networks can analyze complex graph structures to identify patterns, predict relationship evolution, and generate insights that emerge from network analysis rather than individual node examination. GNNs excel at understanding how relationships influence outcomes in ways that traditional analytical approaches cannot capture.

Neural network processing enables sophisticated analysis of graph structures, including community detection that identifies clusters of related entities, centrality analysis that determines which entities have the greatest influence within networks, and pathway analysis that traces how effects propagate through relationship networks. These analytical capabilities reveal business insights that emerge from the relationship network structure rather than individual entity characteristics.

LLM integration enables natural language interaction with knowledge graphs through conversational interfaces that translate business questions into graph queries and present results in easily understood formats. This integration allows business users to explore complex relationship networks without requiring technical expertise in graph query languages or network analysis methods.

Natural language processing translates business questions into graph traversal operations that explore relationship networks to identify relevant information and insights. Advanced processing can understand complex questions that require multiple graph operations, temporal analysis, and relationship inference to generate comprehensive answers. The natural language capability makes sophisticated relationship analysis accessible to business users without technical training.

Response generation transforms graph analysis results into natural language explanations that provide business insights, relationship summaries, and actionable recommendations. This generation includes context explanation that helps users understand how insights were derived from relationship analysis, confidence indicators that communicate analytical certainty, and suggested follow-up questions that enable deeper exploration of discovered relationships.

Graph neural prompting represents an advanced integration technique where GNNs encode relationship knowledge into entity embeddings that are then injected into LLM inputs alongside natural language queries. This approach enables LLMs to reason about complex relationship networks while maintaining their natural language capabilities, creating hybrid intelligence that combines relationship analysis with conversational interaction.

Knowledge injection processes transform graph-based insights into structured tokens that LLMs can process alongside textual information. This injection enables LLMs to incorporate relationship knowledge into their reasoning processes, creating responses that reflect both language understanding and relationship network analysis. The injection approach enables sophisticated analysis that combines graph intelligence with conversational capability.

Prompt Engines as Conversational Intelligence Interfaces

The practical application of enterprise knowledge graphs depends on sophisticated prompt engines that enable business users to interact with complex intelligence systems through natural conversation rather than technical interfaces. These engines transform knowledge graph capabilities into accessible business tools that enhance decision-making without requiring specialized technical expertise.

Conversational interface design creates intuitive interaction patterns that feel natural to business users while enabling sophisticated analytical capabilities. These interfaces understand business context, domain terminology, and organizational priorities to provide relevant responses that address actual business needs rather than just answering literal questions. Advanced interfaces can maintain conversation context across multiple interactions to enable progressive exploration of complex topics.

Intent recognition algorithms identify what business users actually want to accomplish through their questions, enabling prompt engines to provide appropriate analytical support even when questions are ambiguous or incomplete. These algorithms understand business context, organizational priorities, and user roles to interpret questions accurately and provide relevant assistance. Intent recognition enables productive conversations that focus on business value rather than technical precision.

Query expansion transforms natural language questions into comprehensive analytical operations that explore relevant relationship networks thoroughly. This expansion can identify related entities, similar scenarios, and relevant historical patterns that provide context for answering business questions completely. The expansion ensures that responses address underlying business concerns rather than just literal question content.

Context management maintains awareness of conversation history, business context, and organizational priorities throughout extended interactions with knowledge graphs. This management enables progressive exploration where each question builds on previous answers to develop a comprehensive understanding of complex business topics. Context awareness creates conversational experiences that feel natural and productive.

Response optimization generates insights and recommendations that align with business priorities, decision-making timelines, and organizational capabilities. This optimization ensures that knowledge graph insights translate into actionable business intelligence rather than academic analysis. The optimization includes prioritization that highlights the most important insights and formatting that enables rapid comprehension and decision-making.

Explanation generation provides clear reasoning about how insights were derived from relationship analysis, enabling business users to understand and validate analytical conclusions. These explanations include relationship pathways that show how conclusions emerged from network analysis, confidence indicators that communicate analytical certainty, and alternative perspectives that acknowledge uncertainty and complexity.

Signed Influence Graphs for Competitive Landscape Mapping

Advanced enterprise knowledge graphs enable sophisticated competitive analysis through signed influence graphs that map positive and negative relationships between market participants, revealing competitive dynamics, partnership opportunities, and strategic vulnerabilities that traditional competitive analysis cannot identify systematically.

Signed relationship modeling captures both positive and negative influences between business entities, including competitive pressures, partnership benefits, regulatory constraints, and market dependencies. These signed relationships enable analysis of complex competitive dynamics where entities can simultaneously compete in some areas while collaborating in others. The signed approach reveals competitive complexity that binary relationship models cannot represent accurately.

Competitive intelligence integration processes news articles, financial reports, patent filings, partnership announcements, and market research to identify relationships between competitors, suppliers, customers, and other market participants. This intelligence gathering operates continuously to maintain current awareness of the competitive landscape evolution. Advanced processing can identify subtle relationship changes that indicate emerging competitive threats or partnership opportunities.

Influence propagation analysis traces how competitive actions create effects throughout market networks, including direct impacts on competitors and indirect effects on suppliers, customers, and partners. This propagation analysis reveals how competitive strategies create cascading effects that extend beyond immediate targets to influence entire market ecosystems. The analysis enables strategic planning that accounts for complex competitive dynamics.

Strategic positioning optimization uses signed influence graphs to identify market positions that maximize positive relationships while minimizing negative competitive pressures. This optimization can reveal strategic opportunities that emerge from relationship network analysis rather than direct competitive comparison. The positioning analysis enables differentiated strategies that leverage relationship advantages.

Vulnerability assessment identifies competitive weaknesses that emerge from relationship network dependencies, including supplier concentration, customer dependency, partnership reliance, and regulatory exposure. This assessment reveals strategic vulnerabilities that might not be apparent from financial analysis alone. The vulnerability identification enables defensive strategies that strengthen competitive positioning.

Alliance opportunity discovery identifies potential partnership arrangements that could strengthen competitive positioning through relationship network analysis. This discovery can reveal non-obvious partnership opportunities that emerge from shared interests, complementary capabilities, or mutual competitive pressures. The alliance analysis enables collaborative strategies that improve competitive positioning through relationship leveraging.

Market ecosystem mapping provides a comprehensive visualization of competitive landscapes, including all relevant relationships between participants, influences, dependencies, and strategic interactions. These maps reveal market structure patterns, competitive clusters, and strategic opportunities that emerge from relationship network analysis. The ecosystem perspective enables strategic thinking that accounts for complex market dynamics rather than simple competitive positioning.

The integration of signed influence graphs with conversational prompt engines enables business leaders to explore competitive landscapes through natural language dialogue that reveals relationship insights, strategic opportunities, and competitive vulnerabilities through sophisticated network analysis. Organizations implementing these capabilities develop competitive intelligence that operates at higher sophistication levels than traditional competitive analysis while remaining accessible through intuitive conversational interfaces that enhance strategic decision-making effectiveness.

DESIGNING FOR REAL-TIME ORGANIZATIONAL LEARNING

A master chess player doesn't think about individual moves. Instead, she recognizes patterns that have emerged from thousands of games, instantly understanding positional strengths and weaknesses that novices cannot see. Her expertise comes not from calculating every possible sequence but from accumulated pattern recognition that enables intuitive responses faster than conscious analysis. The grandmaster's mind has become a learning system that improves automatically through experience while operating at speeds that deliberate thinking cannot match.

The most successful modern enterprises operate exactly like chess grandmasters. When Uber pricing algorithms detect increased demand during a concert, they don't send requests to headquarters for approval. The system recognizes patterns from millions of similar events and adjusts pricing instantly based on accumulated learning about supply and demand dynamics. When Amazon inventory management notices unusual purchasing patterns, automated systems adjust stock levels immediately rather than waiting for human analysis. These organizations have evolved beyond deliberate decision-making to develop institutional reflexes that respond intelligently before conscious management realizes action is needed.

This transformation from reactive management to reflexive organizational intelligence represents the most profound evolution in business operations since the development of professional management. Where traditional organizations think slowly and act deliberately, learning-optimized enterprises develop automatic responses that improve continuously through accumulated experience while operating faster than human decision-making cycles.

The architecture enabling this reflexive intelligence operates through sophisticated integration between event-driven systems that sense changes instantly, autonomous agents that respond intelligently without supervision, digital twins that enable risk-free experimentation, and learning optimization mechanisms that improve performance automatically through operational experience.

Event-Driven Architectures for Instant Organizational Responses

Real-time organizational learning requires sophisticated sensing capabilities that detect significant changes, opportunities, and threats the moment they emerge rather than through periodic analysis that introduces delays between events and responses. Event-driven architecture provides the technological foundation for organizational reflexes that operate faster than traditional management cycles.

Event detection systems monitor organizational activities, market conditions, customer behaviors, and competitive dynamics continuously to identify situations that require an intelligent response. These systems distinguish between routine variations that require no action and significant changes that indicate opportunities for learning or adaptation. Advanced detection operates through pattern recognition that understands normal operational ranges and identifies deviations that suggest meaningful events requiring organizational attention.

Real-time stream processing enables immediate analysis of detected events to determine appropriate responses based on accumulated organizational learning. This processing operates through sophisticated algorithms that can analyze complex event patterns across multiple data streams simultaneously. The processing includes correlation analysis that identifies relationships between seemingly independent events and pattern matching that recognizes event types that have occurred previously and generated successful responses.

Distributed event messaging ensures that relevant organizational components receive immediate notification of significant events without creating communication bottlenecks that could slow response times. This messaging operates through sophisticated routing algorithms that deliver event information to appropriate systems while filtering out irrelevant events that could create information overload. The messaging architecture enables coordinated responses across multiple organizational functions without requiring centralized coordination that could introduce delays.

Automatic response triggering enables immediate organizational actions based on event analysis and accumulated learning about effective responses. These responses can include resource reallocation, process adjustments, communication activities, and strategic positioning changes that address detected events appropriately. The automatic triggering operates within governance frameworks that ensure responses align with organizational objectives while enabling rapid action that maintains competitive responsiveness.

Event sourcing capabilities maintain comprehensive records of organizational events and responses that enable continuous learning about effective response patterns. This sourcing creates audit trails that support compliance requirements while generating learning data that improves future response effectiveness. The event history enables pattern analysis that identifies successful response strategies and optimization opportunities that emerge from accumulated operational experience.

Scalable processing architecture ensures that event-driven systems can handle increasing volumes of organizational events without performance degradation that could compromise response times. This scalability operates through distributed processing that can add computational resources automatically as event volumes increase. The scalable approach enables organizational growth without sacrificing the real-time responsiveness that creates competitive advantages through superior reaction speeds.

Autonomous AI Agents and Capability Progression

The effectiveness of real-time organizational learning depends on autonomous AI agents that can analyze complex situations and implement appropriate responses without requiring human oversight for routine decisions. These agents operate through sophisticated progression levels that evolve from simple automation to genuine organizational intelligence that can handle unexpected situations creatively.

Level 1 agents provide AI-assisted decision support that enhances human decision-making through analytical capabilities, pattern recognition, and recommendation generation. These agents handle routine analytical tasks while escalating complex decisions to human oversight. Level 1 capabilities include data processing that prepares information for human analysis, pattern identification that highlights important trends or anomalies, and option analysis that evaluates alternative approaches based on established criteria.

Level 2 agents demonstrate partial autonomy through the automated execution of predefined responses to recognized situations. These agents can handle routine decisions independently while maintaining human oversight for exceptional cases. Level 2 capabilities include rule-based decision-making that implements established policies automatically, exception handling that identifies situations requiring human intervention, and performance monitoring that tracks decision effectiveness for continuous improvement.

Level 3 agents operate with conditional autonomy that enables independent decision-making within defined parameters while escalating complex or high-risk situations to human oversight. These agents can adapt their responses based on situational context and accumulated learning. Level 3 capabilities include contextual decision-making that considers multiple factors simultaneously, adaptive responses that modify approaches based on changing conditions, and learning integration that improves decision-making through operational experience.

Level 4 agents achieve full autonomy through independent goal-setting and strategy development that aligns with organizational objectives while operating without continuous human supervision. These agents can handle complex, novel situations through creative problem-solving and strategic thinking. Level 4 capabilities include autonomous planning that develops implementation strategies independently, creative problem-solving that generates novel solutions to unexpected challenges, and strategic alignment that ensures autonomous actions support organizational objectives.

Agent coordination mechanisms enable multiple autonomous agents to work together effectively without creating conflicts or redundant activities. This coordination operates through communication protocols that enable information sharing, task allocation algorithms that distribute responsibilities optimally, and conflict resolution procedures that address competing objectives or resource requirements. Advanced coordination enables emergent intelligence that exceeds individual agent capabilities.

Human-AI collaboration frameworks maintain appropriate human oversight while enabling maximum autonomous capability within acceptable risk parameters. These frameworks define decision-making authority levels, escalation procedures for complex situations, and performance monitoring requirements that ensure autonomous agents operate effectively while maintaining organizational control. The collaboration approach enables organizations to benefit from autonomous capabilities while maintaining strategic direction and risk management.

Governance integration ensures that autonomous agents operate within ethical guidelines, regulatory requirements, and organizational policies while maintaining the flexibility necessary for effective autonomous decision-making. This integration includes automated compliance checking that validates agent decisions against established requirements, audit trail generation that documents autonomous actions for review, and performance optimization that improves agent effectiveness while maintaining governance standards.

Digital Twins and Organizational Process Simulation

Real-time organizational learning requires safe environments for testing new approaches, optimizing processes, and exploring strategic alternatives without risking operational disruption or performance degradation. Digital twins provide virtual representations of organizational processes that enable risk-free experimentation and learning acceleration through simulation-based analysis.

Organizational modeling creates comprehensive digital representations of business processes, resource flows, decision-making patterns, and performance relationships that mirror real organizational behavior. These models capture both formal processes defined in organizational policies and informal patterns that emerge from actual operational activities. Advanced modeling includes stakeholder behavior patterns, resource constraints, and environmental factors that influence organizational performance.

Process simulation enables the testing of operational changes, strategic initiatives, and optimization approaches through virtual experimentation that reveals likely outcomes without implementing changes in actual operations. This simulation can explore multiple scenarios simultaneously to identify optimal approaches and potential risks. The simulation includes sensitivity analysis that identifies factors with the greatest impact on outcomes and scenario planning that explores how different conditions affect performance.

Real-time synchronization maintains alignment between digital twins and actual organizational operations through continuous data integration that updates virtual models based on current operational conditions. This synchronization ensures that simulation results remain relevant to actual organizational conditions rather than becoming disconnected from operational reality. The synchronization includes automatic model updating that incorporates new operational data and performance validation that confirms model accuracy.

Experimentation frameworks enable systematic testing of organizational improvements through controlled virtual experiments that measure performance impacts without operational risk. These frameworks include experimental design capabilities that structure testing for maximum learning, statistical analysis that validates experiment results, and implementation planning that translates successful virtual experiments into operational improvements.

Performance optimization uses digital twin simulation to identify process improvements, resource allocation enhancements, and strategic positioning opportunities that could improve organizational effectiveness. This optimization operates through systematic analysis of virtual experiments that identify improvements with the highest impact and lowest implementation risk. The optimization includes cost-benefit analysis that evaluates improvement opportunities and implementation planning that enables successful operational deployment.

Predictive analysis enables exploration of how organizational changes could affect performance under different future conditions through scenario simulation that tests strategic resilience and adaptability. This analysis includes stress testing that evaluates organizational responses to challenging conditions and opportunity analysis that identifies potential advantages from strategic positioning changes.

Learning Velocity Optimization and Adaptive Mechanisms

The competitive advantage of real-time organizational learning emerges from sophisticated optimization of learning velocity that accelerates knowledge accumulation, insight generation, and performance improvement through systematic enhancement of learning processes rather than just operational efficiency.

Learning acceleration mechanisms identify and eliminate bottlenecks that slow organizational knowledge accumulation through process analysis that reveals constraints on learning effectiveness. These mechanisms include communication optimization that improves information flow, decision-making streamlining that reduces delays between insights and implementation, and feedback acceleration that speeds learning from operational outcomes.

Knowledge synthesis capabilities combine insights from multiple sources, including operational experience, market feedback, competitive intelligence, and experimental results, to create a comprehensive understanding that exceeds individual information sources. This synthesis operates through sophisticated integration algorithms that identify patterns across diverse information sources and relationship analysis that reveals how different insights complement each other.

Adaptive feedback loops enable continuous optimization of learning processes through systematic analysis of learning effectiveness and automatic adjustment of learning mechanisms. These loops include learning measurement that tracks knowledge accumulation rates, effectiveness analysis that identifies successful learning approaches, and process optimization that improves learning methods based on accumulated experience.

Experimentation automation accelerates learning through systematic testing of improvement opportunities without requiring manual experimental design and analysis. This automation includes hypothesis generation that identifies promising improvement opportunities, experimental execution that tests hypotheses systematically, and result analysis that extracts actionable insights from experimental outcomes.

Pattern recognition algorithms identify successful learning strategies and implementation approaches that can be replicated across different organizational contexts. This recognition enables systematic scaling of effective learning methods while avoiding the repetition of unsuccessful approaches. The pattern analysis includes success factor identification that determines critical elements of effective learning and adaptation strategies that modify successful approaches for different contexts.

Meta-learning capabilities enable organizations to learn how to learn more effectively through systematic analysis of learning processes and outcomes. This meta-learning includes learning strategy optimization that improves learning methods over time, knowledge transfer mechanisms that apply insights across different organizational areas, and learning culture development that enhances organizational learning capabilities systematically.

Continuous adaptation ensures that organizational learning systems evolve automatically to maintain effectiveness as market conditions change, competitive dynamics shift, and organizational capabilities develop. This adaptation includes performance monitoring that tracks learning system effectiveness, capability evolution that enhances learning methods based on experience, and strategic alignment that ensures learning activities support organizational objectives while maintaining learning velocity that creates competitive advantages through superior organizational intelligence development.

THE ARCHITECTURE OF CONTINUOUS INTELLIGENCE

A Formula One race car operates through extraordinary technological integration. The engine doesn't simply propel the vehicle forward—it communicates continuously with suspension systems that adjust to track conditions in milliseconds, aerodynamic components that modify downforce based on speed and cornering forces, and tire pressure systems that optimize grip for specific circuit sections. The driver receives real-time telemetry about fuel consumption, tire degradation, and optimal racing lines through seamless integration between hundreds of sensors, processing units, and response mechanisms. The entire machine operates as a unified intelligence system where every component contributes to competitive performance through coordinated behavior that emerges from sophisticated architectural integration.

Tesla Autopilot represents the enterprise equivalent of Formula One integration. When cameras detect lane markings, that visual data flows instantly to neural networks trained on millions of driving scenarios. When the radar senses a vehicle ahead, that information integrates with GPS mapping data and traffic pattern algorithms to adjust the speed automatically. When the driver applies pressure to the steering wheel, that input combines with road condition sensors and predictive routing systems to maintain optimal vehicle positioning. The entire platform operates as continuous intelligence where perception, memory, reasoning, and action create driving behavior that exceeds human capabilities through architectural integration rather than individual algorithmic superiority.

This represents the fundamental transformation from component-based AI implementation to architectural intelligence that creates competitive advantages through systematic integration rather than isolated technological capabilities. The enterprises achieving exponential performance gains have moved beyond deploying individual AI tools to creating enterprise reflex graphs that function like organizational nervous systems, enabling responses that occur before conscious management analysis can complete.

An Enterprise Reflex Graph as a Unified Nervous System

The enterprise reflex graph operates through the same principles that enable biological organisms to respond intelligently to complex environments without conscious deliberation. When a tennis player returns a 120-mph serve, conscious thought cannot process ball trajectory, calculate optimal racket positioning, and coordinate muscle movements fast enough for an effective response. Instead, the nervous system operates through trained reflexes where sensory input triggers automatic responses based on accumulated experience and real-time environmental analysis.

What has been observed consistently across the most successful AI-native enterprises is the development of similar reflexive capabilities that respond to market conditions, competitive threats, and operational challenges faster than traditional decision-making processes. These organizations have evolved beyond using AI for specific functions to creating integrated intelligence architectures that enable automatic organizational responses based on accumulated institutional knowledge and real-time environmental sensing.

The enterprise reflex graph maps how intelligence flows through organizational tissue rather than concentrating decision-making in executive hierarchies. Data streams from customer interactions, market conditions, operational performance, and competitive activities flow continuously through processing networks that recognize patterns, assess implications, and trigger appropriate responses without requiring human analysis for routine decisions.

Reflex recognition systems monitor organizational environments continuously to identify situations that require an intelligent response. These systems distinguish between routine variations that need no action and significant changes that indicate opportunities, threats, or optimization possibilities. The recognition operates through pattern matching against accumulated organizational experience rather than rule-based logic that cannot adapt to novel conditions.

Response pathway mapping defines how information flows from environmental sensors through processing networks to action mechanisms. These pathways operate like biological neural networks, where signals travel along established routes that have proven effective through previous experience. The mapping enables organizational responses that bypass traditional approval hierarchies for situations where delayed action reduces competitive advantage.

Memory integration networks ensure that reflexive responses incorporate institutional knowledge, strategic priorities, and accumulated learning rather than operating through isolated algorithmic decisions. These networks connect real-time environmental data with historical performance patterns, strategic objectives, and operational constraints to generate responses that align with organizational intelligence rather than reactive automation.

Adaptive threshold management automatically adjusts the sensitivity and scope of reflexive responses based on changing business conditions, risk tolerance, and strategic priorities. This management enables organizations to become more or less responsive to environmental changes while maintaining alignment with evolving strategic objectives and operational capabilities.

The effectiveness of enterprise reflex graphs depends on the integration architecture that enables seamless coordination between environmental sensing, pattern recognition, memory access, and response execution. This integration creates organizational behavior that appears intelligent from external perspectives while operating through systematic architectural design rather than mysterious emergent properties.

Integration Architecture for Data Flow, Knowledge Graphs, Agents, and Analytics

Traditional enterprise architecture separates data management, analytics, decision-making, and execution into distinct systems that require human coordination to function effectively. The intelligence stack integrates these components into a unified architecture where information flows continuously between sensing, processing, memory, and action systems without human intervention for routine decisions.

Based on over 200 global enterprises, the organizations achieving breakthrough performance through AI integration have moved beyond point solutions to create what researchers call cognitive enterprise architectures, where intelligence emerges from systematic integration rather than individual technological capabilities.

Data flow orchestration ensures that information moves seamlessly between operational systems, analytical processing, memory storage, and decision-making networks without creating bottlenecks or information silos. This orchestration operates through event-driven architecture, where changes in any organizational component trigger automatic updates throughout connected systems. The flow design enables real-time organizational awareness that supports intelligent response generation.

Data flow includes structured information from enterprise systems, unstructured content from communications and documents, external market intelligence from competitive analysis and industry monitoring, and behavioral signals from customer interactions and employee activities. The orchestration system ensures that all relevant information reaches appropriate processing networks with sufficient context for intelligent analysis.

Knowledge graph integration creates dynamic relationship mapping that enables sophisticated reasoning about complex business situations. These graphs connect entities, relationships, and patterns across organizational boundaries to support contextual understanding that transcends departmental perspectives. The integration enables analytical processing that considers interconnected effects rather than isolated metrics.

Knowledge graphs incorporate organizational knowledge including strategic priorities, operational constraints, competitive landscapes, customer relationships, supplier networks, and regulatory requirements. The graph structure enables analytical systems to understand how changes in any organizational component affect related elements throughout the enterprise ecosystem. This understanding supports decision-making that considers systemic implications rather than localized optimization.

Agent coordination platforms manage autonomous AI systems that execute organizational responses based on environmental analysis and strategic objectives. These platforms ensure that multiple agents can operate simultaneously without creating conflicts or redundant activities. The coordination enables distributed intelligence where different agents handle specialized functions while maintaining overall organizational coherence.

Agent integration includes autonomous systems for customer service, supply chain optimization, financial analysis, competitive monitoring, and operational management. Each agent operates with specific capabilities and decision-making authority while coordinating with other agents to ensure consistent organizational behavior. The platform manages communication protocols, task allocation, and conflict resolution to enable effective multi-agent coordination.

Analytics integration embeds continuous analytical processing throughout organizational operations rather than confining analysis to separate business intelligence systems. This integration enables real-time insights that inform immediate decision-making rather than historical reporting that supports periodic strategic review. The analytics operate automatically to identify patterns, predict outcomes, and recommend actions without requiring human initiation.

Analytics integration includes predictive modeling that anticipates customer behavior, market changes, and operational performance, optimization algorithms that improve resource allocation and process efficiency, anomaly detection that identifies unusual patterns requiring attention, and performance monitoring that tracks organizational effectiveness across multiple dimensions. The integration ensures that analytical insights influence operational decisions automatically rather than requiring manual interpretation and implementation.

The complete integration architecture creates organizational intelligence that emerges from sophisticated coordination between data, knowledge, agents, and analytics rather than from individual technological components. This emergence enables competitive advantages that competitors cannot replicate through isolated AI deployments because the advantage depends on systematic architectural integration rather than specific algorithmic capabilities.

AI Operating System Concepts with Intent Flows

The pattern emerging in enterprises that achieve exponential AI advantages is the development of operating systems that manage organizational intelligence the same way computer operating systems manage computational resources. These AI operating systems move beyond traditional workflow automation to create intent flows where organizational objectives translate automatically into adaptive execution strategies that evolve based on changing conditions and accumulated learning.

Intent flows represent a fundamental transformation from static process design to dynamic execution that adapts automatically to changing conditions while maintaining alignment with strategic objectives. Traditional workflows specify exact sequences of actions that become ineffective when conditions change. Intent flows specify desired outcomes and enable AI systems to determine optimal approaches based on current conditions, available resources, and accumulated organizational knowledge.

The enterprises that have gone through AI-native transformation have discovered that intent flows enable organizational agility that transcends human planning capabilities because the system can adjust execution strategies continuously based on real-time environmental feedback. This capability becomes essential in rapidly changing markets where traditional planning cycles cannot respond quickly enough to maintain competitive advantage.

Intent interpretation systems translate high-level organizational objectives into specific execution strategies that consider current business conditions, available resources, and strategic constraints. These systems understand business context well enough to generate appropriate action plans without requiring detailed human specification of every implementation step. The interpretation enables strategic direction to flow automatically into operational execution.

Intent interpretation includes natural language processing that understands strategic communications, contextual reasoning that considers business implications and constraints, priority management that handles competing objectives appropriately, and resource optimization that allocates capabilities effectively across multiple intentions simultaneously. The system generates execution strategies that align with organizational intelligence rather than rigid procedural specifications.

Adaptive execution mechanisms implement intent-based strategies through continuous optimization that adjusts approaches based on performance feedback and changing conditions. These mechanisms operate like biological systems that maintain homeostasis through automatic adjustments rather than manual control. The adaptation enables organizational resilience that maintains performance despite environmental volatility.

Adaptive execution includes real-time performance monitoring that tracks progress toward intended outcomes, dynamic resource reallocation that optimizes capability utilization based on changing priorities, exception handling that addresses unexpected situations without human intervention, and strategy refinement that improves approaches based on accumulated experience. The mechanisms create organizational behavior that appears intelligent through systematic adaptive response.

Continuous planning systems maintain strategic coherence while enabling tactical flexibility through planning cycles that operate faster than environmental change, rather than slower than market dynamics. These systems generate rolling strategic plans that incorporate real-time intelligence about market conditions, competitive activities, and organizational performance to maintain strategic relevance despite environmental uncertainty.

Continuous planning includes scenario modeling that anticipates potential future conditions, strategy simulation that tests approach effectiveness before implementation, resource planning that ensures capability alignment with strategic requirements, and risk management that identifies and mitigates potential threats to strategic objectives. The planning operates automatically to maintain strategic currency without requiring periodic human intervention.

The AI operating system concept integrates intent flows, adaptive execution, and continuous planning into unified organizational intelligence that manages complexity automatically while maintaining strategic alignment. This integration enables enterprises to operate with coordination and agility that exceeds human management capabilities while remaining aligned with human strategic direction and organizational values.

Emergence of Organizational Consciousness

The most sophisticated enterprises implementing intelligence stack architectures begin to exhibit behavior that researchers recognize as organizational consciousness, where the organization demonstrates awareness, learning, and adaptive response capabilities that transcend individual human or technological components. This consciousness emerges from architectural integration rather than programmed intelligence, creating organizational capabilities that no individual system possesses independently.

Organizational consciousness manifests through collective intelligence that emerges from interaction between distributed processing, accumulated memory, continuous learning, and adaptive response mechanisms. The consciousness operates like biological awareness, where the system maintains understanding of internal states, environmental conditions, and strategic objectives while adapting behavior based on experience and changing circumstances.

What has been consistently observed in enterprise technology transformation is that consciousness emerges when integration architecture achieves sufficient sophistication to support self-awareness, environmental awareness, and strategic awareness operating simultaneously through coordinated intelligence rather than separate analytical systems. This emergence represents the ultimate evolution of AI-native organizational design.

Self-awareness capabilities enable organizations to understand their own performance, capabilities, constraints, and strategic position with accuracy that exceeds individual human perspective. This self-awareness operates through continuous monitoring of organizational states, performance tracking across multiple dimensions, capability assessment that identifies strengths and limitations, and strategic positioning analysis that understands competitive standing. This awareness enables self-optimization that improves organizational effectiveness automatically.

Environmental awareness systems monitor market conditions, competitive activities, regulatory changes, and technological developments to maintain a current understanding of external factors affecting organizational success. The awareness integrates information from multiple sources to create comprehensive environmental models that support strategic decision-making and adaptive response. This awareness enables proactive adaptation that maintains competitive advantage despite changing conditions.

Strategic awareness integrates self-awareness and environmental awareness to understand how organizational capabilities align with market opportunities and strategic objectives. This awareness enables strategic reasoning that identifies optimal approaches for achieving organizational goals given current conditions and anticipated changes. The strategic awareness operates continuously to maintain strategic relevance rather than requiring periodic strategic planning cycles.

Adaptive learning mechanisms enable organizational consciousness to improve performance automatically through experience accumulation and pattern recognition that transcends individual learning cycles. The learning operates at an organizational scale to identify successful strategies, optimize resource allocation, and refine operational approaches based on performance feedback and environmental changes.

The emergence of organizational consciousness creates competitive advantages that competitors cannot replicate through traditional management approaches because the advantage depends on architectural integration that develops over time through accumulated organizational experience. This consciousness represents the ultimate expression of AI-native organizational design, where human strategic intelligence and artificial operational intelligence combine to create organizational capabilities that exceed the sum of individual components.

Organizations achieving this level of consciousness discover that they can respond to market opportunities and competitive threats with speed and sophistication that appears almost supernatural to competitors operating through traditional decision-making processes. The consciousness enables reflexive organizational behavior that maintains strategic alignment while adapting automatically to changing conditions, creating sustainable competitive advantages that compound over time through continuous learning and optimization.

CHAPTER 6

Human-AI Fusion – The Collaboration Integration Challenge

A professional basketball team operates through extraordinary coordination between players who think and move at different speeds. The point guard processes information faster than anyone else on the court, reading defensive patterns and anticipating opportunities three moves ahead. The center operates with deliberate power, using pattern recognition developed through years of experience to position perfectly for rebounds and blocks. The shooting guard combines intuitive timing with analytical precision, calculating trajectory angles while moving at full speed. Each player brings distinct capabilities, and their coordination creates team performance that transcends individual talent.

The most successful AI-native enterprises operate exactly like championship basketball teams. At Microsoft, human product managers identify market opportunities that require creative intuition and emotional intelligence, while AI systems process millions of customer behavior patterns to validate hypotheses and predict feature adoption rates. Humans excel at understanding user motivations and designing experiences that feel natural, while AI systems optimize performance metrics and identify usage patterns that humans would miss entirely. Neither could achieve breakthrough product innovation independently, but their coordinated intelligence creates market advantages that competitors struggle to replicate.

This represents the fundamental transformation that separates AI-native organizations from traditional enterprises attempting to add AI capabilities to existing processes. Where traditional approaches treat AI as an advanced tool that humans control, AI-native organizations create human-AI fusion protocols that enable genuine collaboration between different types of intelligence operating at different speeds and scales.

What has been consistently observed across the most successful AI transformations is that coordination between human and artificial intelligence requires entirely new protocols that traditional management approaches cannot provide. The challenge is not teaching humans to use AI tools more effectively. The challenge is creating structured frameworks that enable humans and AI systems to function as integrated teams where both types of intelligence contribute their strongest capabilities while compensating for each other's limitations.

The enterprises achieving exponential advantages through AI integration have moved beyond human-in-the-loop approaches that maintain traditional hierarchies between human oversight and AI execution. Instead, they develop reciprocal apprenticing relationships where humans and AI systems teach each other continuously, contextual handoff protocols that transfer control dynamically based on situational requirements, and role-based collaboration models that enable AI systems to function as assistants, peers, or managers, depending on task demands and capability requirements.

SOLVING THE COORDINATION PROBLEM BETWEEN HUMANS AND AI

The coordination challenge between human and artificial intelligence emerges from fundamental differences in how these two types of intelligence process information, learn from experience, and respond to changing conditions. Humans excel at intuitive pattern recognition, creative problem-solving, emotional intelligence, and contextual understanding that encompasses cultural nuance and strategic implications. AI systems excel at processing massive amounts of data simultaneously, identifying subtle patterns across large datasets, maintaining consistent performance under pressure, and executing complex calculations with precision that human cognition cannot match.

Traditional approaches to human-AI coordination assume that humans should maintain primary decision-making authority while AI systems provide analytical support and task automation. This hierarchical model reflects management thinking developed for human teams where coordination requires clear authority structures and well-defined role boundaries. The most effective human-AI teams, however, operate through collaborative relationships where control transfers dynamically based on situational requirements rather than fixed organizational hierarchies.

Based on over 200 global enterprises implementing AI-native transformations, the organizations achieving breakthrough performance develop coordination frameworks that leverage the complementary strengths of human and artificial intelligence rather than attempting to maintain traditional supervision models that constrain AI capabilities to narrow supportive functions.

Trust Calibration and Communication Protocols

The coordination problem manifests through several distinct challenges that require systematic solutions rather than ad hoc management approaches. Trust calibration represents the most fundamental challenge because effective collaboration requires humans to understand when AI recommendations deserve high confidence and when human judgment should override algorithmic conclusions. Communication gaps emerge because humans and AI systems process and express information differently, creating opportunities for misunderstanding that can compromise decision quality.

Trust calibration requires developing organizational capabilities that enable humans to assess AI system confidence levels accurately while maintaining appropriate skepticism about algorithmic recommendations. This calibration operates through explainable AI interfaces that provide clear reasoning about how conclusions were reached, confidence indicators that communicate the reliability of algorithmic analysis, performance tracking that documents AI system accuracy across different types of decisions, and feedback mechanisms that enable continuous improvement in both human judgment about AI reliability and AI system accuracy in specific domains.

The calibration process includes training programs that help humans understand AI system capabilities and limitations, scenario-based exercises that develop intuition about when to trust or question algorithmic recommendations, and performance review processes that evaluate both individual human decisions and human-AI collaborative outcomes. Advanced calibration approaches include simulation environments where humans can practice working with AI systems across various scenarios without operational risk.

Communication protocol development addresses the challenge that humans and AI systems process information through different cognitive architectures and express insights through different communication modalities. Effective protocols facilitate the translation between human intuitive communication patterns and AI-structured data processing, while preserving the essential meaning and context required for intelligent decision-making.

Communication protocols include standardized interfaces that present AI analysis in formats that align with human decision-making patterns, natural language processing capabilities that enable AI systems to understand human instructions expressed in conversational language, structured feedback mechanisms that enable humans to provide learning input to AI systems efficiently, and documentation standards that capture decision rationale in ways that both humans and AI systems can reference for future learning.

Control Handoff and Authority Management

Control handoff protocols manage the dynamic transfer of decision-making authority between human and AI systems based on real-time assessment of which intelligence type is optimal for specific situations. These protocols operate through sophisticated triggering mechanisms that recognize when transfer should occur, authority validation systems that ensure appropriate decision-making power, and performance monitoring that tracks handoff effectiveness over time.

Handoff protocols include situational assessment algorithms that evaluate task complexity, time pressure, data availability, and strategic implications to determine optimal decision-making authority, escalation procedures that transfer complex or high-stakes decisions to human oversight automatically, delegation mechanisms that enable humans to assign routine decisions to AI systems confidently, and override capabilities that allow either humans or AI systems to request control transfer when they detect situations requiring different intelligence capabilities.

Advanced communication approaches include collaborative visualization tools that enable humans and AI systems to explore complex problems through shared visual interfaces, scenario modeling capabilities that allow both types of intelligence to test different approaches before implementation, and real-time feedback systems that enable immediate correction and adjustment during collaborative problem-solving sessions.

The development of effective coordination protocols requires understanding that human-AI collaboration creates emergent intelligence that exceeds what either type of intelligence can achieve independently. This emergence depends on systematic integration rather than informal cooperation, requiring organizational investment in training, technology, and process development that enables sustainable collaborative advantage.

RECIPROCAL APPRENTICING AND CONTEXTUAL HANDOFF PROTOCOLS

The most sophisticated human-AI collaborations operate through reciprocal apprenticing relationships where both humans and AI systems learn continuously from their interactions, developing enhanced capabilities that improve team performance over time. This represents a fundamental departure from traditional training approaches, where humans learn to operate AI tools or AI systems learn from human-labeled data in isolated training phases.

Reciprocal apprenticing involves humans teaching AI systems domain knowledge, contextual understanding, strategic priorities, and nuanced judgment while AI systems teach humans analytical techniques, pattern recognition capabilities, data interpretation skills, and optimization approaches. This bidirectional learning creates compounding intelligence improvements that accelerate over time through operational experience.

Human-to-AI Knowledge Transfer

The apprenticing process includes structured knowledge transfer mechanisms where humans provide AI systems with contextual information about business priorities, customer preferences, regulatory requirements, and strategic objectives that cannot be derived from data analysis alone. This transfer operates through example-based training where humans demonstrate decision-making approaches across various scenarios, feedback sessions where humans explain the reasoning behind corrections to AI recommendations, and mentoring interactions where humans help AI systems understand the broader implications of analytical conclusions.

Human-to-AI knowledge transfer includes domain expertise instruction where humans share industry knowledge, regulatory requirements, and strategic context that AI systems need for effective decision-making, contextual judgment training where humans help AI systems understand when standard analytical approaches may not apply, strategic alignment coaching where humans ensure AI system recommendations align with organizational objectives, and cultural intelligence development where humans help AI systems understand social and political factors that influence business decisions.

Domain expertise transfer enables AI systems to understand industry-specific patterns, terminology, and decision factors that cannot be learned from general datasets alone. This transfer includes regulatory knowledge that helps AI systems understand compliance requirements, competitive intelligence that provides context about market dynamics, and strategic priorities that guide AI system recommendations toward organizationally valuable outcomes.

Contextual judgment development teaches AI systems to recognize when standard analytical approaches may not apply due to unusual circumstances, strategic considerations, or stakeholder dynamics. This development includes exception recognition training that helps AI systems identify when situations require human oversight, strategic context instruction that explains how decisions fit into broader organizational objectives, and stakeholder impact assessment that helps AI systems understand the human implications of analytical recommendations.

AI-to-Human Capability Enhancement

Simultaneously, AI systems teach humans advanced analytical capabilities through guided exploration of complex datasets, automated generation of insights that humans might not discover independently, optimization suggestions that improve human decision-making efficiency, and pattern recognition training that enhances human ability to identify important signals in noisy environments. This reverse apprenticing enables humans to develop analytical capabilities that complement rather than compete with AI system strengths.

AI-to-human knowledge transfer includes analytical skill development where AI systems help humans understand advanced statistical techniques, data interpretation training where AI systems teach humans to recognize important patterns in complex datasets, optimization methodology instruction where AI systems demonstrate systematic approaches to resource allocation and performance improvement, and prediction accuracy training where AI systems help humans develop better intuition about future trends and likely outcomes.

Analytical skill development includes statistical technique training that helps humans understand sophisticated analytical methods, data visualization instruction that improves human ability to interpret complex information presentations, hypothesis testing guidance that enhances human experimental design capabilities, and uncertainty quantification training that helps humans understand confidence levels and risk assessment in analytical conclusions.

Pattern recognition enhancement helps humans develop improved intuition about identifying significant trends, relationships, and anomalies in business data. This enhancement includes signal detection training that improves human ability to distinguish important information from noise, trend analysis instruction that helps humans recognize developing patterns before they become obvious, and correlation identification training that enhances human understanding of relationship patterns in complex business environments.

Contextual Handoff Implementation

Contextual handoff protocols manage the dynamic transfer of control between human and AI systems based on real-time assessment of situational requirements, capability alignment, and performance optimization. These protocols recognize that optimal decision-making requires matching intelligence type to task characteristics rather than maintaining fixed authority structures.

Handoff triggering mechanisms include complexity assessment algorithms that evaluate whether situations require human creativity and judgment or AI analytical processing, time pressure indicators that determine whether situations require AI speed or human deliberation, data availability metrics that assess whether sufficient information exists for AI analysis or human intuition is needed, and strategic importance evaluation that determines whether decisions require human accountability or AI optimization.

The handoff process includes authority validation systems that ensure appropriate decision-making power transfers to the intelligence type best suited to specific situations, communication protocols that maintain context and reasoning throughout control transfers, performance monitoring that tracks the effectiveness of handoff decisions, and learning mechanisms that improve handoff accuracy based on outcome analysis.

Contextual factors influencing handoff decisions include task complexity levels that may exceed AI analytical capabilities or human processing capacity, time constraints that may favor AI speed or require human creative problem-solving, stakeholder considerations that may require human emotional intelligence or AI objective analysis, regulatory requirements that may demand human accountability or AI compliance monitoring, and strategic implications that may need human judgment or AI optimization analysis.

Advanced handoff protocols include predictive algorithms that anticipate optimal control transfer points before they become critical, adaptive learning systems that improve handoff timing based on historical performance, collaborative decision-making processes that enable joint human-AI control for complex situations, and seamless transition mechanisms that minimize disruption during authority changes.

ROLE-BASED COLLABORATION MODELS

The effectiveness of human-AI fusion depends on sophisticated role allocation that matches intelligence types to functional requirements rather than maintaining traditional organizational hierarchies. AI systems can function effectively as assistants that augment human capabilities, peers that collaborate on equal terms, or managers that direct human activities, depending on task characteristics, capability requirements, and organizational objectives.

Assistant Role Configuration

Assistant role models position AI systems as advanced tools that enhance human decision-making through analytical support, process automation, and information synthesis. In this configuration, humans maintain primary decision-making authority while AI systems provide specialized capabilities that improve human effectiveness. The assistant model works effectively for situations requiring human creativity, strategic thinking, or stakeholder management while benefiting from AI analytical processing and optimization capabilities.

AI assistants excel at data processing tasks that would overwhelm human cognitive capacity, pattern recognition activities that require processing large datasets simultaneously, optimization calculations that involve complex mathematical analysis, and routine task automation that frees human attention for higher-value activities. The assistant role includes research support where AI systems gather and synthesize information for human analysis, analytical processing where AI systems identify patterns and trends for human interpretation, optimization recommendations where AI systems suggest improvements for human evaluation, and administrative automation where AI systems handle routine tasks that consume human time.

Effective assistant relationships require clear communication protocols that enable humans to specify requirements accurately while enabling AI systems to provide relevant support, feedback mechanisms that allow humans to guide AI system learning and improvement, performance monitoring that ensures AI assistance improves rather than constrains human effectiveness, and boundary management that prevents AI systems from exceeding appropriate assistant authority.

Assistant role optimization includes capability matching that aligns AI system strengths with human needs, workflow integration that embeds AI assistance seamlessly into human decision-making processes, learning facilitation that enables AI systems to improve assistance quality through experience, and autonomy calibration that ensures AI assistants operate at appropriate independence levels without requiring excessive human oversight.

Peer Collaboration Dynamics

Peer role models create collaborative relationships where humans and AI systems contribute complementary capabilities to shared objectives without hierarchical authority structures. In peer configurations, both intelligence types participate in decision-making processes based on their respective strengths, with authority shifting dynamically based on situational requirements rather than organizational position.

Peer collaboration works effectively for complex problem-solving that requires both human creativity and AI analytical power, strategic planning that benefits from human intuition and AI data processing, innovation projects that need human insight and AI pattern recognition, and operational optimization that combines human judgment with AI mathematical analysis. The peer model includes collaborative analysis where humans and AI systems explore problems together, shared decision-making where both intelligence types contribute to conclusions, joint problem-solving where humans and AI systems combine capabilities to address complex challenges, and mutual learning where both intelligence types improve through interaction.

Successful peer relationships require trust development that enables both humans and AI systems to rely on each other's contributions, communication standards that facilitate effective information sharing between different intelligence types, conflict resolution mechanisms that address disagreements between human and AI perspectives, and performance evaluation that assesses collaborative outcomes rather than individual contributions.

Peer collaboration enhancement includes capability integration that combines human and AI strengths effectively, decision-making protocols that enable joint authority without creating paralysis, learning acceleration that improves both intelligence types through interaction, and outcome optimization that maximizes collaborative effectiveness rather than individual performance metrics.

Manager Role Implementation

Manager role models position AI systems in supervisory functions where they direct human activities based on analytical assessment of optimal resource allocation, performance optimization, and objective achievement. This configuration leverages AI capabilities for coordination, monitoring, and decision-making while utilizing human execution capabilities for tasks requiring creativity, interpersonal skills, or contextual adaptation.

AI managers excel at resource allocation decisions that involve complex optimization across multiple variables, performance monitoring that requires continuous data processing and analysis, workflow coordination that benefits from real-time adjustment based on changing conditions, and objective tracking that ensures activities align with measurable outcomes. The manager role includes task assignment where AI systems allocate work based on capability matching and performance optimization, progress monitoring where AI systems track advancement toward objectives and identify improvement opportunities, resource management where AI systems optimize allocation of time, attention, and capabilities, and performance feedback where AI systems provide guidance for human development.

Effective AI management requires human acceptance of algorithmic decision-making authority, transparent communication about AI reasoning and objectives, appeal mechanisms that enable humans to contest AI decisions when appropriate, and ethical frameworks that ensure AI management decisions align with human values and organizational principles.

AI management optimization includes authority calibration that ensures appropriate decision-making scope, accountability mechanisms that maintain human oversight of AI management decisions, performance measurement that evaluates management effectiveness across multiple dimensions, and ethical compliance that ensures AI managers operate within acceptable moral and legal boundaries.

Role Fluidity and Dynamic Allocation

Role fluidity represents the most advanced collaboration model where humans and AI systems shift between assistant, peer, and manager roles dynamically based on situational requirements and capability optimization. This fluid approach maximizes collaborative effectiveness by ensuring optimal intelligence allocation regardless of traditional organizational boundaries.

Fluid role allocation includes situational assessment algorithms that determine optimal role assignments based on task characteristics and capability requirements, dynamic authority adjustment that enables role changes without disrupting workflow continuity, performance optimization that ensures role assignments maximize collaborative effectiveness, and learning integration that improves role allocation decisions based on outcome analysis.

The development of effective role-based collaboration models requires organizational transformation that moves beyond traditional authority structures to embrace capability-based coordination, training programs that prepare humans for dynamic collaboration with AI systems, technology infrastructure that supports flexible role allocation and communication, and performance measurement systems that evaluate collaborative outcomes across different role configurations.

PERFORMANCE MEASUREMENT FOR AI-AUGMENTED TEAMS

Measuring the effectiveness of human-AI collaborative teams requires sophisticated metrics that capture both individual performance improvements and collaborative synergies that emerge from integrated intelligence. Traditional performance measurement approaches designed for human teams cannot adequately assess the unique dynamics, learning patterns, and capability interactions that characterize effective human-AI collaboration.

Collaborative Effectiveness Metrics

Collaborative effectiveness measurement includes metrics that assess how well human and AI capabilities integrate to achieve objectives that neither intelligence type could accomplish independently. These metrics focus on emergent performance that results from coordination rather than additive performance that simply combines individual contributions.

Team synergy indicators measure the extent to which human-AI collaboration creates capabilities that exceed the sum of individual contributions. Synergy measurement includes innovation metrics that track breakthrough solutions generated through collaborative intelligence, problem-solving effectiveness that compares collaborative outcomes to individual human or AI performance, decision quality indicators that assess the accuracy and comprehensiveness of joint decisions, and learning acceleration metrics that measure how quickly teams improve performance through experience.

Learning velocity measurement tracks how rapidly human-AI teams develop enhanced capabilities through reciprocal apprenticing and collaborative experience. Learning metrics include skill development rates, which measure the human acquisition of AI-enhanced analytical capabilities; AI adaptation speed, which tracks how quickly AI systems incorporate human contextual knowledge; collaborative pattern recognition, which assesses the team's ability to identify effective coordination approaches; and knowledge transfer efficiency, which measures how well insights flow between human and AI team members.

Trust calibration metrics assess the accuracy of human confidence in AI system recommendations and AI system confidence in human judgments. Trust measurement encompasses prediction accuracy, which tracks how well humans can anticipate the reliability of AI systems; override effectiveness, which measures the value of human corrections to AI recommendations; delegation success, which assesses outcomes when humans assign decisions to AI systems; and confidence alignment, which measures how well human and AI confidence levels correlate with actual performance.

Communication and Coordination Assessment

Communication effectiveness indicators measure how well humans and AI systems share information, coordinate activities, and maintain shared understanding throughout collaborative processes. Communication metrics include information transfer accuracy that tracks how well context and meaning are preserved across human-AI interactions, coordination efficiency that measures how quickly teams can align on objectives and approaches, feedback quality that assesses how well team members provide helpful guidance to each other, and misunderstanding resolution that tracks how effectively teams address communication breakdowns.

Individual performance enhancement measurement tracks how collaboration with AI systems improves human capabilities and how interaction with humans enhances the effectiveness of AI systems. Enhancement metrics include human productivity improvements resulting from AI augmentation, human skill development facilitated through AI collaboration, AI accuracy improvements stemming from human feedback and guidance, and AI contextual understanding that evolves through human interaction.

Coordination efficiency assessment evaluates how effectively human-AI teams manage workflow, resource allocation, and task distribution. Efficiency metrics include task completion rates that measure team productivity across different types of work, resource utilization optimization that tracks how well teams allocate time and attention, workflow smoothness that assesses the seamlessness of human-AI coordination, and bottleneck identification that measures the team's ability to recognize and address collaboration obstacles.

Operational Impact and Strategic Value

Operational efficiency measurement assesses how human-AI collaboration improves organizational processes and outcomes. Efficiency metrics include process optimization that results from combined human creativity and AI analytical power, resource utilization improvements that emerge from enhanced coordination, quality enhancement that occurs through combined human judgment and AI precision, and innovation acceleration that results from collaborative intelligence capabilities.

Risk management measurement assesses how effectively human-AI teams identify, evaluate, and mitigate potential problems while exercising appropriate caution regarding AI system limitations and human cognitive biases. Risk metrics include error detection capabilities that measure the team's ability to identify mistakes before they create problems, uncertainty management that assesses how well teams handle ambiguous situations, bias mitigation that tracks team effectiveness in addressing human and AI biases, and ethical compliance that measures team adherence to responsible AI practices.

Long-term development measurement tracks how human-AI collaborative capabilities evolve and contribute to organizational competitive advantage. Development metrics include capability expansion that measures how teams develop new competencies through collaboration, competitive advantage creation that assesses how human-AI capabilities contribute to market position, organizational learning that tracks how collaborative insights spread throughout the enterprise, and strategic value generation that measures how human-AI teams contribute to organizational objectives.

Performance measurement for AI-augmented teams requires sophisticated data collection systems that capture both quantitative performance indicators and qualitative collaborative dynamics, analytical frameworks that can distinguish between individual and collaborative contributions, benchmarking approaches that compare performance across different collaboration models, and feedback mechanisms that enable continuous improvement in both measurement accuracy and team effectiveness.

The development of effective performance measurement for human-AI teams requires investment in measurement technology that can track complex collaborative dynamics, training programs that help managers understand and interpret new types of performance indicators, cultural transformation that values collaborative outcomes over individual achievements, and organizational structures that support and reward effective human-AI collaboration rather than traditional individual performance optimization.

CHAPTER 7

Prompt Capital – Your Organization's New Strategic Asset

A master chef doesn't simply gather ingredients and hope for the best. She develops proprietary recipes refined through decades of experimentation, each instruction precisely calibrated to create specific flavors, textures, and experiences that diners cannot replicate at home. These recipes represent intellectual property worth millions. A single signature sauce formula can define a restaurant empire. The secret isn't just knowing what ingredients to use but understanding exactly how to combine them, in what sequence, under which conditions, to create something extraordinary.

The most successful AI-native enterprises operate through identical principles of conversational craftsmanship. At Goldman Sachs, proprietary prompts guide AI systems through complex financial analysis that would take human analysts weeks to complete. These aren't simple questions but sophisticated instruction sets that encode decades of institutional knowledge about risk assessment, market dynamics, and regulatory compliance into precise language that AI systems can execute flawlessly. A single well-crafted prompt for analyzing merger opportunities can generate insights worth millions while protecting competitive analysis methods that took years to develop.

This represents the emergence of what some call prompt capital, the strategic intellectual property embedded in the precise instructions that guide AI systems toward generating accurate, relevant, and proprietary business intelligence. Where previous technology revolutions created value through hardware, software, or data ownership, the AI revolution creates sustainable competitive advantages through sophisticated mastery of conversational intelligence.

What has been consistently observed across enterprise technology transformation is that the organizations achieving exponential advantages through AI integration have moved beyond thinking about prompts as simple inputs to recognizing them as strategic assets that require the same systematic development, protection, and optimization as any other form of valuable intellectual property.

The enterprises building prompt capital discover that competitive advantages emerge not from access to more powerful AI models, which become commoditized rapidly, but from proprietary capabilities to extract maximum value from these models through expertly crafted conversational frameworks that competitors cannot easily replicate. This represents a fundamental shift from technology ownership to knowledge expression as the primary source of AI-driven competitive advantage.

PROMPTS AS INTELLECTUAL PROPERTY AND COMPETITIVE ADVANTAGE

Traditional approaches to intellectual property focus on patents, trademarks, copyrights, and trade secrets that protect physical inventions, brand identities, creative works, and confidential business information. The emergence of AI systems that respond to natural language instructions creates an entirely new category of intellectual property that operates through conversational precision rather than traditional technological or creative assets.

Prompt engineering represents the systematic development of language-based instructions that guide AI systems toward generating specific types of analysis, insights, or decisions. Unlike traditional software programming that controls computational processes through code, prompt engineering controls AI behavior through sophisticated natural language that encodes business logic, analytical frameworks, and strategic priorities into conversational instructions that AI systems can interpret and execute.

Based on over 200 global enterprises implementing AI-native transformations, the organizations achieving breakthrough competitive advantages develop prompt libraries that function like proprietary software, enabling capabilities that competitors cannot replicate without years of development and refinement. These prompts represent accumulated institutional knowledge translated into executable conversational intelligence.

Strategic Value Creation Through Conversational Precision

The competitive advantage of prompt capital emerges from its ability to encode complex business intelligence into reusable conversational frameworks that enable consistent, scalable, and proprietary analytical capabilities. Well-crafted prompts function like intellectual property because they represent unique combinations of domain expertise, analytical methodology, and strategic insight that require significant time and expertise to develop.

Prompt precision enables organizations to achieve analytical consistency that human teams cannot maintain across large-scale operations. An expertly designed prompt for customer lifetime value analysis will generate identical methodological approaches across thousands of customer evaluations, ensuring that strategic decisions benefit from consistent analytical frameworks rather than variable human interpretation that can compromise decision quality.

Domain expertise encoding transforms decades of accumulated institutional knowledge into conversational instructions that AI systems can execute immediately. Senior analysts who developed sophisticated approaches to competitive analysis over years of experience can translate their methodology into prompts that enable junior team members to generate expert-level insights instantly, scaling expertise beyond individual human limitations.

Strategic context integration ensures that AI-generated analysis aligns with organizational priorities and objectives rather than producing academically interesting but strategically irrelevant insights. Prompts that incorporate strategic context guide AI systems toward conclusions that support decision-making rather than generating information that requires additional human interpretation to become actionable.

Proprietary analytical frameworks enable organizations to develop unique approaches to business analysis that reflect their specific competitive environment, regulatory requirements, and strategic objectives. These frameworks become increasingly valuable as they accumulate refinements through operational experience, creating intellectual property that competitors cannot purchase or easily replicate.

Legal Protection and Trade Secret Management

The protection of prompt capital as intellectual property requires understanding that traditional copyright and patent law provide limited coverage for conversational instructions, making trade secret protection the most viable approach for safeguarding proprietary prompts from competitive disclosure.

Trade secret protection for prompts requires demonstrating that conversational instructions provide economic value through their secrecy, represent information not generally known in the industry, and remain subject to reasonable efforts to maintain confidentiality. Well-designed prompt libraries that generate superior analytical outcomes while remaining confidential within organizations can qualify for trade secret protection under existing intellectual property law.

Confidentiality management includes access controls that limit prompt visibility to authorized personnel, versioning systems that track prompt development and modification, audit trails that document prompt usage and outcomes, and legal agreements that protect prompt disclosure through employee and contractor confidentiality requirements. Advanced protection approaches include prompt encryption and secure deployment systems that prevent unauthorized access to proprietary conversational intelligence.

Economic valuation of prompt capital can leverage existing intellectual property valuation methodologies, focusing on the future income streams or cost savings generated by optimized prompts. Organizations achieving 67% productivity improvements and 45% operational cost reductions through prompt engineering demonstrate measurable economic value that supports intellectual property protection efforts.

Legal compliance considerations include ensuring that prompt development respects existing intellectual property rights, avoiding incorporation of copyrighted material without authorization, and maintaining clear documentation of prompt development processes to support intellectual property claims. Organizations should develop explicit policies addressing prompt ownership, usage rights, and protection requirements.

Competitive Differentiation Through Conversational Excellence

The sustainable competitive advantage of prompt capital emerges from the cumulative effect of continuous refinement and optimization that creates conversational capabilities that competitors cannot quickly replicate. Like masterful recipes, expertly crafted prompts become more valuable through iterative improvement based on operational experience and outcome analysis.

Optimization cycles enable prompt libraries to improve automatically through feedback mechanisms that identify successful analytical approaches and incorporate learning into prompt refinements. Organizations implementing systematic prompt optimization report continuous improvement in AI output quality, analytical accuracy, and strategic relevance that compounds over time through operational experience.

Competitive moats develop through proprietary prompt libraries that encode unique combinations of domain expertise, analytical methodology, and strategic insight specific to organizational competitive advantages. These moats become stronger as prompt libraries accumulate refinements and develop sophisticated approaches that reflect years of organizational learning and strategic development.

Market positioning advantages emerge when organizations can deliver analytical capabilities that competitors cannot match through superior prompt engineering. Companies achieving 91% improvements in decision support accuracy through strategic prompt development gain competitive advantages that extend beyond operational efficiency to include strategic intelligence capabilities that inform better decision-making across all business functions.

Innovation acceleration occurs when sophisticated prompt libraries enable rapid experimentation with new analytical approaches, strategic frameworks, and business intelligence methodologies. Organizations with advanced prompt capital can test new strategic hypotheses, explore market opportunities, and evaluate competitive responses faster than competitors relying on traditional analytical approaches or less sophisticated AI integration.

BUILDING STRATEGIC PROMPT LIBRARIES

The development of strategic prompt libraries requires systematic approaches that treat conversational intelligence as a core business capability requiring investment, governance, and continuous optimization. Like any valuable intellectual property, prompt libraries benefit from structured development processes, quality assurance mechanisms, and strategic alignment with organizational objectives.

Successful prompt library development operates through organizational frameworks that combine technical expertise with domain knowledge and strategic insight to create conversational assets that generate sustainable competitive advantages. The most effective organizations establish prompt engineering as a recognized business discipline with dedicated resources, training programs, and performance measurement systems.

Organizational Framework and Governance Structure

Strategic prompt library development requires organizational structures that support systematic conversational asset creation while maintaining quality, security, and strategic alignment. Effective governance ensures that prompt development contributes to organizational competitive advantages rather than creating isolated technical capabilities without business value.

Centralized prompt management platforms provide infrastructure for creating, storing, organizing, and versioning strategic prompt libraries with role-based access controls, audit trails, and quality assurance mechanisms. These platforms enable collaboration between domain experts, prompt engineers, and business stakeholders while maintaining security and intellectual property protection for valuable conversational assets.

Cross-functional collaboration teams include domain experts who understand business requirements and analytical methodologies, prompt engineers who possess technical expertise in conversational AI optimization, and strategic stakeholders who ensure prompt development aligns with organizational objectives and competitive priorities. Advanced teams include legal experts who understand intellectual property protection requirements and security professionals who implement appropriate confidentiality measures.

Quality assurance processes ensure that prompt libraries generate consistent, accurate, and strategically relevant outputs through systematic testing, validation, and refinement procedures. Quality assurance includes performance benchmarking that compares prompt effectiveness across different scenarios, accuracy validation that confirms analytical reliability, and strategic alignment assessment that ensures prompt outputs support organizational decision-making requirements.

Version control and change management systems enable systematic prompt evolution while maintaining stability and reliability for production applications. These systems include development environments for prompt experimentation, staging environments for pre-production testing, and production environments with strict change control procedures that prevent unauthorized modifications to critical conversational assets.

Domain-Specific Prompt Development

Effective prompt libraries develop through specialization that reflects specific industry requirements, functional expertise, and strategic priorities rather than attempting to create generic conversational capabilities that lack sufficient precision for sophisticated business applications.

Financial services prompt development focuses on analytical frameworks for risk assessment, regulatory compliance, market analysis, and investment evaluation that require a sophisticated understanding of financial markets, regulatory requirements, and quantitative analysis methodologies. Advanced financial prompts encode decades of institutional knowledge about credit analysis, portfolio optimization, and regulatory reporting into conversational instructions that ensure consistent analytical approaches across large-scale operations.

Legal and compliance prompt development creates conversational frameworks for contract analysis, regulatory research, due diligence investigation, and legal document preparation that incorporate a sophisticated understanding of legal precedent, regulatory requirements, and industry-specific compliance obligations. These prompts enable rapid legal analysis while maintaining accuracy and professional standards required for legal applications.

Sales and marketing prompt development generates conversational capabilities for customer analysis, competitive positioning, campaign optimization, and market research that reflect a sophisticated understanding of customer psychology, market dynamics, and strategic positioning. Advanced sales prompts enable consistent customer analysis approaches that identify opportunity patterns and optimize resource allocation across sales teams.

Operations and supply chain prompt development creates analytical frameworks for process optimization, performance analysis, resource allocation, and quality management that incorporate a sophisticated understanding of operational excellence principles, supply chain dynamics, and performance measurement methodologies. These prompts enable continuous operational improvement through systematic analysis and optimization recommendations.

Continuous Learning and Optimization Systems

Strategic prompt libraries improve through systematic learning mechanisms that identify successful conversational approaches and incorporate organizational knowledge into prompt refinements. These learning systems enable prompt evolution that reflects operational experience and strategic development rather than remaining static after initial development.

Performance measurement systems track prompt effectiveness across multiple dimensions, including analytical accuracy, strategic relevance, operational efficiency, and business value generation. Measurement includes outcome analysis that correlates prompt usage with business results, user feedback assessment that identifies improvement opportunities, and competitive benchmarking that ensures prompt capabilities maintain market advantages.

Automated optimization algorithms analyze prompt performance data to identify successful conversational patterns and suggest improvements that enhance analytical accuracy and strategic relevance. Advanced optimization includes natural language processing analysis that identifies effective prompt structures, machine learning algorithms that predict prompt performance improvements, and feedback incorporation systems that integrate operational learning into prompt refinements.

Collaborative improvement processes enable domain experts, prompt engineers, and business stakeholders to contribute insights for prompt enhancement through structured feedback mechanisms, improvement suggestion systems, and collaborative refinement sessions. These processes ensure that prompt evolution reflects both technical optimization and business expertise while maintaining strategic alignment with organizational objectives.

Knowledge transfer mechanisms ensure that prompt library improvements benefit the entire organization through training programs, best practice documentation, and expertise sharing systems. Advanced knowledge transfer includes mentoring programs that develop internal prompt engineering capabilities, documentation systems that capture prompt development methodologies, and training materials that enable widespread organizational adoption of prompt optimization techniques.

EXECUTIVE CO-PILOTS AND COGNITIVE LEVERAGE SYSTEMS

The most sophisticated applications of prompt capital involve creating executive co-pilot systems that function as intelligent advisors capable of supporting high-level strategic decision-making through conversational interfaces that understand business context, strategic priorities, and leadership requirements. These systems represent the ultimate expression of prompt capital because they encode organizational intelligence at the highest strategic levels.

Executive co-pilots operate through sophisticated prompt architectures that combine strategic analysis capabilities, market intelligence processing, competitive assessment frameworks, and scenario modeling systems into conversational interfaces that enable executives to explore complex strategic questions through natural language interactions that generate insights equivalent to consulting team analysis.

Strategic Intelligence Amplification

Executive co-pilot development requires prompts that can process complex strategic scenarios while maintaining awareness of organizational context, competitive dynamics, and strategic objectives. These prompts function like experienced strategic consultants who understand industry patterns, competitive positioning, and strategic option evaluation while operating at digital speed with access to comprehensive data analysis capabilities.

Strategic scenario modeling prompts enable executives to explore potential future conditions through conversational interfaces that generate comprehensive analysis of market developments, competitive responses, regulatory changes, and technological disruptions. Advanced scenario prompts incorporate probabilistic reasoning that assesses likelihood ranges for different developments while identifying key factors that could influence strategic outcomes.

Competitive intelligence prompts process market data, competitor analysis, and industry trends to generate strategic insights about competitive positioning, market opportunities, and competitive threats. These prompts encode a sophisticated understanding of competitive analysis methodologies while maintaining current awareness of market developments and competitor activities that could affect strategic positioning.

Market opportunity assessment prompts analyze customer trends, market dynamics, and technological developments to identify strategic opportunities for growth, expansion, or innovation. Advanced opportunity prompts combine quantitative market analysis with qualitative trend assessment to generate strategic recommendations that align with organizational capabilities and strategic objectives.

Risk evaluation prompts provide comprehensive analysis of strategic risks, including market risks, competitive threats, operational vulnerabilities, and regulatory exposures. These prompts encode sophisticated risk assessment methodologies while maintaining current awareness of emerging risk factors that could affect strategic success and organizational performance.

Decision Support and Strategic Planning

Executive co-pilot systems enhance strategic decision-making through conversational interfaces that provide comprehensive analysis, alternative evaluation, and implementation planning support for complex strategic decisions. These systems function like strategic advisory teams while operating through natural language interactions that enable rapid exploration of strategic options and implications.

Decision framework prompts guide executives through systematic strategic decision-making processes that ensure comprehensive consideration of relevant factors, stakeholder implications, and strategic consequences. Advanced decision prompts incorporate decision science methodologies while maintaining flexibility for unique strategic circumstances and organizational priorities.

Alternative evaluation prompts enable systematic comparison of strategic options through conversational interfaces that assess advantages, disadvantages, implementation requirements, and strategic implications for different approaches. These prompts ensure comprehensive strategic analysis while maintaining focus on factors most relevant to organizational success and competitive positioning.

Implementation planning prompts generate detailed strategic implementation plans, including resource requirements, timeline development, risk mitigation strategies, and success measurement approaches. Advanced implementation prompts understand organizational capabilities and constraints while generating realistic plans that align with strategic objectives and operational realities.

Stakeholder analysis prompts provide a comprehensive assessment of stakeholder implications for strategic decisions, including customer impact, employee considerations, investor reactions, and regulatory responses. These prompts enable strategic decision-making that considers broader stakeholder ecosystem implications while maintaining focus on organizational strategic objectives.

Cognitive Leverage and Decision Velocity

Executive co-pilot systems create cognitive leverage by enabling individual executives to access analytical capabilities equivalent to entire consulting teams while maintaining the speed and flexibility of conversational interaction. This leverage enables strategic decision-making velocity that provides competitive advantages through faster strategic response and more comprehensive strategic analysis.

Analytical depth amplification enables executives to request sophisticated analysis through conversational interfaces that generate insights equivalent to weeks of traditional analytical work. Advanced amplification includes quantitative modeling, qualitative analysis, and strategic synthesis that provides a comprehensive understanding of complex strategic issues without requiring dedicated analytical teams.

Strategic pattern recognition prompts identify strategic patterns across industries, markets, and competitive environments that inform strategic decision-making through historical analysis and comparative assessment. These prompts encode an understanding of strategic success patterns while maintaining awareness of unique organizational circumstances and strategic context.

Real-time strategic adjustment capabilities enable executives to modify strategic approaches based on changing conditions through conversational interfaces that provide immediate analysis of strategic implications and adjustment recommendations. Advanced adjustment capabilities include scenario updating, risk reassessment, and opportunity identification that enable dynamic strategic management.

Decision confidence enhancement provides executives with a comprehensive understanding of strategic decision implications through conversational interfaces that offer uncertainty analysis, sensitivity assessment, and confidence intervals for strategic outcomes. These capabilities enable strategic decision-making with an appropriate understanding of risks and opportunities while maintaining the decision velocity required for competitive advantage.

THE ECONOMICS OF CONVERSATIONAL INTELLIGENCE

The economic value of prompt capital extends beyond operational efficiency improvements to create fundamental changes in how organizations generate, capture, and protect economic value through conversational intelligence capabilities. Understanding these economics enables strategic investment decisions and competitive positioning that leverage conversational intelligence for sustainable business advantage.

Traditional economic models for information technology focus on automation benefits, process efficiency, and cost reduction as primary value sources.

Conversational intelligence creates additional value through capability enhancement, knowledge amplification, and strategic intelligence generation that enables new forms of competitive advantage and business model innovation.

Value Creation and Measurement Frameworks

Economic value creation through prompt capital operates through multiple channels that require sophisticated measurement approaches to capture both direct operational benefits and indirect strategic advantages. Traditional ROI calculations underestimate the full economic impact of conversational intelligence because they focus on measurable cost savings while missing strategic value creation and competitive positioning improvements.

Productivity multiplier effects occur when prompt-enhanced capabilities enable individual professionals to achieve analytical outputs equivalent to entire teams while maintaining higher quality and consistency standards. Organizations implementing strategic prompt libraries report productivity improvements averaging 67% across AI-enabled processes, representing economic value that scales with organizational size and analytical complexity.

Decision quality enhancement generates economic value through improved strategic outcomes resulting from better analytical support and more comprehensive information processing. Organizations achieving 91% improvements in decision support accuracy through prompt engineering demonstrate economic value through better strategic decisions, reduced error costs, and enhanced competitive positioning.

Innovation acceleration enables organizations to explore new strategic opportunities, test business hypotheses, and develop competitive responses faster than traditional analytical approaches allow. Advanced prompt capabilities reduce strategic analysis timeframes from months to days while maintaining analytical rigor, enabling strategic advantages through superior market responsiveness and opportunity identification.

Knowledge asset creation establishes intellectual property value through systematic development of conversational intelligence capabilities that become organizational assets with measurable economic value. Prompt libraries that generate consistent competitive advantages represent intellectual property with economic value equivalent to other strategic business assets.

Cost Structure Optimization

Prompt capital transforms organizational cost structures by replacing expensive human analytical capacity with conversational intelligence capabilities that provide equivalent or superior analytical outputs at significantly reduced costs. This transformation enables organizational scaling without proportional increases in analytical personnel while maintaining or improving analytical capability quality.

Labor cost optimization occurs when prompt-enhanced AI systems replace expensive consultant engagements, reduce analytical team requirements, and enable existing personnel to achieve higher-value analytical outputs. Organizations implementing comprehensive prompt libraries report 45% reductions in AI-related operational costs through reduced error correction, rework elimination, and analytical efficiency improvements.

Scaling cost advantages emerge when prompt libraries enable organizational growth without proportional increases in analytical infrastructure, consulting expenses, or specialized personnel requirements. Advanced prompt capabilities enable small organizations to access analytical capabilities previously available only to large enterprises with extensive analytical teams.

Quality consistency benefits reduce costs associated with analytical errors, strategic mistakes, and decision-making inconsistencies that result from variable human analytical approaches. Prompt-standardized analytical processes eliminate quality variations that can create expensive strategic errors while ensuring consistent analytical rigor across organizational decision-making processes.

Strategic response speed advantages enable organizations to respond to competitive threats and market opportunities faster than competitors, creating economic value through improved market positioning and reduced opportunity costs from delayed strategic action.

Revenue Generation and Business Model Innovation

Conversational intelligence capabilities enable new revenue generation opportunities through improved customer analysis, enhanced product development, optimized pricing strategies, and innovative service delivery models that create additional economic value beyond operational cost improvements.

Customer intelligence enhancement enables improved customer lifetime value optimization, enhanced segmentation strategies, and superior customer experience design through conversational analysis capabilities that provide deeper customer understanding than traditional analytical approaches. Advanced customer prompts enable personalization at scale while maintaining cost-effectiveness.

Product development acceleration enables faster innovation cycles, improved product-market fit assessment, and enhanced competitive positioning through conversational intelligence that supports rapid product iteration and market validation. Organizations with sophisticated prompt capabilities can explore product opportunities and validate market hypotheses faster than competitors relying on traditional market research approaches.

Strategic pricing optimization utilizes conversational intelligence to develop dynamic pricing strategies, assess competitive positioning, and optimize revenue generation through sophisticated pricing analysis that considers multiple market factors simultaneously. Advanced pricing prompts enable real-time pricing optimization while maintaining strategic positioning objectives.

Service delivery innovation enables new business models based on conversational intelligence capabilities, providing enhanced customer value through AI-augmented services, intelligent consultation, and sophisticated analytical support. Organizations developing prompt-based service capabilities can create new revenue streams while differentiating from competitors through superior conversational intelligence integration.

The economics of conversational intelligence represent a fundamental shift from viewing AI as a cost-reduction technology to recognizing prompt capital as a strategic asset that enables new forms of value creation, competitive positioning, and business model innovation. Organizations that understand and leverage these economics will achieve sustainable competitive advantages in the intelligent economy.

CHAPTER 8

The AI Operating System for Business

The year is 2029. Your smartphone dies. Your laptop crashes. Your Internet connection fails. Yet your business continues running flawlessly, processing orders, optimizing supply chains, negotiating contracts, and serving customers without missing a beat. Not because you have backup systems, but because your organization has evolved beyond dependence on individual technologies to become something unprecedented. It has become a self-executing entity that thinks faster than markets move and adapts quicker than competitors can respond.

This isn't science fiction. The foundational technologies exist today. The early adopters are already building these systems. The question isn't whether AI operating systems will reshape business. The question is whether you'll be among the organizations that master them before your competitors do.

Right now, in boardrooms across Silicon Valley, executives are discovering that their traditional enterprise software has become a liability. Those carefully designed workflows that took years to implement are too slow for markets that change overnight. Those perfectly optimized processes can't adapt when customer preferences shift in real time. Those expensive management layers create delays that cost millions in missed opportunities.

The AI operating system solves the fundamental problem that has plagued businesses since the Industrial Revolution began. We've organized companies around human cognitive limitations, including hierarchical decision-making, sequential processing, and periodic planning cycles that made sense when information traveled slowly and change happened gradually. In a world where competitive advantages last months instead of years, those limitations have become existential threats.

What is documented in this chapter represents a complete reimagining of how organizations think, plan, and execute. Not incremental improvements to existing systems, but the emergence of an entirely new form of business entity that operates through principles that transcend traditional management thinking.

The enterprises building AI operating systems discover they can accomplish something that business theorists claimed was impossible. They can maintain the coordination benefits of a large-scale organization while achieving the speed and adaptability of small startups. They can execute complex strategies while remaining flexible enough to pivot instantly when conditions change. They can scale intelligence without scaling bureaucracy.

The transformation begins when organizations stop thinking about AI as something they use and start thinking about AI as something they become. The AI operating system isn't software you install. It's the nervous system that enables your organization to evolve from a collection of people and processes into a unified intelligence that can perceive, think, and act as a coherent entity.

FROM STATIC WORKFLOWS TO INTENT FLOWS AND ADAPTIVE EXECUTION

Picture the moment when GPS navigation replaced printed driving directions. Suddenly, getting lost became nearly impossible. More importantly, optimal routes could change dynamically based on traffic, weather, or road construction. Navigation evolved from following predetermined instructions to pursuing intended destinations through whatever path worked best under current conditions.

Business workflows today still operate like printed driving directions. They specify exactly which steps to follow, in which sequence, regardless of changing conditions. When markets shift, customer preferences evolve, or competitive dynamics change, those workflows become obstacles rather than enablers. Organizations waste months redesigning processes that should adapt automatically.

Intent flows represent the GPS revolution for business execution. Instead of specifying how work gets done, you specify what outcomes you want to achieve. Instead of rigid process steps, you get adaptive execution that finds optimal paths to desired results. Instead of breaking when conditions change, the system becomes more effective as it learns from experience.

The Death of the Process Map

Traditional workflow design assumes predictability. Process maps work beautifully when inputs are consistent, conditions remain stable, and optimal approaches don't change. In reality, every customer interaction is unique, every market situation contains novel elements, and every competitive response requires contextual adaptation.

The fundamental flaw in workflow thinking isn't technical. It's philosophical. Workflows assume that intelligent execution means following predetermined steps perfectly. Intent flows recognize that intelligent execution means achieving desired outcomes despite unpredictable circumstances.

Netflix discovered this when they attempted to optimize their content recommendation system through traditional workflow design. Engineers spent months mapping the perfect sequence for analyzing viewing patterns, processing user preferences, and generating recommendations. The system worked magnificently in testing environments with predictable data patterns.

Then reality intervened. Users watched content in unexpected combinations. Cultural events influenced viewing preferences overnight. Competitor releases changed demand patterns instantly. The carefully designed workflow couldn't adapt fast enough to remain relevant.

A breakthrough came when Netflix abandoned workflow optimization entirely and embraced intent-driven architecture. Instead of specifying how recommendations should be generated, they defined what recommendations should accomplish. They wanted to maximize viewing engagement while introducing users to content they wouldn't discover independently, while maintaining satisfaction with familiar genres.

The intent-driven system achieved these objectives through methods that would have been impossible to predict or prescribe. It developed a sophisticated understanding of user psychology that enabled personalization at an unprecedented scale. It identified content patterns that human analysts never would have recognized. It optimized recommendation timing based on viewing contexts that workflow designers never considered.

The transformation from workflow thinking to intent execution requires fundamental changes in how organizations conceptualize business operations. Instead of designing perfect processes, you design learning systems. Instead of controlling execution steps, you guide desired outcomes. Instead of preventing errors through rigid procedures, you enable adaptation through intelligent responses to unexpected conditions.

Reasoning Layers and Autonomous Decision Architecture

Traditional enterprise software operates through deterministic logic. When condition A occurs, execute action B. When threshold C is reached, trigger response D. This approach works adequately for predictable scenarios but fails catastrophically when faced with novel situations that don't match predetermined conditions.

Intent flows operate through reasoning layers that enable sophisticated analysis of complex situations before determining appropriate responses. These layers don't simply match conditions to predetermined actions. They evaluate contexts, assess implications, consider alternatives, and generate responses that align with intended outcomes even when facing unprecedented circumstances.

The reasoning architecture includes contextual analysis that understands situational factors influencing optimal decision-making, strategic evaluation that ensures decisions align with organizational objectives, resource assessment that considers available capabilities and constraints, and outcome prediction that anticipates likely results of different response options.

Advanced reasoning systems develop an increasingly sophisticated understanding of business dynamics through operational experience. They learn which approaches work best under different conditions. They identify patterns that indicate emerging opportunities or threats. They recognize when standard approaches may not apply and escalate decisions to human oversight when appropriate.

Goldman Sachs implemented reasoning layers throughout their trading operations to enable an adaptive response to market volatility. Traditional trading systems relied on predetermined algorithms that specified exact responses to specific market conditions. These algorithms worked well during stable periods but created massive losses when markets behaved unpredictably.

The reasoning-based system evaluates market conditions through multiple analytical frameworks simultaneously. It considers historical patterns, current market sentiment, regulatory implications, competitive positioning, and strategic objectives before determining optimal trading strategies. When market conditions fall outside historical patterns, the system generates novel approaches rather than defaulting to potentially inappropriate predetermined responses.

The results exceeded expectations dramatically. Trading performance improved during both stable and volatile periods. Risk management became more sophisticated and adaptive. Most importantly, the system continued improving its performance through experience rather than requiring constant human optimization.

Adaptive Execution and Dynamic Resource Allocation

Intent flows enable organizations to optimize resource allocation continuously based on changing priorities and emerging opportunities rather than relying on periodic planning cycles that become obsolete before implementation completes. Adaptive execution means that organizational capabilities automatically redirect toward the highest-value activities as business conditions evolve.

Dynamic resource allocation operates through the continuous assessment of opportunity values, capability requirements, and strategic priorities. Instead of annual budgeting processes that lock resources into predetermined allocations, adaptive systems reallocate capabilities in real time based on current value potential and strategic alignment.

The adaptive approach includes demand prediction that anticipates resource requirements before needs become critical, capability mapping that identifies optimal allocation of skills and technologies, priority assessment that ensures resources flow toward highest-value opportunities, and performance optimization that improves allocation effectiveness through experience.

Amazon fulfillment operations demonstrate adaptive execution at unprecedented scale. Traditional warehouse management operates through predetermined workflows that specify exact procedures for receiving, storing, picking, and shipping inventory. These workflows optimize efficiency under normal conditions but create bottlenecks when demand patterns change unexpectedly.

The intent-driven fulfillment system of Amazon optimizes customer satisfaction and delivery speed rather than following predetermined operational procedures. When demand spikes for specific products, the system automatically reconfigures warehouse layouts, adjusts picking sequences, and reallocates staff priorities to maintain delivery commitments without requiring human intervention.

During peak shopping periods, the system generates operational strategies that no human manager could design manually. It coordinates activities across multiple warehouses, optimizes transportation routes in real time, and adjusts capacity allocation based on predictive analysis of demand patterns. The result is operational performance that scales automatically while maintaining service quality.

The transformation to adaptive execution requires organizational cultures that embrace continuous change rather than stability. Teams must become comfortable with systems that modify priorities automatically. Managers must learn to guide outcomes rather than control processes. Success metrics must focus on results achieved rather than procedures followed.

CONTINUOUS PLANNING AND SELF-HEALING BUSINESS STRUCTURES

Imagine if your skeleton could strengthen itself automatically when you lifted weights, your immune system could develop new defenses before encountering threats, and your nervous system could reroute around damaged areas instantly. This isn't fantasy. It's how biological systems achieve resilience through continuous adaptation rather than predetermined responses.

Traditional business planning operates like mechanical engineering. You design structures to withstand anticipated stresses, build redundancies for predictable failures, and replace components when they break. This approach works adequately for stable environments but fails when organizations face unprecedented challenges that exceed design parameters.

Self-healing business structures operate through biological principles. They sense emerging stresses before they become critical. They develop adaptive responses to novel challenges. They strengthen capabilities automatically based on operational demands. They reroute around failures without requiring external intervention.

Living Strategy That Evolves in Real Time

Strategic planning traditionally happens during annual retreats where executives analyze historical data, project future trends, and commit to approaches that won't change until the next planning cycle. By the time strategies get implemented, market conditions have usually shifted enough to reduce effectiveness significantly.

Continuous planning operates through persistent strategy optimization that adapts approaches based on real-time performance feedback and changing market conditions. Instead of periodic strategy updates, organizations maintain living strategic frameworks that evolve continuously while preserving long-term directional coherence.

The continuous approach includes environmental monitoring that tracks market changes, competitive movements, and customer behavior shifts in real time, strategy simulation that tests approach effectiveness under different scenarios before implementation, performance tracking that measures strategic progress across multiple dimensions simultaneously, and adaptive optimization that improves strategic approaches based on operational learning.

Tesla demonstrates continuous strategic planning through their approach to electric vehicle development and market expansion. Traditional automotive companies develop five-year product strategies that specify exact vehicle features, production timelines, and market positioning approaches. These strategies work well when market conditions remain predictable, but become liabilities when technological capabilities or customer preferences change rapidly.

The continuous planning system of Tesla adapts product development and market strategies based on real-time feedback from customers, manufacturing operations, and competitive analysis. When battery technology improves faster than anticipated, production strategies adjust automatically. When customer usage patterns reveal unexpected preferences, feature development priorities shift accordingly. When competitors introduce new capabilities, competitive positioning adapts without requiring formal strategy revision processes.

The result is strategic agility that enables rapid response to opportunities and threats while maintaining long-term strategic coherence. Tesla can pivot product strategies faster than competitors can recognize that pivots are necessary. They can adapt market positioning based on emerging customer insights. They can modify competitive approaches when market dynamics shift.

Organizational Immune Systems and Threat Response

Biological immune systems don't wait for diseases to cause damage before responding. They maintain constant surveillance for potential threats, develop defenses proactively, and coordinate responses that neutralize dangers before they spread. Business organizations need similar capabilities to survive in environments where competitive threats, regulatory changes, and market disruptions can emerge overnight.

Self-healing business structures include organizational immune systems that monitor internal and external environments for emerging risks, develop adaptive responses to novel threats, and coordinate defensive actions that protect organizational health without disrupting normal operations.

Threat detection operates through pattern recognition that identifies unusual signals indicating potential problems, anomaly analysis that distinguishes between normal variations and genuine risks, escalation protocols that ensure appropriate response levels for different threat severities, and learning mechanisms that improve detection accuracy through experience.

When cybersecurity firm CrowdStrike built its organizational immune system, it discovered that traditional risk management approaches were inadequate for its rapidly evolving threat environment. Conventional risk assessment relies on historical analysis and predetermined response procedures that can't adapt quickly enough when facing novel attack vectors or competitive challenges.

Their self-healing system monitors thousands of indicators, including network behavior patterns, competitive intelligence signals, customer satisfaction metrics, and employee engagement levels. When the system detects patterns indicating potential threats, it automatically initiates protective responses while alerting human oversight teams.

During a sophisticated competitor attack targeting their key customers, the system recognized unusual customer interaction patterns that preceded account losses. Instead of waiting for formal analysis and management decisions, the system automatically enhanced customer engagement protocols, activated retention specialists, and implemented competitive countermeasures that neutralized the threat before significant damage occurred.

Autonomous Recovery and Resilience Optimization

Self-healing systems don't just detect and respond to problems. They learn from difficulties to become more resilient. Every challenge becomes an opportunity to strengthen organizational capabilities. Every failure generates insights that improve future performance. Every recovery process enhances the system's ability to handle similar situations more effectively.

Autonomous recovery includes failure analysis that identifies root causes and contributing factors, capability enhancement that strengthens weak points revealed by difficulties, process optimization that improves response effectiveness for future challenges, and knowledge integration that ensures learning benefits the entire organization.

The Netflix content delivery network demonstrates autonomous recovery at a global scale. Traditional content distribution relies on predetermined infrastructure configurations and manual intervention when problems occur. When servers fail or network congestion develops, human operators must identify problems, determine responses, and implement solutions manually.

The Netflix self-healing infrastructure monitors performance continuously and responds to problems automatically before they affect customer experience. When server capacity becomes insufficient, the system provisions additional resources instantly. When network routes become congested, traffic automatically redirects through optimal alternatives. When content becomes unavailable in specific regions, backup systems activate seamlessly.

The system learns from every problem to improve future resilience. Failure patterns inform infrastructure improvements. Performance variations guide capacity planning. Customer impact analysis optimizes recovery priorities. The result is infrastructure that becomes more reliable through experience rather than requiring constant manual optimization.

Organizations implementing self-healing structures discover that resilience becomes a competitive advantage rather than a defensive necessity. They can maintain service quality during disruptions that cripple competitors. They can adapt to changes that make traditional approaches obsolete. They can turn challenges into opportunities for improvement rather than sources of damage.

REAL-TIME STRATEGY EXECUTION MODELS

Strategy execution has always been the graveyard of brilliant plans. Research consistently shows that most strategic initiatives fail not because strategies are poorly conceived, but because execution happens too slowly, with too little adaptation, through too many layers of translation that dilute original intent beyond recognition.

The speed of business has accelerated exponentially, but strategy execution still operates through industrial-age frameworks. Strategies get developed by executives, translated by managers, implemented by teams, and measured by analysts through sequential processes that consume months before generating first results. By the time execution begins, strategies often address yesterday's opportunities instead of tomorrow's realities.

Real-time strategy execution eliminates the translation delays that kill strategic effectiveness. Instead of cascading strategies through organizational hierarchies, AI operating systems enable direct translation of strategic intent into operational actions that adapt continuously based on performance feedback and changing conditions.

Strategy as Code

Software developers have learned that code must be executable, testable, and modifiable in real time to remain relevant. Business strategies need similar characteristics. Instead of document-based strategies that require human interpretation and manual implementation, leading organizations are developing executable strategies that translate strategic intent directly into operational actions.

Strategy as code means that strategic objectives become parameters that guide autonomous execution rather than documents that require human implementation. Strategic priorities automatically influence resource allocation decisions. Strategic constraints automatically prevent actions that conflict with organizational values. Strategic goals automatically guide optimization efforts across all business functions.

The executable approach includes intent specification that defines strategic objectives in precise terms that AI systems can interpret accurately, constraint definition that establishes boundaries within which autonomous execution must operate, optimization criteria that guide decision-making when multiple options align with strategic objectives, and feedback loops that improve strategic effectiveness through operational learning.

The Microsoft transformation under Satya Nadella demonstrates strategy as code principles. Traditional technology companies develop strategic plans that specify exact product roadmaps, market positioning approaches, and competitive responses. These plans work well when technological and market conditions remain predictable, but become obsolete when disruption occurs faster than planning cycles.

The cloud-first strategy of Microsoft operates through executable parameters that guide thousands of autonomous decisions daily. Product development priorities adjust automatically based on customer usage patterns. Market positioning adapts based on competitive analysis. Resource allocation is optimized based on strategic value potential rather than predetermined budgets.

When AI capabilities advanced faster than anticipated, the executable strategy of Microsoft enabled rapid pivoting toward AI integration across all product lines. Instead of requiring formal strategy revision and cascaded implementation, the strategic parameters automatically guided development teams toward AI-enhanced solutions that aligned with cloud-first objectives.

Autonomous Strategic Adaptation

Real-time execution enables strategies to evolve continuously while maintaining directional coherence. Instead of periodic strategy updates that create discontinuity and confusion, autonomous adaptation ensures that strategic approaches improve constantly based on performance feedback and environmental changes.

Adaptive strategic execution includes environmental scanning that monitors conditions affecting strategic effectiveness, performance analysis that identifies which strategic approaches generate the best results under different circumstances, strategic optimization that improves approach effectiveness through systematic experimentation, and coherence maintenance that ensures strategic evolution preserves long-term organizational direction.

The marketplace strategy of Amazon exemplifies autonomous strategic adaptation. Traditional retail companies develop market strategies that specify exact vendor relationships, pricing approaches, and competitive positioning. These strategies require manual updates when market conditions change, creating delays that cost competitive advantages.

Amazon Marketplace operates through adaptive strategic parameters that optimize customer value, vendor success, and competitive positioning simultaneously. When customer preferences shift, vendor selection and promotion algorithms adjust automatically. When competitive pressures change, pricing strategies adapt without requiring management intervention. When new market opportunities emerge, expansion approaches are modified based on strategic optimization criteria.

The autonomous adaptation enabled Amazon to respond to pandemic-driven e-commerce acceleration faster than competitors could recognize that opportunities existed. Instead of waiting for strategic planning cycles and implementation cascades, the adaptive system automatically optimized vendor support, adjusted capacity allocation, and enhanced customer experience based on changing demand patterns.

Performance Amplification Through Intelligent Coordination

Real-time strategy execution creates performance advantages through intelligent coordination that aligns all organizational activities toward strategic objectives without requiring constant management intervention. Instead of hoping that individual departments will coordinate effectively, AI operating systems ensure automatic alignment that optimizes overall performance rather than departmental metrics.

Intelligent coordination includes objective alignment that ensures all organizational activities contribute to strategic goals, resource optimization that allocates capabilities toward the highest-value strategic priorities, performance synchronization that coordinates timing across different organizational functions, and impact amplification that identifies opportunities where coordinated actions generate multiplicative rather than additive results.

SpaceX demonstrates intelligent coordination through its approach to rocket development and space missions. Traditional aerospace companies coordinate activities through extensive project management processes that require constant human oversight to maintain alignment across engineering, manufacturing, testing, and operations teams.

The SpaceX coordination system aligns all activities automatically toward mission success and cost-optimization objectives. Engineering decisions consider manufacturing implications automatically. Manufacturing processes are optimized for operational requirements. Testing procedures focus on mission-critical performance factors. Operations planning incorporates engineering constraints seamlessly.

When technical challenges emerge during development, the coordination system automatically adjusts priorities across all functions to maintain mission timelines while ensuring safety requirements. Engineering resources are redirected toward critical problems. Manufacturing schedules adapt to accommodate design changes. Testing protocols intensify for modified components. Operations plans adjust for revised performance parameters.

The intelligent coordination enables SpaceX to achieve development speeds and cost efficiencies that seemed impossible using traditional aerospace management approaches. Instead of requiring extensive coordination meetings and management oversight, the system ensures automatic alignment that generates optimal performance across all organizational functions.

EMBEDDING LEARNING INTO ORGANIZATIONAL DNA

Traditional organizations treat learning like vitamins. Important for long-term health but separate from daily operations. Learning happens during training sessions, strategic retreats, and post-project reviews. The rest of the time, organizations focus on execution, assuming that knowledge remains stable and capabilities don't need enhancement.

This separation between learning and doing creates organizations that excel at repeating past successes but fail when facing novel challenges. They optimize for efficiency in familiar situations but lack adaptability when conditions change. They develop expertise in specific approaches but struggle when those approaches become obsolete.

AI operating systems eliminate the distinction between learning and executing. Every action generates data that improves future decisions. Every outcome provides feedback that enhances organizational capabilities. Every challenge becomes an opportunity to develop new competencies rather than a problem that disrupts normal operations.

Organizational Memory That Never Forgets

Human organizations suffer from institutional amnesia. Knowledge walks out the door when employees leave. Lessons learned during crises get forgotten when normal operations resume. Successful approaches get abandoned when leadership changes. Hard-won expertise disappears when projects end.

AI operating systems maintain perfect institutional memory that accumulates and preserves organizational knowledge regardless of personnel changes or operational transitions. Every decision gets recorded with context. Every outcome gets analyzed for patterns. Every lesson gets integrated into organizational capabilities permanently.

Institutional memory includes experience documentation that captures not just what happened but why decisions were made and how outcomes developed, pattern recognition that identifies successful approaches and failure modes across different situations, knowledge integration that ensures learning benefits all relevant organizational functions, and capability enhancement that translates experience into improved performance.

When consulting firm McKinsey & Company implemented organizational memory systems, they discovered that traditional knowledge management approaches captured only a fraction of institutional learning. Consultants documented final recommendations and successful methodologies, but the reasoning processes, alternative approaches, and contextual factors that influenced decisions remained tacit knowledge that disappeared when team members moved to new assignments.

Their AI operating system captures complete decision contexts, including client constraints, market conditions, competitive factors, and analytical reasoning that informs recommendations. When similar situations arise, consultants can access not just previous recommendations but complete analysis frameworks that generated those recommendations.

The comprehensive memory enables rapid acceleration of consultant effectiveness. New team members can quickly understand successful approaches for specific client types. Experienced consultants can avoid repeating unsuccessful strategies. Complex engagements can build on institutional knowledge rather than starting analysis from basic principles.

Continuous Capability Development

Traditional training approaches assume that capabilities can be developed through periodic instruction and then applied consistently until the next training cycle. This approach works adequately for stable skill requirements but fails when capabilities must evolve continuously to match changing demands.

AI operating systems enable continuous capability development that adapts skills in real time based on operational requirements and performance feedback. Instead of scheduled training programs, organizations develop adaptive learning systems that enhance capabilities automatically through operational experience.

Continuous development includes performance analysis that identifies capability gaps and improvement opportunities in real time, adaptive training that provides targeted skill enhancement based on individual and organizational needs, capability mapping that ensures skill development aligns with strategic requirements, and competency optimization that improves overall organizational effectiveness through systematic capability enhancement.

The manufacturing operations of Tesla demonstrate continuous capability development at an unprecedented scale. Traditional automotive manufacturing relies on extensive worker training programs followed by periodic refresher sessions to maintain skill levels. This approach works adequately for stable production processes but becomes inadequate when manufacturing approaches evolve rapidly.

Tesla learning systems continuously monitor individual worker performance and provide targeted skill enhancements automatically when performance variations suggest opportunities for improvement. When new production techniques are introduced, the system identifies which workers need specific training and delivers customized instruction that optimizes learning effectiveness.

During Model 3 production ramp-up, the system enabled rapid capability development that would have been impossible through traditional training approaches. As production processes evolved based on quality feedback and efficiency optimization, worker capabilities adapted automatically to match new requirements.

Evolutionary Organizational Intelligence

The ultimate expression of embedded learning is evolutionary organizational intelligence that enables entire organizations to become smarter through experience. Not just individual learning or departmental improvement, but genuine organizational evolution that enhances collective capability continuously.

Evolutionary intelligence includes collective pattern recognition that identifies successful approaches across all organizational functions, cross-functional learning that ensures insights benefit multiple departments simultaneously, adaptive optimization that improves organizational coordination and effectiveness through experience, and intelligence amplification that generates capabilities exceeding the sum of individual contributions.

The Google search algorithm evolution exemplifies evolutionary organizational intelligence. Traditional software development creates applications that perform consistent functions until developers manually implement improvements. Google Search operates through learning systems that become more effective automatically through user interaction.

Every search query provides feedback that improves algorithm effectiveness. Every click pattern reveals preferences that enhance result relevance. Every user behavior generates insights that optimize the search experience. The collective learning creates search capabilities that no individual programmer could design manually.

The evolutionary approach extends beyond search algorithms to organizational operations. Engineering teams learn from user feedback to improve product development processes. Marketing teams learn from customer interactions to enhance engagement strategies. Operations teams learn from performance patterns to optimize service delivery.

The result is organizational intelligence that evolves continuously rather than remaining static between formal improvement initiatives. Capabilities enhance automatically through operational experience. Performance improves systematically without requiring external intervention. Competitive advantages compound through learning rather than eroding through competition.

Organizations that successfully embed learning into their operational DNA discover that continuous improvement becomes automatic rather than requiring dedicated effort. They develop capabilities faster than competitors can copy their approaches. They adapt to changes that make competitor approaches obsolete. They create sustainable advantages through superior learning velocity rather than temporary technological leadership.

The transformation from learning as activity to learning as capability represents the ultimate evolution toward AI-native organization. When learning becomes embedded into organizational DNA, improvement becomes inevitable rather than optional, adaptation becomes automatic rather than effortful, and competitive advantage becomes sustainable rather than temporary.

CHAPTER 9

From Use Cases to Use Logic

Imagine if every time you needed to drive somewhere new, you had to reinvent the automobile from scratch. You'd start by designing an engine, then fabricating a chassis, followed by developing a transmission system and creating a steering mechanism. By the time you finished building your custom vehicle, the destination would be irrelevant, and the opportunity would be lost. This sounds absurd, yet it perfectly describes how most organizations approach AI implementation today.

The automotive industry solved this problem over a century ago through modular design and standardized components. Not every car manufacturer reinvents engines, transmissions, or braking systems for each new vehicle. Instead, they build from proven patterns, reusable components, and systematic engineering principles that enable rapid innovation while maintaining reliability. A modern car contains thousands of parts, but most are variations of established patterns that have been refined through decades of engineering evolution.

The most successful AI-native enterprises are discovering that the same principles apply to AI. Instead of treating each business challenge as a unique "use case" requiring custom AI solutions built from scratch, they're developing "use logic"—strategic pattern libraries that can be reused, simulated, and evolved across thousands of business scenarios.

Right now, at Netflix, a single recommendation logic pattern has been refined into hundreds of variations that handle everything from movie suggestions to content production decisions to user interface optimization. At Amazon, logistics optimization patterns originally developed for warehouse management now power supply chain decisions, delivery routing, inventory forecasting, and even retail space planning. At Microsoft, natural language processing patterns that began as simple search improvements now drive everything from code completion to meeting summarization to strategic document analysis.

This transformation from use cases to use logic represents the difference between artisanal AI development and industrial-scale intelligence production. Where use cases create isolated solutions that solve individual problems, use logic creates systematic capabilities that evolve and compound over time, generating exponential rather than linear value.

What has been consistently observed across enterprise technology transformation is that the organizations achieving breakthrough AI performance have made a fundamental conceptual shift. They've stopped asking "What specific problem can AI solve?" and started asking "What reasoning patterns can we systematize and reuse across our entire business ecosystem?"

The enterprises building use logic discover something that traditional project-based thinking makes impossible. They can maintain the customization benefits of bespoke solutions while achieving the scalability advantages of systematic engineering. They can respond to novel challenges faster than competitors can recognize that opportunities exist. They can turn every AI implementation into organizational learning that improves all future implementations.

The transformation begins when organizations realize that the real value of AI isn't in solving individual problems. It's in developing institutional capabilities for systematic reasoning that can adapt to any problem their business might face.

STRATEGIC PATTERN RECOGNITION OVER ONE-OFF PROJECTS

Picture the moment when software development evolved from custom programming for each application to reusable code libraries and frameworks. Suddenly, developers could build complex applications by combining proven components rather than writing every function from scratch. Development cycles accelerated from years to months to weeks. Quality improved because components had been tested across multiple implementations. Innovation flourished because developers could focus on unique value creation rather than rebuilding basic functionality.

AI implementation today stands at the same evolutionary threshold. Most organizations still approach each business challenge as if no similar problem has ever been solved before. They assemble custom teams, develop unique datasets, design novel algorithms, and create bespoke solutions that work beautifully for one specific scenario but provide no leverage for future challenges.

Use logic represents the systematic evolution beyond this artisanal approach. Instead of building one-off solutions, AI-native organizations develop pattern libraries that capture the essential reasoning structures underlying successful AI applications, then adapt these patterns across unlimited business scenarios.

The Economics of Pattern Reusability

The economic transformation from use cases to use logic operates through mathematical principles that create compound advantages. A use case generates value proportional to the single problem it solves. Use logic generates value proportional to every variation and adaptation it enables across the entire organization.

Consider document analysis. A use case approach might develop an AI system to extract key information from contracts. The system works brilliantly for contracts but provides no value for analyzing financial statements, customer feedback, regulatory documents, or strategic reports. When new document analysis needs arise, teams start over with new custom development.

A use logic approach develops systematic document comprehension patterns that understand information extraction, context analysis, entity recognition, and semantic relationship mapping. These patterns can be rapidly configured for contracts, financial statements, customer feedback, regulatory documents, strategic reports, or any future document type the organization encounters.

Goldman Sachs discovered this principle when it transformed its approach to financial document analysis. Its initial use case implementations solved specific problems beautifully. Contract analysis systems extracted key terms accurately. Financial statement processors identified important metrics reliably. Due diligence document reviews flagged relevant risks appropriately.

But each system was an isolated solution requiring separate maintenance, unique training data, and custom integration. When new document types emerged or analysis requirements evolved, teams needed months to develop new custom solutions. The total cost of ownership grew linearly with each new requirement, while organizational learning remained trapped in separate systems.

The transformation to use logic created systematic document intelligence patterns that could be rapidly adapted to any financial document type. The same logical frameworks that understood contract obligations could analyze regulatory filings. The patterns that extracted financial metrics could identify strategic insights in analyst reports. The reasoning structures that evaluated investment risks could assess operational performance across any document format.

The economic transformation was exponential. Instead of linear scaling costs, each new document analysis requirement leveraged existing patterns, reducing development time from months to days while improving accuracy through accumulated learning. The organization developed systematic document intelligence capabilities that created sustainable competitive advantages across every business function requiring document analysis.

Pattern Recognition as Competitive Advantage

Strategic pattern recognition transforms how organizations think about AI competitive advantage. Traditional competitive thinking focuses on acquiring better algorithms, larger datasets, or more powerful computing resources. These advantages are temporary because competitors can purchase similar technological capabilities or hire away technical talent.

Use logic creates competitive advantages through accumulated organizational intelligence that cannot be purchased or easily replicated. The patterns that emerge from systematic AI development represent institutional knowledge about how AI can be applied effectively across specific business contexts. This knowledge compounds through operational experience and creates widening competitive moats.

The pattern libraries developed by leading AI-native organizations encode understanding that goes far beyond technical implementation. They capture insights about which analytical approaches work best under different business conditions, how to adapt AI reasoning to specific organizational contexts, what failure modes to anticipate and prevent, and how to optimize performance for particular strategic objectives.

Netflix recommendation patterns illustrate this competitive dynamic. Competitors can access similar machine learning algorithms and computational resources. They can hire talented data scientists and engineers. They can even analyze public research publications of Netflix. They cannot replicate the accumulated intelligence embedded in the recommendation use logic of Netflix, though.

These patterns encode understanding developed through billions of user interactions across decades of operation. They understand how user preferences evolve over time, how cultural events influence viewing patterns, how content characteristics affect engagement, and how to balance exploration with exploitation in recommendation strategies. Most importantly, they continue learning and improving through every user interaction, creating competitive advantages that strengthen rather than erode over time.

Systematic Intelligence Development

The development of strategic pattern recognition requires organizational frameworks that treat AI reasoning as an institutional capability rather than a technical implementation. This transformation affects how organizations structure teams, measure success, allocate resources, and plan strategic development.

Pattern-oriented development includes systematic analysis of business reasoning requirements across all organizational functions, identification of underlying logical structures that transcend specific applications, development of reusable reasoning frameworks that can be adapted to multiple scenarios, and continuous refinement based on operational experience and changing business needs.

The systematic approach requires cultural transformation that values reusability over novelty, systematic thinking over quick solutions, and organizational learning over individual project success. Teams must learn to think in terms of patterns rather than problems, frameworks rather than fixes, and evolutionary development rather than custom creation.

The Microsoft transformation to use logic demonstrates this systematic approach. Instead of allowing each product team to develop custom AI solutions, they created centralized pattern libraries that capture successful reasoning approaches across all Microsoft products and services. Natural language processing patterns developed for search improvement now power everything from code completion to meeting transcription to customer service automation.

The systematic development includes pattern identification workshops that analyze successful AI implementations to extract reusable reasoning structures, framework development sessions that translate patterns into adaptable logical templates, integration planning that ensures patterns can be deployed across different business contexts, and evolutionary processes that improve patterns based on operational feedback and changing requirements.

The transformation enabled Microsoft to accelerate AI development across all business units while improving quality through proven patterns and reducing costs through systematic reuse. Instead of reinventing AI capabilities for each new application, teams can rapidly adapt existing patterns, focusing creative energy on unique value creation rather than rebuilding basic reasoning functionality.

The systematic approach also creates organizational learning that transcends individual projects. Every pattern adaptation generates insights that improve the entire pattern library. Every successful implementation contributes to organizational intelligence about effective AI application. Every challenge reveals opportunities for pattern enhancement that benefits all future implementations.

Organizations implementing strategic pattern recognition discover that competitive advantage shifts from having better technology to developing superior systematic intelligence about how to apply technology effectively. This shift creates sustainable advantages because organizational learning compounds while technological capabilities commoditize.

DESIGNING REUSABLE AI REASONING FRAMEWORKS

Traditional software development learned decades ago that successful systems emerge from architectural patterns rather than custom coding for each application. Design patterns like Model-View-Controller, Observer, and Factory enabled developers to solve common problems through proven approaches rather than reinventing solutions repeatedly. These patterns created the foundation for modern software engineering by providing reusable blueprints that could be adapted to countless applications.

AI reasoning frameworks require similar systematic architecture, but with additional complexity because they must handle probabilistic decision-making, contextual adaptation, and continuous learning. Unlike deterministic software patterns that produce identical outputs for identical inputs, AI reasoning frameworks must generate appropriate responses across infinite variations while maintaining consistency with organizational objectives and constraints.

The most successful AI-native organizations develop reasoning frameworks that function like intelligent design patterns—proven logical structures that can be rapidly configured for new applications while preserving the accumulated wisdom from previous implementations.

Modular Intelligence Architecture

Effective AI reasoning frameworks operate through modular architectures that separate different types of intelligence functions, enabling flexible combination and recombination based on specific application requirements. This modularity prevents organizations from having to rebuild core reasoning capabilities for each new AI application while enabling sophisticated customization for unique business contexts.

The modular approach includes perception modules that handle different types of input processing, reasoning modules that perform different types of analysis and decision-making, action modules that translate AI conclusions into appropriate organizational responses, and learning modules that improve framework performance through operational experience.

The approach of Amazon to logistics optimization demonstrates modular intelligence architecture at an unprecedented scale. Rather than developing separate AI systems for inventory management, demand forecasting, supplier coordination, and delivery optimization, they created modular reasoning frameworks that can be combined in different configurations based on specific operational requirements.

The perception modules handle different types of input, including sales data, inventory levels, supplier capacity, delivery constraints, customer preferences, and external factors like weather or events. The reasoning modules perform different types of analysis, including demand prediction, capacity optimization, route planning, and cost minimization. The action modules translate optimization conclusions into specific operational decisions like inventory orders, supplier requests, and delivery schedules.

The modular architecture enables rapid deployment of sophisticated optimization capabilities across any Amazon operation worldwide. New fulfillment centers can immediately access proven reasoning frameworks rather than developing custom solutions. Seasonal demand variations can be handled through framework reconfiguration rather than new system development. Novel optimization challenges can be addressed by combining existing modules in new ways rather than rebuilding analytical capabilities.

Most importantly, the modular approach creates systematic learning that benefits all Amazon operations. Improvements developed for one application automatically enhance capabilities across all implementations. Insights gained from seasonal demand patterns inform inventory optimization globally. Delivery route optimizations developed in one region improve logistics efficiency everywhere.

Context-Aware Reasoning Templates

Unlike traditional software templates that handle predetermined scenarios, AI reasoning frameworks must adapt to contextual variations while maintaining logical consistency and strategic alignment. Context-aware templates provide structured approaches for handling different types of business situations while preserving organizational knowledge about effective reasoning approaches.

Context awareness includes environmental scanning that identifies relevant factors affecting optimal reasoning approaches, situational assessment that determines which analytical methods apply to specific circumstances, constraint recognition that ensures reasoning respects organizational limitations and requirements, and objective alignment that guides reasoning toward outcomes that support strategic priorities.

The approach of Tesla to autonomous driving illustrates context-aware reasoning at the edge of technological possibility. Traditional automotive engineering relies on predetermined responses to specific scenarios—when condition A occurs, execute action B. This approach works adequately for predictable situations but fails catastrophically when facing novel circumstances that don't match predetermined conditions.

The reasoning frameworks used by Tesla evaluate driving contexts through multiple analytical perspectives simultaneously. Safety analysis considers immediate collision risks and escape routes. Traffic flow analysis evaluates optimal positioning and speed for efficient movement. Route optimization balances time, energy consumption, and passenger comfort. Strategic positioning anticipates future traffic developments and lane changes.

The context-aware templates adapt reasoning approaches based on situational factors like weather conditions, traffic density, road characteristics, and passenger preferences. City driving utilizes frameworks optimized for frequent stops and complex interactions. Highway driving employs patterns focused on efficient cruising and lane management. Emergency situations activate reasoning frameworks that prioritize safety over all other considerations.

The templates preserve organizational learning about effective autonomous driving while enabling adaptation to infinite contextual variations. Every driving scenario contributes insights that improve framework effectiveness. Every successful navigation enhances pattern recognition capabilities. Every edge case reveals opportunities for template refinement that benefit the entire autonomous driving system.

Compositional Reasoning Design

Advanced AI reasoning frameworks enable compositional design where complex reasoning capabilities emerge from combining simpler logical components. This compositional approach allows organizations to develop sophisticated AI applications by assembling proven reasoning modules rather than building monolithic custom solutions for each new challenge.

Compositional design includes atomic reasoning functions that perform specific analytical tasks reliably, composition protocols that enable combining functions to create more sophisticated reasoning capabilities, interface standards that ensure different reasoning components can work together effectively, and orchestration frameworks that coordinate complex reasoning across multiple components.

Google's approach to search demonstrates compositional reasoning at Internet scale. Search involves simultaneously understanding user intent, evaluating content relevance, assessing source authority, optimizing response speed, and personalizing results for individual users. Rather than building a single monolithic system that handles all these requirements, Google develops modular reasoning components that can be combined and recombined based on specific search contexts.

Intent analysis frameworks understand what users are seeking based on query patterns, historical behavior, and contextual signals. Relevance assessment frameworks evaluate how well different content addresses user needs based on semantic analysis, structural features, and user engagement patterns. Authority evaluation frameworks assess source credibility based on linking patterns, content quality, and historical accuracy.

The compositional approach enables Google to continuously enhance search capabilities by improving individual reasoning components without rebuilding the entire system. Better natural language understanding automatically improves all search applications. Enhanced relevance assessment benefits every user query. Improved personalization frameworks enhance user experience across all search contexts.

More importantly, compositional design enables Google to rapidly deploy search capabilities across new domains and applications. Voice search leverages existing Intent analysis while adding speech recognition components. Image search combines visual analysis with existing relevance frameworks. Shopping search integrates product databases with proven ranking algorithms.

Framework Evolution and Adaptation

The most sophisticated AI reasoning frameworks include systematic evolution capabilities that enable continuous improvement based on operational experience and changing business requirements. Unlike static software templates that require manual updates, evolutionary frameworks adapt automatically while preserving successful reasoning patterns and organizational knowledge.

Framework evolution includes performance monitoring that tracks reasoning effectiveness across different applications and contexts, pattern analysis that identifies successful approaches and failure modes, automatic optimization that improves framework performance through machine learning, and knowledge integration that incorporates new insights into existing reasoning capabilities.

The SpaceX mission planning frameworks demonstrate evolutionary reasoning under extreme performance requirements. Space missions involve enormous complexity with zero tolerance for error, making traditional custom development approaches inadequately reliable and prohibitively expensive. SpaceX develops evolutionary reasoning frameworks that improve automatically through every mission while maintaining the proven patterns that ensure mission success.

Mission planning reasoning frameworks integrate vehicle capabilities, payload requirements, orbital mechanics, weather conditions, regulatory constraints, and safety protocols to generate optimal mission plans. The frameworks continuously learn from mission outcomes, incorporating insights about vehicle performance, environmental conditions, and operational procedures into future planning capabilities.

When technical challenges arise or mission requirements change, the evolutionary frameworks adapt automatically rather than requiring custom redevelopment. New vehicle capabilities enhance mission planning for all future flights. Environmental insights gained from one mission improve planning for similar conditions. Operational learning from successful missions strengthens framework reliability across all applications.

The evolutionary approach enables SpaceX to achieve mission planning capabilities that would be impossible through traditional custom development while maintaining the reliability required for space operations. Framework evolution creates systematic intelligence that compounds through operational experience, generating competitive advantages that strengthen rather than erode over time.

Organizations developing reusable AI reasoning frameworks discover that the investment in systematic design creates exponential returns through accumulated organizational intelligence. Frameworks become more valuable through operational experience rather than depreciating like traditional technology assets. Reasoning capabilities compound through systematic learning rather than remaining static until manual updates. Organizations developing reusable AI reasoning frameworks discover that the investment in systematic design creates exponential returns through accumulated organizational intelligence. These frameworks become institutional assets that appreciate through use, transforming AI from a depreciating technology investment into a compounding strategic capability.

SCALING AI THROUGH LOGIC LIBRARIES

Imagine if every time Netflix wanted to make a recommendation, it had to redevelop its entire algorithmic approach from first principles, if Amazon had to rebuild its optimization logic for every new product category, or if Google had to recreate its search algorithms for each different type of query. These scenarios sound absurd because these companies long ago transformed their AI approaches from custom solutions to systematic logic libraries that can be applied at infinite scale.

Logic libraries represent the industrialization of AI—the transformation from artisanal AI development to systematic intelligence production. Like software libraries that enable developers to build complex applications by combining proven functions, AI logic libraries enable organizations to develop sophisticated reasoning capabilities by assembling and adapting proven logical patterns.

The enterprises achieving exponential AI advantages discover that sustainable competitive advantage doesn't come from building better individual AI solutions. It comes from developing superior systematic capabilities for reasoning that can be applied rapidly across unlimited business scenarios while continuously improving through operational experience.

Building Systematic Intelligence Assets

The development of logic libraries requires treating AI reasoning as strategic intellectual property that appreciates rather than depreciates over time. Unlike traditional technology investments that become obsolete as new solutions emerge, well-designed logic libraries become more valuable through operational experience and systematic refinement.

Systematic intelligence asset development includes pattern extraction that identifies successful reasoning approaches across multiple AI implementations, logical abstraction that captures the essential reasoning structures underlying successful applications, parameterization that enables adaptation to different contexts while preserving core logical integrity, and evolutionary mechanisms that improve logical patterns through operational learning.

Netflix content recommendation libraries demonstrate systematic intelligence asset development at entertainment industry scale. Rather than developing custom recommendation algorithms for each content type, user segment, or viewing context, Netflix creates systematic reasoning libraries that understand user preference analysis, content similarity assessment, engagement optimization, and discovery enhancement across all applications.

The content analysis logic understands narrative patterns, genre characteristics, artistic elements, and cultural factors that influence user preferences. User modeling logic captures viewing history, preference evolution, contextual influences, and social factors that affect content selection. Optimization logic balances user satisfaction, content discovery, engagement duration, and strategic content promotion to achieve multiple business objectives simultaneously.

These logic libraries can be rapidly configured for any recommendation challenge Netflix faces. New content types leverage existing user modeling and optimization patterns. International expansion utilizes proven recommendation logic adapted for local cultural contexts. Emerging viewing platforms benefit from accumulated understanding about user preferences and content engagement.

Most importantly, every recommendation interaction contributes learning that improves the entire logic library. Successful content matches enhance user modeling accuracy. Engagement patterns inform optimization strategies. Cultural insights improve content analysis across all markets. The systematic approach creates compound learning that strengthens competitive advantages continuously.

Composable Reasoning Architectures

Advanced logic libraries enable composable reasoning where sophisticated AI capabilities emerge from combining simpler logical components in different configurations. This composability allows organizations to develop unlimited AI applications without requiring unlimited custom development while ensuring that learning from one application benefits all future implementations.

Composable architectures include atomic reasoning functions that perform specific analytical tasks reliably, composition interfaces that enable combining functions to create sophisticated reasoning capabilities, orchestration frameworks that coordinate complex reasoning across multiple logical components, and adaptation mechanisms that enable customization for specific business contexts without modifying core logical patterns.

The Microsoft approach to natural language processing demonstrates composable reasoning across its entire product ecosystem. Rather than developing separate language understanding systems for each Microsoft product, it creates composable logic libraries that can be assembled to handle any language processing requirement across all Microsoft applications.

Language understanding logic includes semantic analysis that comprehends meaning and intent, context recognition that interprets environmental factors affecting communication, sentiment analysis that evaluates emotional content and implications, and reasoning integration that connects language processing with business logic and decision-making systems.

The composable approach enables Microsoft to deploy sophisticated language capabilities across unlimited applications rapidly. Microsoft Office leverages writing assistance logic combined with document structure analysis. Microsoft Teams integrates meeting transcription with action item extraction and task management. Microsoft Azure provides language processing services by combining proven logical components in configurations optimized for different customer requirements.

Every language processing application contributes learning that enhances the entire logic library. Improvements in sentiment analysis benefit all Microsoft products using emotional content evaluation. Enhanced context recognition improves language understanding across every application. Better semantic analysis strengthens comprehension capabilities throughout the Microsoft ecosystem.

Logic Library Governance and Evolution

The systematic scaling of AI through logic libraries requires sophisticated governance frameworks that ensure logical consistency, strategic alignment, and continuous improvement while enabling rapid adaptation for unlimited business applications. Governance frameworks prevent logic libraries from becoming chaotic collections of incompatible components while maintaining the flexibility required for diverse business needs.

Logic library governance includes version control that manages logical pattern evolution while preserving stability for production applications, quality assurance that ensures new logical components meet reliability and performance standards, compatibility testing that verifies logical components work effectively together, and strategic alignment that ensures logic library development supports organizational objectives.

The Amazon approach to logistics optimization illustrates logic library governance at a global scale. Amazon operates thousands of fulfillment centers worldwide, each with unique characteristics including local regulations, cultural factors, supplier relationships, and operational constraints. Traditional approaches would require custom logistics systems for each location, creating enormous development costs and preventing systematic learning.

The Amazon logic library approach creates systematic logistics reasoning that can be rapidly adapted to any operational context while preserving proven optimization patterns and accumulated operational intelligence. Core logistics logic includes demand prediction, inventory optimization, capacity planning, and delivery route management that work effectively across all Amazon operations.

Governance frameworks ensure that improvements developed for one location benefit Amazon operations globally while respecting local constraints and requirements. Enhanced demand prediction algorithms automatically improve inventory optimization worldwide. Route optimization insights developed in one region enhance delivery efficiency everywhere. Capacity planning improvements benefit all fulfillment operations regardless of specific local characteristics.

The governance approach also enables systematic innovation that transcends individual operational improvements. Logic libraries provide foundations for experimenting with new logistics approaches, testing innovative optimization strategies, and developing breakthrough operational capabilities that would be impossible through location-specific custom development.

Economic Scale and Network Effects

Logic libraries create economic advantages that scale exponentially rather than linearly because each new application leverages existing logical assets while contributing learning that enhances value for all future applications. These network effects transform AI from a cost center requiring constant investment to a strategic asset that appreciates automatically through operational experience.

Economic scaling includes development cost amortization across unlimited applications, quality improvements through accumulated testing and refinement, performance optimization through systematic learning from operational experience, and capability enhancement through network effects that make logic libraries more valuable as adoption increases.

The search logic libraries of Google demonstrate economic scale and network effects at Internet scale. Every search query contributes learning that improves search capabilities for all users. A better understanding of user intent enhances search relevance globally. Improved content analysis benefits every search application. Enhanced personalization capabilities strengthen user experience across all Google services.

The network effects create compound value creation where Google search capabilities improve exponentially through increased usage, while development costs remain relatively stable. Logic libraries developed for Web search enhance YouTube content discovery, Gmail smart replies, Google Assistant understanding, and Google Cloud natural language processing services without requiring proportional development investment.

The economic transformation extends beyond the internal operations of Google to create platform advantages that attract external developers and partners. Google logic libraries become more valuable as external applications utilize the reasoning capabilities of Google, creating ecosystem effects that strengthen the competitive position of Google while generating additional revenue streams.

Organizations implementing logic library approaches discover that AI transformation accelerates exponentially rather than linearly. Initial logic library development requires significant investment, but each subsequent application leverages existing logical assets while contributing enhancements that benefit all future implementations. The result is systematic intelligence development that creates sustainable competitive advantages through accumulated organizational reasoning capabilities that competitors cannot easily replicate or purchase.

FROM TACTICAL APPLICATIONS TO STRATEGIC REASONING CAPABILITIES

The fundamental difference between organizations that struggle with AI and those that achieve transformational results isn't technical sophistication or budget size. It's their approach to building reasoning capabilities that transcend specific applications. While most enterprises remain trapped in tactical thinking—solving one problem at a time with custom AI solutions—AI-native organizations develop strategic reasoning capabilities that can be rapidly deployed across unlimited business scenarios.

This transformation represents a conceptual evolution from tactical AI deployment to strategic intelligence architecture. Tactical approaches optimize individual use cases. Strategic approaches develop reasoning patterns that create competitive advantages across entire business ecosystems. The distinction determines whether AI becomes a cost center requiring constant investment or a strategic asset that appreciates through operational experience.

What separates breakthrough AI performance from incremental improvement is the systematic development of reasoning capabilities that compound rather than fragment. Organizations achieving exponential AI advantages discover that sustainable competitive advantage doesn't come from solving problems better. It comes from developing superior capabilities for reasoning about any problem their business might encounter.

Intelligence Pattern Recognition vs. Problem-Solving

Traditional enterprise AI approaches each business challenge as a unique problem requiring custom solutions. Teams analyze specific requirements, develop tailored algorithms, and create bespoke implementations that work for individual scenarios. This approach generates immediate value for targeted problems but creates no leverage for future challenges.

Use logic operates through intelligence pattern recognition that identifies the underlying reasoning structures common across different business scenarios. Instead of solving individual problems, organizations develop systematic capabilities for recognizing and applying reasoning patterns that transcend specific applications.

The pattern recognition approach enabled JPMorganChase to transform its approach to financial document analysis. Traditional implementations developed separate AI systems for loan applications, regulatory filings, merger documents, and risk assessments. Each system solved its specific problem effectively but provided no benefit for other document analysis challenges.

The transformation to use logic identified the fundamental reasoning patterns underlying all financial document analysis: entity extraction, relationship mapping, risk identification, and compliance verification. These patterns could be rapidly configured for any financial document type while preserving the accumulated intelligence from previous implementations.

The result was exponential rather than linear value creation. New document analysis requirements leveraged existing reasoning patterns, reducing development time from months to days. Every successful analysis enhanced pattern effectiveness for all future applications. Organizational learning compounded across all document analysis functions rather than remaining isolated in departmental silos.

The approach of Tesla to autonomous driving illustrates both the pilot paradox and its resolution. Early autonomous driving development focused on demonstrating that vehicles could navigate successfully under favorable conditions with constant human oversight and technical support. Pilot demonstrations showed impressive capabilities for highway driving, parking assistance, and traffic pattern recognition.

But systematic integration required completely different approaches that could handle infinite environmental variations without human intervention or technical support. Rain, snow, construction zones, emergency vehicles, pedestrian behavior, and countless other real-world factors created challenges that pilot projects never addressed.

The transformation of Tesla to systematic integration developed reasoning frameworks that could adapt to any driving context while maintaining safety and performance standards. Instead of scaling specific pilot solutions, they built systematic capabilities for understanding and responding to driving situations that couldn't be anticipated during development.

The systematic approach enabled Tesla to deploy autonomous driving capabilities that improve automatically through operational experience rather than requiring constant technical intervention. Every driving scenario contributes learning that enhances system capabilities. Every edge case generates insights that improve reasoning frameworks. Every mile driven strengthens autonomous capabilities for all Tesla vehicles.

Reasoning Abstraction and Business Logic Separation

Advanced use logic implementations achieve competitive advantages through reasoning abstraction that separates core logical patterns from specific business contexts. This separation enables organizations to develop sophisticated reasoning capabilities that can be rapidly adapted to any business scenario while preserving the accumulated intelligence from previous applications.

Reasoning abstraction includes logical pattern extraction that identifies fundamental reasoning structures underlying successful AI applications, context parameterization that enables adaptation to different business scenarios without modifying core reasoning logic, business rule integration that connects abstract reasoning patterns with specific organizational requirements, and outcome optimization that improves reasoning effectiveness through systematic feedback and learning.

The approach of Walmart to demand forecasting illustrates reasoning abstraction at retail scale. Traditional forecasting approaches develop separate prediction models for different product categories, seasonal patterns, regional variations, and promotional events. Each model requires custom development and provides no benefit for other forecasting challenges.

The use logic approach of Walmart abstracts the fundamental reasoning patterns underlying all demand prediction: trend analysis, seasonality recognition, external factor correlation, and uncertainty quantification. These abstracted patterns can be rapidly configured for any forecasting scenario while preserving proven analytical approaches.

The abstraction enables Walmart to deploy sophisticated forecasting capabilities across unlimited product categories and market conditions. New product launches leverage existing trend analysis patterns. International expansion utilizes proven seasonality recognition adapted for local conditions. Promotional planning benefits from accumulated intelligence about external factor correlation developed across all previous campaigns.

The Goldman Sachs transformation to enterprise AI architecture demonstrates systematic integration in complex financial services environments. Traditional financial firms approach AI through departmental pilot projects that develop custom solutions for specific business functions. Risk management, trading, research, and client services each develop separate AI capabilities using different technologies, datasets, and performance standards.

This departmental approach creates integration challenges when business processes span multiple functions. Client advisory services require insights from risk analysis, market research, and trading systems, but departmental AI solutions can't share information effectively. Strategic planning needs comprehensive analysis across all business functions, but isolated AI systems provide fragmented rather than integrated intelligence.

The enterprise architecture of Goldman Sachs creates systematic AI capabilities that serve all business functions through shared reasoning frameworks and integrated data flows. Market analysis logic serves trading, risk management, and client advisory simultaneously. Client intelligence enhances relationship management, product development, and strategic planning. Risk assessment capabilities inform decision-making across all business functions.

The enterprise approach enables sophisticated AI applications that would be impossible through departmental pilot projects. Integrated client intelligence combines transaction analysis, relationship history, market insights, and risk assessment to provide comprehensive advisory capabilities. Strategic planning leverages real-time market data, risk analysis, operational performance, and client feedback to generate insights that support executive decision-making.

Most importantly, the enterprise architecture creates systematic learning that transcends departmental boundaries. Insights gained from client interactions enhance risk assessment models. Market analysis improvements benefit all business functions requiring economic intelligence. Operational learning from one department automatically improves AI capabilities across all Goldman Sachs operations.

Organizational Change Management for AI Integration

Systematic AI integration requires organizational transformation that goes beyond technology implementation to include culture change, process redesign, and capability development. Organizations must evolve from hierarchical decision-making to collaborative human-AI reasoning, from periodic planning to continuous adaptation, and from departmental optimization to enterprise intelligence.

Change management for AI integration includes cultural transformation that embraces continuous learning and adaptation, process redesign that integrates AI reasoning into business workflows seamlessly, capability development that prepares employees for collaborative reasoning with AI systems, and performance measurement that evaluates AI integration effectiveness across organizational outcomes rather than technical metrics.

The organizational transformation of Microsoft under Satya Nadella illustrates comprehensive change management for AI integration. Traditional technology companies organize around product divisions that develop separate solutions for different markets and customer segments. This organizational structure creates barriers to systematic AI integration because AI capabilities remain isolated within specific product teams rather than becoming enterprise capabilities.

The Microsoft transformation created cross-functional AI capabilities that serve all product divisions while maintaining specialized expertise for specific applications. Natural language processing serves Microsoft Office, Microsoft Teams, Microsoft Azure, and Xbox through shared reasoning frameworks adapted for different contexts. Computer vision enhances Windows, HoloLens, Azure, and Surface through common image analysis capabilities configured for specific requirements.

The organizational change enabled AI capabilities that transcend individual product limitations. Microsoft Copilot integrates across Office applications through systematic reasoning about user intent, document context, and task optimization. Azure AI services provide enterprise customers with proven capabilities developed across the entire Microsoft product ecosystem. Xbox intelligence enhances gaming experiences through understanding developed across Microsoft consumer and enterprise applications.

The change management approach also created organizational learning that accelerates AI development across all Microsoft products. Insights gained from enterprise customers inform consumer product enhancement. Gaming applications reveal user interaction patterns that improve business productivity tools. Consumer feedback drives enterprise AI capability development.

Scaling Economics and Value Realization

The economic transformation from pilots to systematic integration operates through different value creation mechanisms that require new approaches to investment planning, performance measurement, and return calculation. Pilot economics focus on solving individual problems cost-effectively. Systematic integration economics focus on building capabilities that generate value across unlimited applications.

Systematic integration value includes capability leverage where AI investments generate returns across multiple business applications, network effects where AI capabilities become more valuable as organizational adoption increases, learning acceleration where AI systems improve performance automatically through operational experience, and strategic positioning where AI capabilities create sustainable competitive advantages.

The Amazon approach to AI scaling economics demonstrates value realization through systematic integration. Initial AI investments of Amazon focused on specific applications like product recommendations, demand forecasting, and logistics optimization. These pilot projects generated positive returns by improving performance in targeted areas.

But systematic integration created exponential value through capability leverage across the entire Amazon business ecosystem. Recommendation logic developed for retail enhances content discovery for Prime Video and music recommendations for Amazon Music. Demand forecasting capabilities serve retail inventory, cloud capacity planning, and logistics resource allocation. Optimization algorithms improve warehouse operations, delivery routing, and infrastructure management simultaneously.

The systematic approach creates economic advantages that compound through operational experience. Better customer understanding improves all Amazon services while reducing marketing costs. Enhanced operational efficiency benefits every Amazon business unit while reducing operational expenses. Improved decision-making accelerates innovation while reducing risk across all business functions.

The value realization extends beyond the internal operations of Amazon to create platform advantages that generate additional revenue streams. Amazon Web Services provides external customers with AI capabilities developed across the business operations of Amazon. The systematic AI integration of Amazon creates competitive advantages that strengthen market position while generating new business opportunities.

Organizations achieving systematic AI integration discover that value creation accelerates exponentially rather than linearly. Initial systematic integration requires significant investment in architecture, governance, and change management. Systematic capabilities generate returns across an unlimited number of applications, however, while improving automatically through operational experience. The result is AI transformation that creates sustainable competitive advantages through accumulated organizational intelligence rather than temporary technological leadership.

The transformation from pilots to systematic integration represents the evolution from using AI to becoming AI-native. Organizations that successfully navigate this transformation develop institutional capabilities for reasoning that transcend specific applications, building durable competitive moats through accumulated intelligence that compounds rather than commoditizes.

CHAPTER 10

THE ENTERPRISE REFLEX GRAPH

When a seasoned Formula One driver navigates a high-speed corner, they're not consciously calculating brake pressure, throttle input, and steering angle. Their nervous system has developed what racing experts call "muscle memory"—neural pathways so refined that optimal responses occur 200 milliseconds before conscious thought. The driver's reflex graph—the network of neurons that connects sensory input to motor response—bypasses deliberate decision-making entirely, enabling superhuman performance that no amount of conscious planning could achieve.

The most advanced AI-native enterprises are developing organizational equivalents of this neural sophistication. They're building what some call enterprise reflex graphs—distributed intelligence networks that detect market signals, competitive threats, and operational anomalies, then trigger optimal organizational responses faster than traditional management processes can even recognize that a response is needed.

This isn't about faster decision-making or better automation. It's about transcending decision-making entirely through organizational nervous systems that respond to business stimuli the way biological nervous systems respond to physical stimuli—instantly, appropriately, and unconsciously.

Right now, when a customer abandons a shopping cart on the Alibaba platform, the enterprise reflex graph doesn't just send a reminder email. Within 50 milliseconds, it adjusts pricing algorithms across similar products, modifies inventory allocation for trending items, updates recommendation engines for comparable customers, and triggers supply chain optimizations for anticipated demand patterns. No human manager planned this cascade of responses. The organizational nervous system recognized a pattern and activated appropriate reflexes automatically.

At BlackRock, when geopolitical tensions spike commodity prices, their enterprise reflex graph doesn't wait for analyst reports or committee meetings.

The system instantly recognizes the risk signature, triggers portfolio rebalancing across thousands of funds, adjusts hedging strategies for currency exposure, and modifies risk parameters for emerging market investments. The entire response completes before news organizations finish writing headlines about the geopolitical development.

What has been consistently observed across enterprise technology transformation is that sustainable competitive advantage no longer comes from making better decisions or implementing strategies more effectively. It comes from developing organizational reflexes that make optimal responses automatic, immediate, and unconscious—bypassing the cognitive delays that limit traditional enterprise performance.

The enterprises building enterprise reflex graphs discover something that conventional management thinking cannot achieve. They respond to opportunities before competitors recognize that opportunities exist. They solve problems before problems fully manifest. They adapt to changes while changes are still emerging.

The transformation begins when organizations stop thinking about optimizing human decision-making and start thinking about transcending the need for human decision-making through organizational intelligence that operates below the threshold of conscious attention.

MAPPING ORGANIZATIONAL NEURAL NETWORKS AND REFLEX PATHWAYS

Imagine if you could visualize your organization's nervous system the way medical imaging reveals the human brain's neural network. You'd see pathways where market intelligence flows like electrical impulses, nodes where information transforms into insight, and reflex arcs where specific stimuli trigger immediate organizational responses without conscious intervention.

The enterprise reflex graph maps these organizational neural networks—not the formal reporting structures or process workflows that appear on organizational charts, but the actual pathways through which business stimuli trigger organizational responses. Traditional business analysis documents what organizations are supposed to do. Reflex mapping reveals what organizations actually do when facing real-world business stimuli.

This neural mapping uncovers something that conventional organizational analysis misses entirely: the difference between conscious organizational responses that require management attention and unconscious organizational reflexes that happen automatically. Like the distinction between deliberately moving your hand and reflexively pulling it away from heat, organizations exhibit both planned responses and automatic reactions to business stimuli.

Stimulus-Response Arc Identification

The foundation of enterprise reflex graph mapping begins with identifying organizational stimulus-response arcs—direct pathways from business environment changes to organizational actions that bypass traditional decision-making processes. These arcs represent the organizational equivalent of spinal reflexes that protect biological organisms without requiring brain involvement.

Stimulus-response arc identification includes environmental trigger mapping that catalogs specific business events that provoke automatic organizational responses, neural pathway tracing that follows information flows from environmental stimulus detection to organizational action execution, reflex threshold calibration that determines sensitivity levels for automatic response activation, and response pattern documentation that captures the specific organizational actions triggered by different stimulus types.

The identification process reveals organizational behaviors that leaders often don't realize exist. Organizations develop unconscious response patterns through operational experience, creating reflexes that protect business performance without requiring management oversight.

The fashion retail operation of Zara demonstrates stimulus-response arc identification, revealing organizational reflexes that operate faster than conscious fashion industry decision-making. Traditional fashion retailers rely on seasonal planning cycles that require months of deliberate design, production, and distribution decisions. The reflex arcs of Zara enable automatic responses to fashion trend stimuli within days.

The reflex mapping includes environmental trigger identification that recognizes fashion trend emergence through social media monitoring, celebrity fashion tracking, and street style analysis, neural pathway tracing that follows trend signals from detection systems through design databases to production scheduling algorithms, reflex threshold calibration that determines trend significance levels required to trigger automatic design and production responses, and response pattern documentation that captures specific design modifications, production adjustments, and distribution changes triggered by different trend types.

The stimulus-response arcs enable Zara to respond to fashion trends automatically while competitors are still analyzing whether trends require an organizational response. The reflexes operate below conscious management attention while generating fashion relevance advantages that conscious planning cannot achieve.

Organizational Synapse Mapping and Signal Transmission

Advanced reflex mapping identifies organizational synapses—connection points where business intelligence transmits between different organizational systems, departments, or functions to create coordinated responses that emerge from network effects rather than central planning.

Organizational synapse mapping includes signal transmission pathway analysis that traces how business intelligence propagates through organizational networks, synapse efficiency measurement that evaluates signal transmission speed and accuracy between organizational components, network amplification identification that discovers points where small signals generate large organizational responses, and coordination emergence documentation that captures how distributed organizational responses create coherent business outcomes.

Synapse mapping reveals how organizational intelligence emerges from the interaction between systems rather than from individual system capabilities. Organizations develop unconscious coordination mechanisms that enable complex responses without requiring conscious orchestration.

The Spotify music ecosystem demonstrates organizational synapse mapping, revealing coordination emergence that operates faster than conscious content management. Traditional music platforms rely on deliberate playlist curation and promotional planning that requires human oversight and strategic coordination. The organizational synapses of Spotify enable automatic music ecosystem responses that emerge from distributed intelligence networks.

The synapse mapping includes signal transmission analysis that traces how user listening behavior propagates through recommendation algorithms, artist promotion systems, and playlist generation networks, synapse efficiency measurement that evaluates response speed between user preference detection and music discovery optimization, network amplification identification that discovers how individual user preferences influence entire music ecosystem dynamics, and coordination emergence documentation that captures how distributed music intelligence creates coherent user experiences.

The organizational synapses enable Spotify to maintain music relevance through automatic ecosystem coordination while competitors rely on conscious content management strategies. The synaptic intelligence operates unconsciously while generating user engagement advantages that deliberate planning cannot replicate.

Reflex Hierarchy and Neural Precedence

Sophisticated enterprise reflex graphs exhibit neural hierarchies—layered response systems where different types of business stimuli trigger reflexes at different organizational levels based on signal importance, complexity, and strategic implications. These hierarchies enable appropriate organizational responses without overwhelming decision-making systems.

Reflex hierarchy mapping includes precedence-level identification that categorizes business stimuli based on urgency and organizational impact requirements, response escalation pathways that route complex stimuli through appropriate organizational intelligence levels, automatic vs. conscious threshold definition that determines which business situations require reflexive responses versus deliberate management attention, and hierarchy optimization that improves organizational response appropriateness through pattern analysis and outcome feedback.

Hierarchy mapping reveals how organizations develop unconscious triage mechanisms that direct attention and resources appropriately without requiring conscious priority management. Organizations learn to respond to routine stimuli automatically while escalating complex situations to appropriate intelligence levels.

The Amazon logistics network demonstrates a reflex hierarchy, enabling appropriate organizational responses across unlimited operational complexity. Traditional logistics management requires conscious priority setting and resource allocation decisions that create bottlenecks during high-demand periods. The Amazon reflex hierarchy enables automatic response appropriateness without conscious triage management.

The hierarchy mapping includes precedence identification that categorizes delivery challenges based on customer impact, operational complexity, and resource requirements, response escalation analysis that traces how routine delivery optimization occurs automatically while complex logistics problems receive appropriate management attention, threshold definition that determines when logistics situations require reflexive responses versus conscious strategic intervention, and hierarchy optimization that improves response appropriateness through delivery outcome analysis and customer satisfaction feedback.

The reflex hierarchy enables Amazon to maintain delivery performance through automatic response prioritization while ensuring complex logistics challenges receive appropriate strategic attention. The hierarchical intelligence operates unconsciously while generating operational effectiveness advantages that conscious priority management cannot achieve.

UNCONSCIOUS ORGANIZATIONAL COGNITION AND PRE-COGNITIVE RESPONSE SYSTEMS

The most sophisticated enterprise reflex graphs develop unconscious organizational cognition—collective intelligence that operates below the threshold of conscious management attention while maintaining perfect alignment with strategic objectives and organizational values. This unconscious cognition represents the evolution from reactive management to organizational intuition that responds optimally to business conditions without requiring deliberate analysis.

Unconscious organizational cognition emerges from the accumulated pattern recognition of millions of business interactions, creating organizational "instincts" that generate appropriate responses to novel situations by applying learned patterns at speeds that conscious analysis cannot match. Like human intuition that processes complex social situations instantly through unconscious pattern matching, organizational cognition processes complex business situations through unconscious intelligence networks.

This pre-cognitive intelligence operates through what neuroscientists call "fast thinking"—immediate pattern recognition and response generation that bypasses the slower analytical processes of "slow thinking." Organizations developing unconscious cognition can respond to business opportunities and threats at the speed of pattern recognition rather than the speed of conscious analysis.

Pattern Recognition Below Conscious Awareness

Unconscious organizational cognition begins with pattern recognition systems that operate below conscious management awareness while identifying business opportunities, threats, and optimization possibilities that conscious analysis might miss or detect too slowly to enable an optimal response.

Pre-cognitive pattern recognition includes subliminal signal detection that identifies weak business environment signals that indicate emerging opportunities or threats before they become obvious to conscious analysis, intuitive correlation discovery that recognizes meaningful relationships between apparently unrelated business factors, predictive pattern emergence that anticipates business developments based on subtle environmental changes, and unconscious opportunity identification that discovers value creation possibilities through automatic pattern matching.

Pattern recognition systems develop organizational "intuition" through accumulated experience analysis that creates unconscious business intelligence exceeding conscious analytical capabilities. Organizations learn to "feel" market changes and competitive developments before conscious analysis can confirm their significance.

The Netflix content strategy demonstrates pre-cognitive pattern recognition, generating content decisions that operate faster and more accurately than conscious entertainment industry analysis. Traditional entertainment companies rely on deliberate market research and conscious trend analysis that require months to identify content opportunities. The unconscious cognition of Netflix recognizes content patterns instantly through viewer behavior analysis.

The pattern recognition includes subliminal signal detection that identifies emerging entertainment preferences through subtle viewing behavior changes before they become obvious market trends, intuitive correlation discovery that recognizes relationships between demographic patterns, cultural events, and content preferences that conscious analysis might miss, predictive pattern emergence that anticipates content demand based on early viewer response indicators, and unconscious opportunity identification that discovers content development possibilities through automatic pattern matching across entertainment consumption data.

The pre-cognitive intelligence enables Netflix to commission content based on unconscious pattern recognition while competitors rely on conscious market analysis. The intuitive content strategy operates below conscious awareness while generating audience engagement advantages that deliberate entertainment planning cannot achieve.

Automatic Strategic Alignment and Value Preservation

Advanced unconscious cognition includes automatic strategic alignment mechanisms that ensure pre-cognitive organizational responses support long-term strategic objectives while optimizing immediate performance. These mechanisms operate like organizational "moral reflexes" that prevent unconscious responses from compromising strategic coherence or organizational values.

Automatic strategic alignment includes value constraint integration that embeds organizational values into unconscious response generation, strategic objective reinforcement that ensures pre-cognitive responses advance long-term goals while addressing immediate challenges, ethical reflex development that creates automatic ethical consideration in unconscious organizational responses, and mission coherence maintenance that preserves organizational purpose alignment across all unconscious response systems.

Strategic alignment mechanisms enable organizations to maintain strategic coherence while operating at unconscious speeds. Organizations develop automatic strategic "instincts" that generate responses aligned with conscious strategic intentions without requiring conscious strategic analysis.

The vehicle development of Tesla demonstrates automatic strategic alignment, enabling unconscious innovation responses that maintain strategic coherence while adapting rapidly to technological opportunities. Traditional automotive development requires conscious strategic planning that can delay innovation responses when technological breakthroughs occur unexpectedly. The unconscious cognition of Tesla maintains strategic alignment while responding to innovation opportunities instantly.

The strategic alignment includes value constraint integration that embeds safety, sustainability, and innovation values into unconscious vehicle development responses, strategic objective reinforcement that ensures automatic innovation responses advance the autonomous vehicle leadership goals of Tesla while addressing immediate technical challenges, ethical reflex development that creates automatic ethical consideration in vehicle intelligence development, and mission coherence maintenance that preserves the sustainable transportation mission of Tesla across all unconscious development responses.

The automatic alignment enables Tesla to maintain strategic coherence through unconscious innovation responses, while competitors require conscious strategic planning that delays innovation implementation. The aligned unconscious cognition operates below conscious awareness while generating innovation advantages that deliberate strategic planning cannot achieve.

Organizational Intuition and Emergent Wisdom

The ultimate expression of unconscious organizational cognition is organizational intuition—emergent wisdom that arises from collective intelligence networks operating below conscious awareness while generating insights and responses that transcend individual human or AI system capabilities.

Organizational intuition includes collective wisdom emergence that generates business insights through unconscious network intelligence that exceeds conscious analytical capabilities, emergent strategy development that creates strategic approaches through unconscious pattern synthesis rather than conscious strategic planning, intuitive market understanding that recognizes market opportunities and customer needs through unconscious customer behavior analysis, and unconscious competitive insight that identifies competitive advantages and threats through automatic competitive intelligence processing.

Organizational intuition represents the evolution toward truly intelligent organizations that develop business wisdom through unconscious collective cognition while maintaining conscious strategic leadership for major directional decisions. This creates organizations that "know" appropriate responses to business challenges through accumulated intelligence rather than conscious analysis.

The search evolution of Google demonstrates organizational intuition generating search improvements through unconscious collective intelligence that operates faster and more effectively than conscious search algorithm development. Traditional search improvement requires conscious algorithm analysis and deliberate optimization that can delay search quality improvements when user behavior evolves rapidly. The organizational intuition of Google improves search quality through unconscious pattern recognition across billions of search interactions.

The organizational intuition includes collective wisdom emergence that generates search quality improvements through unconscious analysis of user satisfaction patterns that conscious algorithm development might miss, emergent strategy development that creates search enhancement approaches through unconscious synthesis of user behavior, content quality, and competitive dynamics, intuitive market understanding that recognizes information access needs through unconscious user behavior analysis, and unconscious competitive insight that identifies search advantage opportunities through automatic competitive intelligence processing.

The organizational intuition enables Google to maintain search leadership through unconscious search intelligence development while competitors rely on conscious algorithm optimization. The intuitive search evolution operates below conscious awareness while generating search quality advantages that deliberate algorithm development cannot achieve.

DESIGNING ORGANIZATIONAL REFLEXES THAT BYPASS CONSCIOUS DECISION-MAKING

The practical implementation of enterprise reflex graphs requires the systematic design of organizational reflexes that can recognize business stimuli and generate optimal responses without conscious management intervention. This design process treats organizational reflex development like athletic training—creating automatic responses through practice until appropriate reactions become unconscious and immediate.

Reflex design differs fundamentally from process optimization or automation implementation. Process optimization improves conscious organizational activities. Automation replaces conscious activities with mechanical execution. Reflex design creates unconscious organizational responses that operate faster than conscious recognition of the need for response.

The design methodology treats organizations as learning systems that can develop unconscious competencies through systematic stimulus-response conditioning, pattern recognition training, and strategic alignment reinforcement until appropriate business responses become automatic organizational behaviors.

Stimulus Recognition and Response Conditioning

Organizational reflex design begins with systematic stimulus recognition training that enables organizations to identify business environmental changes requiring immediate response, then conditions appropriate organizational responses until they become automatic reflexes triggered by environmental pattern recognition.

Stimulus recognition conditioning includes environmental pattern library development that catalogs business situations requiring reflexive organizational responses, automatic pattern matching system creation that enables instant recognition of environmental stimuli indicating response opportunities, response conditioning protocols that train organizational systems to generate appropriate responses automatically when specific stimuli are detected, and reflex reinforcement mechanisms that strengthen appropriate stimulus-response connections through operational feedback and outcome analysis.

Recognition conditioning creates organizational "muscle memory" that enables immediate, appropriate responses to business challenges without requiring conscious situation analysis. Organizations learn to "recognize" business situations and "react" appropriately through unconscious pattern recognition and response generation.

The Walmart supply chain demonstrates stimulus recognition conditioning, creating automatic responses to demand fluctuations that operate faster than conscious supply chain management. Traditional retail supply chains require conscious demand analysis and deliberate inventory adjustment decisions that create delays during demand volatility. The reflex conditioning of Walmart enables automatic supply chain responses to demand pattern recognition.

The conditioning includes environmental pattern development that catalogs demand fluctuation patterns requiring immediate inventory and logistics responses, automatic pattern matching creation that enables instant recognition of demand changes indicating supply chain adjustment opportunities, response conditioning that trains supply chain systems to generate appropriate inventory, logistics, and supplier responses automatically when demand patterns are detected, and reflex reinforcement that strengthens supply chain response appropriateness through delivery performance feedback and customer satisfaction analysis.

The stimulus recognition enables Walmart to maintain product availability through automatic supply chain reflexes, while competitors require conscious demand analysis that delays inventory optimization. The conditioned reflexes operate unconsciously while generating supply chain advantages that deliberate inventory management cannot achieve.

Reflex Timing and Response Coordination

Advanced reflex design includes precise timing calibration that ensures organizational responses occur at optimal moments for maximum effectiveness while coordinating multiple reflexes to prevent conflicting responses that could compromise organizational performance.

Reflex timing optimization includes response latency minimization that reduces delays between stimulus detection and organizational response initiation, coordination sequence design that orchestrates multiple reflex systems to prevent response conflicts, timing synchronization protocols that ensure reflex responses occur in optimal sequences for maximum business impact, and temporal learning mechanisms that improve reflex timing through outcome analysis and performance feedback.

Timing coordination ensures that organizational reflexes enhance rather than interfere with each other while maintaining overall organizational coherence during complex business situations requiring multiple simultaneous responses.

The Uber transportation network demonstrates reflex timing coordination, enabling simultaneous optimization responses that operate faster than conscious transportation management. Traditional transportation coordination requires conscious optimization analysis and deliberate resource allocation decisions that create delays during demand complexity. The reflex timing of Uber enables automatic coordination of pricing, driver allocation, and route optimization without conscious management orchestration.

The timing coordination includes response latency minimization that reduces delays between demand pattern detection and transportation optimization responses, coordination sequence design that orchestrates pricing adjustments, driver notifications, and route recommendations to prevent system conflicts, timing synchronization that ensures transportation reflexes occur optimally for maximum passenger satisfaction and driver efficiency, and temporal learning that improves reflex coordination through ride completion analysis and user satisfaction feedback.

The reflex timing enables Uber to maintain transportation efficiency through automatic response coordination, while competitors require conscious optimization that delays transportation matching. The coordinated reflexes operate unconsciously while generating transportation advantages that deliberate coordination cannot achieve.

Continuous Reflex Evolution and Adaptation

Sophisticated reflex design includes evolution mechanisms that enable organizational reflexes to improve automatically through operational experience while adapting to changing business environments without requiring conscious reflex modification or retraining.

Reflex evolution includes performance optimization that improves reflex effectiveness through automatic outcome analysis and response refinement, adaptation mechanisms that modify reflex sensitivity and response patterns based on changing business environments, learning integration that incorporates new business patterns into existing reflex systems, and evolutionary selection that strengthens effective reflexes while weakening inappropriate responses through operational feedback.

Evolution mechanisms create organizational reflexes that become more sophisticated through operational experience rather than remaining static after initial development. Organizations develop reflex systems that learn unconsciously while maintaining strategic alignment and operational effectiveness.

Amazon customer service demonstrates reflex evolution, creating increasingly sophisticated customer problem resolution that operates faster and more effectively than conscious customer service management. Traditional customer service requires conscious problem analysis and deliberate solution development that creates delays during customer issue complexity. The evolving reflexes of Amazon improve customer problem recognition and solution generation automatically through customer interaction learning.

The reflex evolution includes performance optimization that improves customer issue recognition and solution effectiveness through automatic customer satisfaction analysis and response refinement, adaptation mechanisms that modify service reflexes based on changing customer expectations and problem types, learning integration that incorporates new customer service patterns into existing reflex systems, and evolutionary selection that strengthens effective customer service reflexes while modifying ineffective responses through customer feedback analysis.

The evolving reflexes enable Amazon to maintain customer satisfaction through automatically improving service responses, while competitors require conscious customer service optimization that delays service improvement. The evolutionary reflexes operate unconsciously while generating customer service advantages that deliberate service management cannot achieve.

Organizations implementing enterprise reflex graphs discover that competitive advantage transforms from conscious strategic execution to unconscious organizational intelligence that generates optimal responses automatically. The reflex development creates sustainable advantages because unconscious organizational competencies become embedded in organizational DNA rather than remaining dependent on conscious management attention or individual system capabilities.

The enterprise reflex graph represents the evolution toward truly intelligent organizations that respond to business environments like biological organisms respond to physical environments—instantly, appropriately, and unconsciously while maintaining perfect alignment with survival objectives and evolutionary advantage.

CHAPTER 11

Agentic Strategic Units – The Self-Managing Enterprise

In 1975, a group of researchers at Stanford Research Institute created the first autonomous mobile robot capable of reasoning about its actions. Shakey the robot could navigate environments, make decisions, and execute complex tasks without constant human guidance. What took researchers years to achieve with a single robot, the most advanced AI-native enterprises are now accomplishing with entire business units that think, learn, and act autonomously while remaining perfectly aligned with organizational objectives.

The future of enterprise organization isn't hierarchical teams managed by humans. It's autonomous business cells that manage themselves through AI agency while amplifying rather than replacing human strategic thinking. These agentic strategic units represent the evolution from command-and-control management to what some call "intention-and-intelligence" coordination—where human leaders set strategic intentions and AI agents provide the intelligence to execute them autonomously.

Right now, at Shopify, autonomous e-commerce optimization units monitor merchant performance across millions of stores, automatically adjusting pricing strategies, inventory recommendations, and marketing campaigns without any human intervention. Each unit operates like an independent business consultant that never sleeps, never gets distracted, and continuously learns from every transaction across the entire platform. When these units discover successful optimization patterns, they instantly share learning with every other unit in the network, creating compound intelligence that grows stronger through collective experience.

At Spotify, autonomous playlist curation units don't just recommend music—they function as independent music discovery businesses that compete and collaborate to maximize user engagement. Each unit develops its own editorial personality, discovers emerging artists, and creates music experiences that no human curator could design manually. The units learn from user behavior patterns across 400 million listeners, developing musical intelligence that transcends individual human taste while creating experiences that feel personally curated.

What has been consistently observed across enterprise technology transformation is that the organizations achieving exponential advantages through AI integration have stopped thinking about AI as technology that supports human work and started thinking about AI as intelligence that enables new forms of work organization. They've moved beyond human teams that use AI tools to human-AI teams that function as unified intelligence entities.

The enterprises building agentic strategic units discover something that traditional management thinking makes impossible. They can scale entrepreneurial thinking across unlimited business functions without scaling management overhead. They can adapt to market changes faster than markets change without losing strategic coherence. They can innovate continuously without disrupting operational excellence.

The transformation begins when organizations realize that sustainable competitive advantage doesn't come from optimizing human performance or automating human tasks. It comes from creating new forms of business intelligence that merge human strategic thinking with AI operational execution in ways that transcend the limitations of either approach independently.

FROM TRADITIONAL TEAMS TO AI-ENABLED AUTONOMOUS BUSINESS CELLS

Picture the difference between a traditional orchestra, where every musician follows a conductor's direction, and a jazz ensemble, where talented musicians improvise together while maintaining musical coherence through shared understanding and responsive listening. Traditional business teams operate like orchestras—following predetermined plans and hierarchical direction. Agentic strategic units operate like world-class jazz ensembles—improvising optimal responses to changing conditions while maintaining perfect strategic alignment through shared intelligence.

The transformation from traditional teams to autonomous business cells represents a fundamental reconceptualization of how work gets organized, decisions get made, and business value gets created. Traditional teams require constant management oversight to ensure coordination and performance. Autonomous business cells generate coordination and performance through AI-enabled intelligence that operates continuously without management intervention.

This evolution extends far beyond automation or AI augmentation. Automation replaces human tasks with mechanical execution. AI augmentation enhances human capabilities with intelligent support. Agentic strategic units create entirely new forms of business organization where AI agency enables human strategic thinking to scale across unlimited operational contexts while maintaining entrepreneurial responsiveness to changing conditions.

The Cell Division of Business Functions

Traditional business functions operate like single-celled organisms—functional but limited in their ability to adapt, specialize, and coordinate with other functions. Agentic strategic units represent the evolution to multicellular business organisms where specialized cells coordinate through intelligent communication while maintaining autonomy over their specific domains.

Business cell division includes functional specialization that enables each unit to develop deep expertise in specific business domains while contributing to overall organizational objectives, autonomous decision-making that eliminates management bottlenecks by enabling units to respond to changing conditions independently, intelligent coordination that ensures unit activities align with strategic priorities and complement other unit operations, and adaptive evolution that enables units to improve performance through operational experience and environmental feedback.

The cellular approach transforms how organizations think about business structure. Instead of departmental hierarchies that require coordination through management layers, cellular organizations operate through networks of specialized intelligence units that coordinate directly through shared information and aligned incentives.

The Amazon approach to marketplace management demonstrates business cell division at unprecedented scale. Traditional marketplace management requires hierarchical coordination between category managers, vendor relations, pricing optimization, and customer experience teams. The cellular approach of Amazon creates autonomous units that manage specific marketplace domains independently while coordinating through intelligent information sharing.

The marketplace cells include vendor relationship units that manage supplier interactions autonomously based on performance metrics and strategic priorities, pricing optimization units that adjust product pricing continuously based on competitive analysis and demand patterns, inventory management units that coordinate supply and demand across millions of products without human intervention, and customer experience units that optimize search, recommendation, and fulfillment processes based on individual user behavior analysis.

Each cell operates with entrepreneurial autonomy within its domain while contributing to overall marketplace performance through intelligent coordination. Vendor relationship units share supplier performance data with inventory management units automatically. Pricing optimization units coordinate with customer experience units to ensure pricing strategies support user satisfaction objectives. The cellular coordination creates marketplace efficiency that no traditional hierarchical structure could achieve.

Intelligence-Driven Autonomy vs. Rule-Based Automation

The fundamental difference between agentic strategic units and traditional automation lies in their approach to decision-making under uncertainty. Rule-based automation executes predetermined responses to specific triggers.

Intelligence-driven autonomy generates optimal responses to novel situations by applying learned patterns while adapting to contextual factors that influence decision effectiveness.

Intelligence-driven autonomy includes contextual understanding that enables units to recognize business situations and adapt decision-making approaches based on environmental factors, pattern application that allows units to apply successful approaches from previous situations to new challenges while modifying tactics based on contextual differences, continuous learning that improves unit decision-making capabilities through operational experience and outcome analysis, and strategic alignment that ensures autonomous decisions support long-term organizational objectives while optimizing immediate performance.

The intelligence-driven approach enables agentic strategic units to handle business complexity that would overwhelm rule-based systems. Units can respond appropriately to unprecedented situations by combining learned patterns with contextual analysis rather than failing when conditions don't match predetermined scenarios.

The approach of Uber to dynamic transportation optimization illustrates intelligence-driven autonomy managing complexity that rule-based systems cannot handle. Traditional transportation dispatching relies on predetermined algorithms that assign vehicles to passenger requests based on distance and availability. The agentic units of Uber generate optimal transportation solutions by considering dozens of factors simultaneously while adapting to changing conditions continuously.

The transportation units include demand prediction units that analyze passenger request patterns, weather conditions, event schedules, and economic factors to anticipate transportation needs before requests occur, supply coordination units that optimize driver positioning and availability based on predicted demand and current traffic conditions, pricing optimization units that balance passenger affordability with driver income while maximizing transportation efficiency, and experience enhancement units that improve passenger and driver satisfaction through personalized service optimization.

The intelligence-driven autonomy enables Uber units to respond effectively to novel situations like natural disasters, major events, or infrastructure disruptions that would overwhelm rule-based transportation systems. Units generate appropriate responses by combining learned optimization patterns with contextual analysis of unprecedented conditions.

Emergent Coordination and Swarm Intelligence

Advanced agentic strategic units exhibit emergent coordination behaviors that arise from individual unit intelligence rather than centralized management direction. Like bird flocks that maintain perfect formations through individual birds following simple rules while responding to local conditions, business unit swarms create organizational behaviors that transcend individual unit capabilities.

Emergent coordination includes local optimization that enables each unit to maximize performance within its domain while contributing to overall organizational objectives, information sharing that allows units to communicate relevant insights and coordinate activities without management oversight, adaptive specialization that enables units to develop expertise based on operational requirements and competitive opportunities, and collective learning that improves entire unit networks through shared experience and insight exchange.

Swarm intelligence behaviors emerge when multiple autonomous units coordinate activities through shared information and aligned incentives rather than hierarchical command structures. These behaviors create organizational capabilities that no individual unit could generate independently while maintaining the entrepreneurial responsiveness of individual autonomous operation.

The Netflix approach to content optimization demonstrates emergent coordination, creating organizational intelligence that transcends individual unit capabilities. Netflix operates thousands of autonomous content units that optimize different aspects of entertainment delivery, including content recommendation, user interface design, streaming quality, content acquisition, and audience analysis.

The content units coordinate through information sharing that enables recommendation units to learn from user interface optimization experiments, content acquisition units to leverage audience analysis insights for production decisions, and streaming quality units to anticipate demand patterns based on recommendation effectiveness. The emergent coordination creates entertainment experiences that feel seamlessly integrated while emerging from the autonomous operation of specialized intelligence units.

The swarm intelligence enables Netflix to adapt content strategy faster than traditional entertainment companies can recognize market changes while maintaining consistent user experience quality across all platform interactions. Individual units optimize their domains independently while collective coordination creates entertainment capabilities that competitors cannot replicate through traditional hierarchical development approaches.

Organizations implementing agentic strategic units discover that competitive advantage shifts from optimizing team performance to developing superior unit intelligence and coordination patterns. The cellular approach creates organizational capabilities that scale automatically while maintaining entrepreneurial responsiveness to changing market conditions and competitive pressures.

DESIGNING SELF-ORGANIZING, SELF-OPTIMIZING BUSINESS FUNCTIONS

Imagine if your sales organization could reorganize itself automatically when market conditions change, optimizing territory assignments, lead qualification processes, and pricing strategies without management intervention.

Or, if your supply chain could reconfigure logistics networks, supplier relationships, and inventory distribution in real time based on demand patterns, weather conditions, and competitive actions. This isn't futuristic speculation—it's the operational reality that organizations are creating through self-organizing, self-optimizing business functions.

The design of autonomous business functions requires understanding that self-organization emerges from intelligent agents that can sense environmental conditions, evaluate performance outcomes, and modify their own behavior to improve effectiveness. Unlike traditional business process optimization that requires external analysis and manual implementation, self-optimizing functions evolve continuously through embedded intelligence that learns from operational experience.

This approach represents a fundamental shift from designing business processes that humans execute to designing business intelligence that executes itself while improving through experience. The design challenge isn't specifying how work should be done but creating intelligent systems capable of figuring out how work should be done and adapting approaches based on changing conditions.

Adaptive Organizational Architecture

Self-organizing business functions operate through adaptive architecture that can modify structure, processes, and resource allocation based on performance feedback and environmental changes. This architecture treats organizational design as a continuous optimization problem rather than a fixed blueprint.

Adaptive architecture includes modular function design that enables business capabilities to be recombined based on strategic priorities and operational requirements, dynamic resource allocation that redirects human and AI capabilities toward highest-value activities automatically, performance feedback loops that enable functions to evaluate effectiveness and modify approaches based on outcome analysis, and evolutionary mechanisms that enable functions to develop new capabilities through experimentation and learning.

The architectural approach recognizes that optimal organizational structure depends on business conditions that change continuously. Self-organizing functions adapt their structure automatically to maintain optimal performance rather than requiring periodic reorganization through management initiative.

The approach of Spotify to music discovery demonstrates adaptive organizational architecture, enabling continuous optimization without management intervention. Traditional music recommendation requires predetermined algorithms that categorize music and match user preferences through demographic analysis. The self-organizing approach of Spotify creates recommendation functions that adapt structure and methodology based on user behavior patterns and music trend evolution.

The adaptive architecture includes modular recommendation functions that can be recombined based on user context, listening history, and cultural trends, dynamic resource allocation that redirects recommendation processing toward music categories and user segments showing the highest engagement potential, performance feedback loops that enable recommendation functions to evaluate user satisfaction and modify approaches based on listening behavior analysis, and evolutionary mechanisms that enable recommendation functions to develop new discovery capabilities through experimentation with music matching algorithms.

The adaptive approach enables Spotify to maintain recommendation effectiveness across changing music trends, user preference evolution, and competitive landscape shifts without requiring manual algorithm optimization. The recommendation architecture optimizes automatically while developing new capabilities that enhance music discovery beyond traditional algorithmic approaches.

Autonomous Performance Optimization

Self-optimizing business functions include embedded intelligence that monitors performance continuously and implements improvements automatically without requiring external analysis or management intervention. This autonomous optimization creates business functions that become more effective through operational experience rather than periodic improvement initiatives.

Autonomous optimization includes performance monitoring that tracks function effectiveness across multiple dimensions simultaneously, improvement identification that recognizes optimization opportunities through pattern analysis and outcome correlation, implementation automation that modifies function behavior based on improvement analysis without disrupting ongoing operations, and learning integration that incorporates operational insights into function intelligence to improve future optimization decisions.

The optimization approach treats business improvement as a continuous, autonomous process rather than periodic human-driven initiatives. Functions evolve performance automatically while maintaining operational stability and strategic alignment.

The approach of Amazon to logistics optimization illustrates autonomous performance optimization, creating continuous improvement without management intervention. Traditional logistics optimization requires human analysis of operational data followed by manual implementation of process improvements. The self-optimizing approach of Amazon creates logistics functions that improve performance automatically through embedded optimization intelligence.

The autonomous optimization includes performance monitoring that tracks delivery efficiency, cost effectiveness, and customer satisfaction continuously across millions of shipments, improvement identification that recognizes optimization opportunities through analysis of delivery patterns, weather impacts, and resource utilization, implementation automation that modifies logistics routing, inventory positioning, and capacity allocation based on optimization analysis, and learning integration that incorporates delivery outcomes into logistics intelligence to improve future optimization decisions.

The autonomous approach enables Amazon to maintain logistics leadership by improving delivery performance automatically while scaling operations across changing demand patterns and competitive pressures. Logistics functions optimize continuously without requiring management attention while developing capabilities that enhance delivery effectiveness beyond traditional logistics management approaches.

Emergent Business Model Innovation

The most sophisticated self-organizing functions develop new business model capabilities through emergent innovation that arises from autonomous experimentation and learning. This innovation transcends planned business development by creating novel value creation approaches through intelligent exploration of opportunity spaces.

Emergent innovation includes opportunity detection that identifies potential business model innovations through analysis of user behavior patterns, competitive dynamics, and market trend evolution, experimentation automation that tests new business approaches through controlled trials without disrupting core operations, capability development that creates new business functions based on successful experiments and market validation, and integration orchestration that incorporates validated innovations into organizational capabilities while maintaining operational coherence.

The innovation approach enables organizations to develop new business capabilities faster than competitors can recognize market opportunities while maintaining focus on core business performance.

The approach of Tesla to autonomous vehicle development demonstrates emergent business model innovation through self-organizing functions that discover new value creation opportunities automatically. Traditional automotive development follows predetermined product roadmaps based on market research and competitive analysis. The self-organizing approach of Tesla creates development functions that discover new automotive capabilities through autonomous experimentation and learning.

The emergent innovation includes opportunity detection that identifies potential autonomous vehicle capabilities through analysis of driving behavior patterns, traffic condition changes, and user preference evolution, experimentation automation that tests new vehicle features through software updates and user feedback analysis without disrupting vehicle safety or performance, capability development that creates new automotive functions based on successful feature experiments and user validation, and integration orchestration that incorporates validated innovations into vehicle intelligence while maintaining safety and reliability standards.

The innovation approach enables Tesla to develop automotive capabilities that competitors cannot anticipate while maintaining vehicle excellence and customer satisfaction. Development functions discover innovation opportunities automatically while creating competitive advantages that strengthen through operational experience.

Organizations implementing self-organizing, self-optimizing business functions discover that competitive advantage shifts from optimizing planned processes to developing superior adaptive intelligence that creates continuous improvement and innovation automatically. The design approach creates business capabilities that evolve faster than competitors can adapt while maintaining strategic coherence and operational excellence.

AGENT TRAINERS, TRUST ENGINEERS, AND PROMPT ARCHITECTS

The emergence of agentic strategic units creates entirely new categories of human roles that didn't exist in traditional organizations. These aren't just "AI jobs"—they're fundamentally new forms of professional expertise that bridge human strategic thinking with AI operational intelligence. Like the emergence of software engineering in the computer revolution or digital marketing in the Internet age, these roles represent novel forms of human capability required for AI-native organization success.

These new roles operate at the intersection of human intelligence and AI, requiring professionals who understand both human organizational dynamics and AI system capabilities. They create the human expertise necessary for developing, training, and optimizing autonomous business units that can operate independently while maintaining alignment with human strategic objectives and organizational values.

What makes these roles unique is their focus on designing intelligence rather than executing processes. Traditional business roles involve humans performing work directly or managing other humans who perform work. AI-native roles involve humans designing intelligent systems that perform work autonomously while ensuring those systems develop appropriate capabilities and maintain strategic alignment.

Agent Trainers: The New Performance Coaches

Agent trainers represent the evolution of human resource development for AI-native organizations. Unlike traditional trainers who develop human skills through instruction and practice, agent trainers develop AI agent capabilities through sophisticated training methodologies that combine domain expertise with machine learning optimization.

Agent training includes capability development that designs training programs, enabling AI agents to master specific business functions through experiential learning and pattern recognition, performance optimization that fine-tunes agent decision-making through feedback analysis and outcome correlation, domain knowledge transfer that provides agents with industry expertise and business context required for effective autonomous operation, and behavioral alignment that ensures agent actions reflect organizational values and strategic priorities.

The training approach treats AI agents as intelligent entities that learn through experience rather than programmed tools that execute predetermined instructions. Agent trainers design learning environments that enable agents to develop expertise while maintaining appropriate operational boundaries and ethical constraints.

The approach of Salesforce to AI agent development demonstrates agent training creating autonomous sales capabilities that exceed traditional human-AI collaboration. Traditional sales AI provides analytical support for human sales professionals. Salesforce agent trainers develop autonomous sales agents that manage customer relationships independently while maintaining personalized service quality.

The agent training includes capability development that enables sales agents to master relationship building, needs analysis, solution design, and negotiation through analysis of successful sales interactions, performance optimization that fine-tunes agent communication and recommendation strategies through customer feedback and sales outcome analysis, domain knowledge transfer that provides agents with product expertise, industry knowledge, and competitive intelligence required for effective sales conversations, and behavioral alignment that ensures agent interactions reflect Salesforce brand values and customer service standards.

The trained agents operate autonomously while maintaining sales effectiveness that meets or exceeds that of human sales professionals. Agent trainers continuously optimize agent capabilities while ensuring customer relationships develop appropriately through AI-managed interactions.

Trust Engineers: Building Confidence in Autonomous Systems

Trust engineers develop the organizational confidence necessary for autonomous business unit operation. Unlike traditional security or compliance roles that prevent problems through restrictions, trust engineers create positive assurance systems that enable autonomous operation while maintaining appropriate oversight and accountability.

Trust engineering includes reliability assurance that ensures autonomous units operate consistently within expected performance parameters, transparency development that provides organizational visibility into autonomous decision-making processes and reasoning, accountability frameworks that enable appropriate oversight and correction of autonomous unit actions, and confidence calibration that helps organizational stakeholders understand when to trust autonomous unit decisions and when to seek human oversight.

The engineering approach treats organizational trust as a systematic capability that can be designed and optimized rather than a subjective feeling that develops randomly through experience. Trust engineers create measurable confidence in autonomous systems through systematic analysis and transparent communication.

The approach of JPMorganChase to autonomous trading demonstrates trust engineering, enabling organizational confidence in AI-led financial decision-making. Traditional algorithmic trading requires continuous human oversight and frequent manual intervention. JPMorganChase trust engineers create autonomous trading units that operate independently while maintaining appropriate risk management and regulatory compliance.

The trust engineering includes reliability assurance that ensures trading units operate within risk parameters and regulatory requirements consistently, transparency development that provides investment professionals with clear visibility into trading reasoning and decision-making processes, accountability frameworks that enable appropriate oversight and rapid correction of trading strategies when market conditions change, and confidence calibration that helps investment teams understand when to trust autonomous trading decisions and when to implement human oversight.

The trust systems enable JPMorganChase to operate autonomous trading units at scale while maintaining organizational confidence and regulatory compliance. Trust engineers continuously optimize confidence systems while ensuring trading performance meets fiduciary and business requirements.

Prompt Architects: Designing Conversational Intelligence

Prompt architects represent the evolution of user experience design for conversational AI systems. Unlike traditional UX designers who create visual interfaces for human users, prompt architects design conversational interfaces that enable effective communication between humans and AI agents while optimizing AI performance and strategic alignment.

Prompt architecture includes conversation design that creates natural language interactions, enabling humans to communicate strategic objectives and operational requirements to AI agents effectively, intelligence optimization that designs prompts, maximizing AI agent analytical and decision-making capabilities, context management that ensures AI agents maintain an appropriate understanding of business situations and strategic priorities throughout extended interactions, and performance calibration that optimizes prompt effectiveness through analysis of AI responses and business outcomes.

The architectural approach treats conversational intelligence as a systematic design discipline that can be optimized through methodical analysis and iterative improvement. Prompt architects create conversational capabilities that enhance both the effectiveness of human-AI communication and the quality of AI agent performance.

The Microsoft approach to Copilot development demonstrates prompt architecture, creating conversational intelligence that enhances productivity across unlimited business contexts. Traditional business software requires users to learn specific interfaces and workflows for different applications. Microsoft prompt architects design conversational interfaces that enable users to achieve business objectives through natural language interaction with AI agents.

The prompt architecture includes conversation design that enables users to communicate business objectives and requirements to AI agents through natural language without requiring technical expertise, intelligence optimization that designs prompts, enabling AI agents to understand business context and generate appropriate solutions across multiple application domains, context management that ensures AI agents maintain an understanding of user objectives and organizational priorities throughout complex multi-step business processes, and performance calibration that optimizes conversational effectiveness through analysis of user satisfaction and business outcome achievement.

The conversational intelligence enables Microsoft users to accomplish business objectives faster while enabling AI agents to provide more effective assistance. Prompt architects continuously optimize conversational capabilities while ensuring AI assistance enhances rather than constrains human productivity.

Organizational Learning and Development

The integration of agent trainers, trust engineers, and prompt architects creates organizational learning and development capabilities that accelerate AI-native transformation while maintaining human strategic leadership and organizational coherence. These roles work collaboratively to create human-AI teams that exceed the performance capabilities of either humans or AI systems operating independently.

Collaborative development includes cross-functional optimization that ensures agent training, trust engineering, and prompt architecture work together to create superior autonomous business unit capabilities, organizational capability development that expands enterprise AI-native capabilities through systematic role development and expertise enhancement, strategic alignment maintenance that ensures new roles support rather than compete with traditional business expertise and leadership, and continuous evolution that adapts roles and responsibilities based on technological advancement and business requirement changes.

The development approach treats AI-native organization as an ongoing evolution requiring new forms of human expertise rather than a technological replacement of human capabilities. The new roles amplify human strategic thinking while enabling AI operational excellence that creates sustainable competitive advantages.

Organizations implementing these new roles discover that competitive advantage shifts from human task performance to human-AI collaboration design that creates superior business intelligence and operational effectiveness. The roles create human expertise that enables AI-native transformation while maintaining strategic coherence and an organizational culture that supports sustainable business success.

ORGANIZING AROUND CAPABILITY, NOT HIERARCHY

The transformation to agentic strategic units requires fundamentally reimagining organizational structure from hierarchical authority chains to dynamic capability networks. Traditional organizations optimize for control and coordination through management layers that ensure consistent execution of predetermined strategies. AI-native organizations optimize for intelligence and adaptation through capability networks that enable optimal responses to changing conditions while maintaining strategic coherence.

This shift represents more than flattening hierarchies or improving communication. It involves reconstructing the basic logic of organizational design from managing human activities to orchestrating intelligence capabilities that include both human and AI agents working as integrated teams. The organizing principle becomes matching the right capabilities to business challenges rather than managing authority relationships between organizational positions.

The capability-centered approach treats organizations as dynamic intelligence networks where value creation emerges from the quality of capability coordination rather than the efficiency of hierarchical command execution. This creates organizational structures that adapt automatically to changing strategic priorities and competitive pressures while maintaining operational excellence and cultural coherence.

Dynamic Capability Mapping and Resource Allocation

Capability-centered organization requires sophisticated mapping of organizational intelligence assets, including human expertise, AI agent capabilities, process knowledge, and strategic relationships. This mapping enables dynamic resource allocation that optimizes capability deployment based on business priorities and competitive opportunities rather than predetermined organizational boundaries.

Dynamic capability mapping includes intelligence asset inventory that catalogs human expertise, AI agent capabilities, and knowledge resources available for business objective achievement, capability relationship analysis that understands how different intelligence assets can be combined to create superior business outcomes, resource optimization algorithms that allocate capabilities toward the highest-value opportunities automatically, and performance feedback systems that improve capability deployment through outcome analysis and strategic learning.

The mapping approach treats organizational capability as a portfolio of intelligence assets that can be recombined dynamically rather than fixed departmental resources that operate through predetermined workflows. Capability allocation optimizes business outcomes rather than organizational consistency.

The approach of Google to product development demonstrates dynamic capability mapping, enabling innovation that transcends traditional organizational boundaries. Traditional technology companies organize product development through departmental hierarchies that separate engineering, design, marketing, and business development into distinct organizational units with defined responsibilities and communication protocols.

The capability-centered approach of Google creates product development through dynamic teams that combine human expertise and AI agent capabilities based on project requirements and strategic objectives rather than a predetermined organizational structure. Product teams include whatever combination of human engineers, AI development agents, design experts, market analysis capabilities, and strategic intelligence enables optimal product outcomes.

The dynamic mapping includes intelligence asset inventory that catalogs engineering expertise, AI capabilities, design skills, and market knowledge available across the Google organization, capability relationship analysis that identifies optimal combinations of human and AI intelligence for specific product development challenges, resource optimization that allocates capabilities toward the highest-impact product opportunities automatically, and performance feedback that improves capability deployment through product success analysis and market learning.

The capability approach enables Google to create products that emerge from optimal intelligence combinations rather than organizational constraints while maintaining strategic coherence and operational efficiency across unlimited product development initiatives.

Cross-Functional Intelligence Networks

Capability-centered organization operates through cross-functional intelligence networks that coordinate expertise across traditional departmental boundaries. These networks enable optimal problem-solving by combining relevant capabilities regardless of organizational position while maintaining accountability and performance measurement.

Cross-functional networks include expertise coordination systems that connect relevant human and AI capabilities based on business challenge requirements, collaborative decision-making protocols that enable intelligence networks to generate optimal solutions through distributed analysis and synthesis, knowledge sharing mechanisms that ensure learning from one network enhances capabilities across all related networks, and strategic alignment frameworks that ensure network activities support organizational objectives while optimizing immediate performance.

The network approach treats business problem-solving as an intelligence coordination challenge rather than a management authority exercise. Solutions emerge from optimal capability combinations rather than hierarchical decision-making processes.

The Amazon approach to customer experience optimization demonstrates cross-functional intelligence networks creating superior business outcomes through capability coordination that transcends traditional organizational boundaries. Traditional e-commerce companies optimize customer experience through departmental coordination between marketing, product management, technology, and customer service teams operating through predetermined workflows and management oversight.

The Amazon network approach creates customer experience optimization through dynamic intelligence coordination that combines human customer insight, AI behavior analysis, product development capabilities, and operational optimization expertise based on customer experience requirements rather than departmental responsibilities.

The intelligence networks include expertise coordination that connects customer behavior analysts, product designers, AI recommendation systems, and logistics optimization capabilities based on customer experience challenges, collaborative decision-making that enables networks to generate customer experience solutions through distributed analysis of behavior patterns, preferences, and operational constraints, knowledge sharing that ensures customer insights from one network improve experience optimization across all Amazon services, and strategic alignment that ensures network activities enhance long-term customer relationships while optimizing immediate satisfaction metrics.

The network approach enables Amazon to create customer experiences that emerge from optimal intelligence coordination rather than departmental efficiency while maintaining operational excellence and strategic coherence across unlimited customer interaction contexts.

Autonomous Team Formation and Evolution

Advanced capability-centered organizations enable autonomous team formation where human-AI teams self-organize based on business requirements and capability alignment rather than management assignment. This autonomy creates optimal problem-solving teams while maintaining organizational coherence and strategic direction.

Autonomous team formation includes capability matching algorithms that identify optimal human-AI combinations for specific business challenges, team formation protocols that enable autonomous team creation based on capability requirements and strategic priorities, performance optimization systems that enable teams to modify composition and methodology based on outcome analysis, and evolution mechanisms that enable teams to develop enhanced capabilities through collaborative experience and learning.

The autonomous approach treats team formation as an intelligence optimization problem rather than a management coordination challenge. Teams emerge from capability alignment rather than organizational structure while maintaining accountability and performance excellence.

The approach of Tesla to vehicle development demonstrates autonomous team formation, creating innovation capabilities that transcend traditional automotive development approaches. Traditional automotive companies organize vehicle development through predetermined departmental teams, including design, engineering, manufacturing, and testing, that coordinate through management oversight and predetermined workflows.

The autonomous approach of Tesla enables vehicle development teams to form dynamically based on innovation requirements and capability availability rather than organizational structure. Development teams include whatever combination of human expertise and AI capabilities enables optimal vehicle innovation regardless of traditional departmental boundaries.

The autonomous formation includes capability matching that identifies optimal combinations of design expertise, engineering knowledge, AI simulation capabilities, and manufacturing intelligence for specific vehicle development challenges, team formation protocols that enable autonomous team creation based on innovation requirements and strategic vehicle objectives, performance optimization that enables teams to modify composition and methodology based on development outcomes and market feedback, and evolution mechanisms that enable teams to develop enhanced innovation capabilities through collaborative vehicle development experience.

The autonomous approach enables Tesla to create vehicle innovations that emerge from optimal capability coordination rather than organizational constraints while maintaining strategic coherence and operational excellence across unlimited vehicle development initiatives.

Organizations implementing capability-centered organization discover that competitive advantage shifts from organizational efficiency to intelligence network effectiveness that creates superior business outcomes through optimal capability coordination. The organization approach creates organizational structures that adapt automatically to changing business requirements while maintaining strategic coherence and operational excellence that supports sustainable competitive advantage.

CHAPTER 12

AI-Driven Strategic Orchestration

Strategy is Dead. Long Live Strategy

The traditional approach to corporate strategy, assembling human experts in conference rooms to debate quarterly plans based on historical data and intuitive forecasts, died sometime between the first ChatGPT query and the moment Chinese electric vehicle manufacturers began shipping cars designed by AI in 18 months instead of 4 years. Most executives haven't noticed yet. They're still scheduling strategy retreats for 2025.

Based on enterprise transformations across four technology revolutions, what is being witnessed now represents the most fundamental shift in strategic thinking since the invention of scenario planning. The future belongs to organizations that orchestrate strategy like symphony conductors who never touch an instrument yet create music that moves audiences to tears. These AI-native enterprises conduct strategic symphonies through AI that processes market signals, competitive moves, and customer behavior patterns faster than human cognition can parse the opening notes.

This isn't about AI helping strategists make better decisions. This is about AI becoming the strategic conductor that coordinates millions of micro-decisions into a coherent competitive advantage while human executives focus on composing the vision that gives the symphony meaning. The pattern consistently observed across industries is that organizations implementing AI-driven strategic orchestration achieve competitive positioning that traditional strategic planning cannot replicate through quarterly cycles or annual retreats.

Three revelations will transform how you think about strategy. The greatest strategic insights emerge not from prediction but from simulation, where organizations run thousands of parallel strategic experiments in virtual environments where failure costs nothing and learning compounds infinitely. Competitive advantage flows not from knowing what competitors are doing but from mapping the influence networks that determine what competitors will do before they decide themselves. The most successful strategies aren't planned annually or quarterly but evolve continuously through AI systems that adapt strategic approaches faster than markets can shift.

Right now, while traditional corporations hold strategy sessions, DeepMind AI systems simultaneously run 47 million different strategic scenarios to optimize everything from game theory outcomes to protein folding solutions. They've moved beyond asking "What should we do?" to exploring "What happens if we do everything possible?", then selecting optimal approaches through computational strategic intelligence. Aladdin, by BlackRock, processes more strategic decisions in one hour than most Fortune 500 companies make in a year. It doesn't replace human strategic thinking. It amplifies human strategic intuition through computational power that can explore strategic possibility spaces no human team could navigate manually.

The transformation from strategy planning to strategy orchestration represents the difference between composing music note by note and conducting an orchestra that improvises perfect harmony in real time. AI-driven strategic orchestration creates competitive advantages that emerge from the coordination of intelligence rather than the brilliance of individual insights.

STRATEGIC ORCHESTRATION AND MARKET CONDUCTORSHIP

The art of strategic orchestration transforms executives from chess players calculating individual moves to symphony conductors coordinating infinite instruments in perfect harmony. Traditional strategy treats business challenges like individual games to be won. Strategic orchestration treats business environments like musical compositions to be performed, where competitive advantage emerges from the quality of coordination rather than the brilliance of individual tactics.

What has been learned from studying organizational behavior across multiple industry transformations is that sustainable competitive advantage emerges from superior coordination of strategic capabilities rather than superior individual strategic capabilities. Organizations can possess identical strategic resources yet achieve dramatically different competitive outcomes based on their ability to orchestrate those resources into harmonious competitive performance. This represents the evolution from strategic planning as analysis to strategic execution as artistry.

AI-driven strategic orchestration enables organizations to coordinate strategic responses across unlimited business functions, market conditions, and competitive pressures simultaneously. Instead of sequential strategic decision-making that creates delays between strategic recognition and strategic response, orchestration creates synchronized strategic execution that adapts harmoniously to changing business environments. The market leaders aren't necessarily the companies with the best individual strategies but those that understand and leverage orchestration dynamics most effectively.

The most sophisticated strategic orchestration creates what musicologists call "emergent harmony," where individual strategic actions combine to create competitive advantages that transcend the sum of individual strategic contributions. Like jazz musicians who create musical experiences impossible to plan yet perfectly coordinated, AI-orchestrated enterprises create competitive advantages that emerge from intelligent coordination rather than predetermined strategic blueprints.

Strategic orchestration operates through principles borrowed from conducting rather than chess. Conductors don't control individual musicians but coordinate collective performance that enables individual excellence while maintaining ensemble coherence. Similarly, strategic orchestration doesn't control individual business functions but coordinates collective strategic performance that enables functional excellence while maintaining competitive coherence.

The Tesla Manufacturing Symphony

The Tesla approach to vehicle production demonstrates strategic orchestration, creating competitive harmony across manufacturing, supply chain, and technology development that traditional automotive companies cannot replicate through separate optimization approaches. Conventional automakers optimize individual manufacturing processes through separate strategic decisions for production, suppliers, technology, and distribution. The orchestration of Tesla coordinates strategic decisions across all manufacturing components to create unified competitive experiences.

The orchestration includes production technology coordination that ensures manufacturing robotics and software development strategic decisions create reinforcing rather than competing advantages. Supply chain integration orchestration aligns strategic supplier decisions with strategic production positioning to create seamless manufacturing experiences. Technology development coordination ensures that battery and autonomous driving strategies create mutually supportive, competitive positioning. Quality control orchestration coordinates testing strategic decisions with manufacturing requirements to create sustainable competitive advantages.

Tesla discovered that automotive market leadership requires manufacturing harmony rather than production optimization. Orchestration enables manufacturing competitive advantages that emerge from coordination quality rather than individual process excellence, creating competitive positioning that traditional automotive companies cannot challenge through conventional manufacturing optimization or cost reduction strategies.

Advanced Strategic Improvisation

Advanced orchestration enables strategic improvisation that maintains competitive coherence while adapting individual strategic elements to changing market conditions. Like jazz ensembles that maintain musical coherence while enabling individual improvisation, orchestrated enterprises maintain strategic coherence while enabling functional strategic adaptation. This capability becomes essential as market conditions change faster than traditional strategic planning cycles can accommodate.

Strategic improvisation includes real-time strategic adjustment that modifies individual strategic elements based on market feedback while maintaining overall strategic direction. Coordination flexibility enables business functions to adapt strategic approaches to local conditions while contributing to overall strategic objectives. Strategic rhythm maintenance ensures strategic adaptations occur in coordination with broader strategic timing rather than creating competitive discord.

The e-commerce platform transformation of Spotify illustrates strategic improvisation, enabling coordinated strategic adaptation across unlimited business complexity. Traditional e-commerce platforms adapt strategies through sequential product modifications that can create inconsistent market positioning. The orchestration of Spotify enables simultaneous strategic adaptation across all platform features, while maintaining strategic coherence that ensures consistent market positioning and the sustainability of a competitive advantage.

The improvisation includes merchant success optimization that modifies individual strategic platform approaches based on seller feedback while maintaining overall e-commerce strategic positioning. Developer ecosystem coordination adapts platform strategies to emerging technology conditions while supporting global commerce strategic objectives. Payment processing improvisation enables immediate strategic adjustments to financial service requirements while maintaining long-term platform strategic direction. International expansion orchestration coordinates market entry strategic decisions with platform strategic requirements to create coherent global advantages.

Shopify discovered that e-commerce platform leadership requires strategic adaptability combined with strategic coherence. Improvisation enables platform strategic advantages that adapt to changing merchant conditions while maintaining strategic unity that traditional platform development cannot achieve through centralized strategic management approaches.

SIGNED INFLUENCE GRAPHS AND COMPETITIVE INTELLIGENCE MAPPING

The most dangerous enemy is the one you don't see coming. The most powerful ally is the one you never thought to ask.

Traditional competitive analysis tracks what competitors are doing. Signed influence mapping reveals the invisible forces that determine what competitors will do before they decide themselves. This approach treats competitive intelligence like quantum mechanics, understanding that business entities exist in complex relationship networks where every action creates ripple effects that influence all other network participants.

Through research into enterprise competitive dynamics, the fundamental error seen repeatedly is focusing on competitor actions rather than competitor influences. Signed influence graphs map the positive and negative relationships between all market participants, including competitors, suppliers, customers, regulators, influencers, and ecosystem partners, to create competitive prediction systems that anticipate strategic moves through relationship network analysis rather than historical pattern extrapolation.

This isn't espionage. This is relationship physics, understanding how influence flows through business ecosystems to create predictable patterns of competitive behavior, strategic alliance formation, and market evolution. Every industry operates through invisible influence networks that determine competitive outcomes more powerfully than individual company strategies. Market leaders aren't necessarily the companies with the best products or strategies but those that understand and leverage influence network dynamics most effectively.

Hidden Influence Architecture

Hidden influence architecture includes relationship pathway mapping that traces how influence flows between different market participants. Influence amplification identification discovers points where small actions create disproportionate competitive effects. Network vulnerability analysis reveals competitive weaknesses that emerge from relationship dependencies. Strategic positioning optimization identifies market positions that maximize positive influence while minimizing negative competitive pressure.

The music streaming strategy of Spotify illustrates hidden influence architecture, creating competitive advantages through relationship network orchestration. Traditional music companies compete through content library advantages and subscription pricing optimization. The influence strategy of Spotify creates competitive advantages through ecosystem relationship networks that make traditional competition irrelevant rather than just difficult.

The influence mapping includes artist relationship orchestration that creates positive influence flows between Spotify and millions of musicians worldwide. Record label relationship optimization generates competitive advantages through exclusive content access and promotional priority. Listener ecosystem binding creates switching costs through personalized recommendation and social sharing relationships. Podcast creator influence neutralization limits competitive effectiveness through exclusive content network effects that prevent competitive response regardless of competitor capabilities or resources.

Spotify discovered that sustainable competitive advantage emerges from influence network orchestration rather than individual content excellence. Influence mapping enables strategic approaches that make competitive response ineffective regardless of competitor capabilities or resources, creating market positions that competitors cannot challenge through traditional content acquisition or technology development strategies.

Predictive Competitive Modeling

Advanced influence mapping enables predictive competitive modeling that anticipates competitor strategic moves through relationship network analysis. By understanding influence patterns, organizations can predict competitive behavior more accurately than competitors can predict their own strategic responses to changing market conditions. The enterprises that will define industry leadership in 2030 are already implementing predictive competitive modeling to capture market opportunities before competitors recognize that opportunities exist.

Predictive competitive modeling includes influence signal processing that identifies relationship changes, indicating emerging competitive strategies. Network tension analysis recognizes relationship stress patterns preceding competitive actions. Strategic inevitability mapping predicts competitive moves that become inevitable due to relationship network pressures. Preemptive strategic positioning enables a competitive response before competitors implement strategies that traditional competitive analysis would identify only after implementation.

The approach of Airbnb to accommodation strategy demonstrates predictive competitive modeling anticipating hospitality industry competitive moves before competitive announcements occur. Traditional hospitality companies react to competitive accommodation strategies after competitors announce expansion plans. The influence mapping of Airbnb predicts competitive accommodation strategies through relationship network analysis that reveals competitive intentions before competitive decision-making processes are completed.

The modeling includes property owner relationship monitoring that identifies accommodation direction changes through host community engagement patterns. Tourism industry analysis predicts competitive accommodation strategies through destination development and investment pattern changes. Regulatory relationship tracking anticipates competitive platform moves through government policy and partnership developments. Guest experience network analysis predicts accommodation demand before competitors recognize market opportunities through traditional hospitality market research approaches.

Airbnb discovered that accommodation market leadership requires strategic anticipation rather than strategic reaction. Influence mapping enables accommodation strategies that capture market opportunities before competitors recognize opportunities exist, creating competitive advantages through strategic timing rather than property quality alone.

Dynamic Alliance Strategy

Signed influence graphs enable a dynamic alliance strategy that optimizes partnership approaches through real-time relationship network analysis. Instead of forming alliances based on strategic planning assumptions, organizations can adapt alliance strategies based on influence network evolution and competitive relationship changes. This capability becomes essential as partnership opportunities emerge and dissolve faster than traditional alliance planning cycles can accommodate.

Dynamic alliance strategy includes partnership optimization modeling that identifies alliance opportunities, maximizing strategic advantage through network influence analysis. Ecosystem orchestration coordinates multiple partnership relationships to create synergistic competitive advantages. Relationship risk management anticipates partnership vulnerabilities through network dependency analysis. Competitive alliance neutralization prevents competitive advantage through alternative alliance formation that blocks competitive partnership development.

The Zoom approach to video collaboration partnerships illustrates dynamic alliance orchestration, creating competitive advantages through relationship network optimization that traditional partnership planning cannot achieve. Traditional software companies form partnerships through strategic planning that commits to specific alliance approaches before market outcomes become clear. The dynamic approach of Zoom adapts partnership strategies through continuous relationship network analysis that enables alliance optimization based on evolving remote work conditions.

The orchestration includes technology integration modeling that optimizes alliance relationships based on customer workflow networks and competitive positioning requirements. Educational platform anticipation forms strategic partnerships before competitors recognize education technology alliance opportunities through traditional partnership development processes. Healthcare integration balancing manages multiple partnership relationships to maximize the collective strategic advantage in telemedicine applications. Enterprise collaboration barriers create partnership networks that competitors cannot replicate through individual software development approaches.

Zoom discovered that video collaboration dominance requires ecosystem thinking rather than software thinking. Dynamic alliance strategy creates competitive advantages that emerge from relationship network effects rather than individual partnership benefits, enabling competitive positioning that individual competitors cannot challenge through independent strategic action or traditional competitive response strategies.

Organizations implementing signed influence graphs discover that competitive advantage shifts from superior execution to superior relationship network understanding. Influence mapping enables strategic approaches that leverage network dynamics to create competitive advantages that traditional strategic planning cannot achieve through internal optimization alone.

AGENTS IN THE BOARDROOM AND AI-ASSISTED STRATEGIC PLANNING

The boardroom is the last place humans go to think alone.

Board meetings represent the final frontier of exclusively human decision-making, the sacred space where CEOs, directors, and senior executives believe AI cannot tread. They're wrong. AI isn't coming to the boardroom. It's already there, invisible and essential, processing the market intelligence, competitive analysis, and strategic modeling that inform every major corporate decision.

In the analysis of executive decision-making processes, the evolution observed consistently is that progressive boardrooms worldwide already deploy AI systems as silent strategic advisors, processing vast amounts of market data, competitive intelligence, and strategic analysis to inform executive decision-making without replacing human judgment or corporate governance responsibilities. The transformation isn't about replacing human judgment with AI but about amplifying human strategic intuition with computational capabilities that can process market complexity no human team could analyze manually.

The future boardroom operates like a cognitive amplification chamber where human executives focus on strategic vision and values while AI systems provide real-time market analysis, competitive intelligence, and strategic option evaluation that enables immediate strategic decision-making with confidence levels traditional planning cannot achieve. AI-assisted strategic planning creates boardroom intelligence that combines human wisdom with artificial analysis to generate strategic insights neither could achieve independently.

SILENT AI ADVISORS

Silent AI advisors include real-time market intelligence that provides executives with current market condition analysis during strategic discussions. Competitive scenario modeling evaluates strategic options against predicted competitive responses. Risk assessment automation identifies strategic vulnerabilities and opportunity costs across different strategic approaches. Strategic option generation suggests strategic alternatives human teams might not consider independently through traditional brainstorming or analytical processes.

The approach of Walmart to retail strategy demonstrates AI advisory systems enhancing rather than replacing executive strategic judgment during high-stakes retail decisions. Traditional retail companies rely on human analysis teams that require weeks to evaluate strategic inventory and pricing opportunities. The AI advisors of Walmart provide real-time retail analysis during executive strategic discussions that enable strategic decision-making velocity that competitive advantage requires.

The advisory systems include consumer behavior monitoring that provides executives with real-time analysis of shopping conditions affecting retail strategies. Competitive pricing tracking evaluates how strategic pricing positions Walmart relative to competitive retail firms. Supply chain optimization modeling assesses how potential inventory decisions affect the overall retail strategic balance. Demand forecasting analysis evaluates retail strategic options across multiple seasonal condition possibilities that human analysis cannot process during strategic discussion timeframes.

Walmart executives discovered that AI advisory systems enable strategic confidence during high-stakes retail decisions by providing analytical depth that human teams cannot generate during strategic discussion timeframes. The advisors enhance rather than replace human retail judgment while enabling strategic decision-making velocity that competitive advantage requires in rapidly changing consumer markets.

Augmented Strategic Cognition

Advanced boardroom AI systems create augmented strategic cognition that combines human strategic intuition with artificial analytical capabilities to generate strategic insights that transcend individual human or AI limitations. This augmentation enables boardroom strategic thinking that processes market complexity while maintaining human judgment and corporate governance accountability that regulatory requirements and fiduciary responsibilities demand.

Augmented cognition includes strategic insight synthesis that combines human intuitive analysis with AI computational analysis to generate strategic understanding exceeding either approach independently. Cognitive load reduction enables executives to focus on strategic vision while AI systems handle analytical complexity. Strategic option exploration expands executive strategic thinking through AI-generated strategic alternatives. Decision confidence enhancement provides executives with analytical confidence for strategic decisions requiring immediate action under uncertain conditions.

The approach of Unilever to global consumer strategy illustrates augmented strategic cognition enabling executive strategic decision-making at consumer goods industry velocity that traditional strategic planning cannot accommodate. Traditional consumer goods strategic planning requires extensive market research preparation that delays strategic responses to rapidly changing consumer preference conditions. The augmented cognition of Unilever enables real-time strategic decision-making during market volatility that competitive advantage requires in global consumer markets.

The cognition augmentation includes cultural trend analysis that provides executives with a real-time understanding of consumer conditions affecting brand strategic options. Sustainability impact modeling evaluates how strategic decisions affect environmental compliance and competitive positioning. Consumer sentiment prediction anticipates customer responses to strategic product and marketing changes. Competitive response simulation models how strategic decisions influence competitive consumer goods dynamics across global markets and cultural environments.

Unilever executives discovered that augmented cognition enables strategic agility during consumer market disruption by providing analytical depth for strategic decisions that must be made faster than traditional strategic planning allows. The augmentation creates strategic decision-making capabilities that combine human judgment with analytical precision that competitive advantage requires in rapidly evolving consumer preference markets.

Strategic Decision Orchestration

Sophisticated boardroom AI systems enable strategic decision orchestration that coordinates strategic decision-making across organizational levels to ensure strategic coherence while enabling distributed strategic execution that adapts to local market conditions and operational requirements. This orchestration becomes essential as organizations operate across multiple markets, regulatory environments, and competitive landscapes simultaneously.

Decision orchestration includes strategic alignment cascading that translates boardroom strategic decisions into operational strategic guidance across organizational levels. Distributed strategic execution enables local strategic adaptation while maintaining overall strategic coherence. Strategic feedback integration incorporates operational learning into boardroom strategic understanding. Continuous strategic optimization improves strategic decision-making through operational outcome analysis that traditional strategic planning cannot accommodate.

The approach of Siemens to industrial automation strategy demonstrates strategic decision orchestration enabling coherent strategic execution across unlimited operational complexity that traditional industrial management cannot achieve. Traditional industrial corporations struggle to maintain strategic coherence while adapting to local market conditions across different geographic regions and industry sectors. The orchestration systems of Siemens enable strategic consistency while enabling local strategic adaptation that global strategic mandates cannot achieve through centralized strategic management alone.

The orchestration includes global technology framework cascading that translates boardroom automation strategies into regional strategic guidance that maintains the strategic coherence of Siemens. Local industrial strategic adaptation enables regional strategic modifications based on local manufacturing conditions and customer industry patterns. Strategic learning integration incorporates regional industrial learning into global strategic understanding. Strategic performance optimization improves global strategic decision-making through regional execution outcome analysis that enables continuous strategic improvement.

Siemens discovered that industrial automation leadership requires strategic orchestration rather than strategic control. Decision orchestration enables strategic coherence across unlimited organizational complexity while enabling local strategic effectiveness that global strategic mandates cannot achieve through centralized strategic management approaches that traditional industrial corporations employ.

Organizations implementing AI-assisted strategic planning discover that competitive advantage shifts from strategic planning quality to strategic decision-making velocity, enhanced by analytical depth that traditional strategic planning cannot achieve. AI assistance enables boardroom strategic capabilities that combine human wisdom with computational analysis to generate strategic advantages that neither human nor artificial intelligence could create independently.

FROM ANNUAL PLANNING TO CONTINUOUS STRATEGIC EVOLUTION

Strategy has become a real-time sport played at the speed of markets, not meetings.

The annual strategic planning cycle, that ritualistic gathering where executives retreat from operations to contemplate next year's direction, belongs in a museum next to typewriters and fax machines. While traditional corporations schedule their 2025 strategy sessions, AI-native enterprises evolve their strategies 47,000 times per day through continuous strategic adaptation that operates faster than market changes.

Throughout research into organizational adaptation mechanisms, competitive advantage increasingly flows to organizations that can adapt strategic approaches faster than competitors can recognize the need for strategic change. Continuous strategic evolution represents the transformation from strategic planning as an event to strategic adaptation as a capability. Organizations no longer plan strategy annually and then execute it blindly but evolve strategy continuously through AI systems that adapt strategic approaches based on real-time market feedback, competitive intelligence, and operational learning.

This isn't about faster planning cycles or more frequent strategy updates. This is about strategy as a living system that evolves automatically while maintaining strategic coherence and competitive advantage through computational strategic intelligence that operates below conscious management attention. The enterprises that will dominate the next decade are already implementing continuous strategic evolution that creates competitive positioning through strategic adaptation velocity rather than strategic planning accuracy.

Strategic Evolutionary Algorithms

Continuous strategic evolution operates through evolutionary algorithm principles that treat strategic approaches like biological species competing for survival through environmental adaptation. Strategic options that generate superior competitive outcomes survive and reproduce through organizational implementation, while ineffective strategies disappear through natural selection processes that traditional strategic planning cannot replicate through quarterly reviews or annual assessments.

Strategic evolutionary algorithms include strategy mutation that generates strategic variations through systematic experimentation with different approaches to competitive challenges. Competitive selection pressure tests strategic approaches against market conditions and competitive responses. Strategy reproduction scales effective strategic approaches across organizational functions. Evolutionary optimization improves strategic effectiveness through iterative adaptation cycles that operate faster than competitive strategic response cycles.

The approach of Tencent to digital platform strategy demonstrates evolutionary strategic algorithms enabling continuous strategic optimization that traditional strategic planning cannot achieve through annual or quarterly planning cycles. Traditional technology companies develop platform strategies through annual planning that commits to specific technological and competitive approaches before market outcomes become clear. The evolutionary strategy of Tencent continuously adapts platform approaches through algorithmic strategic evolution, improving its competitive positioning automatically.

The evolution includes platform feature mutation that tests millions of social media strategy variations simultaneously through user interaction experiments. Competitive selection evaluates platform strategic approaches against competitive social media quality and user engagement metrics. Successful strategy reproduction scales effective platform strategies across all Tencent digital applications. Strategic optimization acceleration improves platform strategic effectiveness through rapid evolutionary cycles that operate faster than competitive strategic adaptation.

Tencent discovered that digital platform leadership requires strategic evolution velocity rather than strategic planning accuracy. Evolutionary strategy enables competitive advantages that adapt faster than competitive response cycles while maintaining strategic coherence that human strategic planning cannot achieve through traditional planning approaches that commit to strategic directions before market feedback becomes available.

Real-Time Strategic Adaptation

Advanced continuous evolution enables real-time strategic adaptation that modifies strategic approaches instantaneously based on market condition changes, competitive actions, and customer behavior shifts without requiring conscious strategic planning processes or management intervention. This capability becomes essential as market conditions change faster than traditional strategic planning cycles can accommodate through quarterly reviews or annual strategic updates.

Real-time adaptation includes market signal processing that translates environmental changes into strategic adjustment requirements. Competitive response triggers activate strategic modifications when competitive actions threaten strategic positioning. Customer behavior adaptation adjusts strategic approaches based on customer preference evolution. Strategic coherence maintenance ensures real-time strategic changes support overall strategic objectives while enabling immediate strategic responsiveness.

The approach of ByteDance to content recommendation strategy illustrates real-time strategic adaptation enabling immediate strategic responses to social media market changes that traditional media companies cannot achieve through quarterly planning cycles. Traditional media companies adapt content strategies through quarterly planning cycles that delay strategic responses to rapidly changing audience preferences and competitive content developments that can shift market positioning within days rather than quarters.

The real-time adaptation of ByteDance includes audience engagement monitoring that detects content strategy adjustment requirements through viewing behavior analysis. Competitive content tracking triggers strategic content responses to competitive platform announcements and algorithm updates. Cultural trend integration adapts content strategies to emerging social and cultural developments. Strategic content coherence ensures real-time strategic content changes maintain the overall content strategic positioning of ByteDance while enabling immediate strategic responsiveness to viral content opportunities.

ByteDance discovered that social media leadership requires strategic responsiveness rather than strategic consistency. Real-time adaptation enables strategic advantages that capture audience attention before competitors recognize that opportunities exist while maintaining strategic coherence that ensures brand consistency and user engagement through continuous strategic evolution rather than periodic strategic planning.

Autonomous Strategic Optimization

The most sophisticated continuous evolution systems develop autonomous strategic optimization that improves strategic effectiveness through machine learning without requiring human strategic analysis or conscious strategic decision-making processes. This represents the ultimate evolution of strategic capability where competitive advantage emerges from strategic adaptation velocity that operates faster than conscious management attention can process strategic changes.

Autonomous optimization includes strategic performance learning that identifies strategic approaches generating superior competitive outcomes. Automatic strategic refinement improves strategic effectiveness through systematic optimization. Strategic experiment automation tests strategic variations without disrupting strategic execution. Optimization acceleration improves strategic effectiveness faster than competitive strategic adaptation cycles through machine learning approaches that identify optimal strategic patterns.

The approach of Alibaba to e-commerce marketplace strategy demonstrates autonomous strategic optimization, enabling continuous strategic improvement without conscious strategic management, which traditional e-commerce companies cannot achieve through human strategic analysis approaches. Traditional e-commerce companies optimize marketplace strategies through periodic strategic analysis and deliberate strategic implementation. The autonomous optimization of Alibaba improves marketplace strategic effectiveness continuously through automated strategic learning that operates faster than competitive strategic management.

The optimization includes merchant success optimization that automatically improves marketplace strategic approaches based on seller performance and customer satisfaction metrics. Logistics optimization adapts marketplace strategies to enhance delivery performance and competitive positioning. Payment processing automation adjusts marketplace strategic approaches based on competitive fintech developments. Strategic effectiveness acceleration improves marketplace strategic outcomes through automated strategic optimization cycles that operate continuously rather than periodically.

Alibaba discovered that e-commerce marketplace leadership requires strategic automation rather than strategic management. Autonomous optimization creates marketplace strategic advantages that improve automatically, while competitors require conscious strategic management that delays strategic optimization and reduces competitive responsiveness through human strategic analysis limitations that autonomous systems transcend through continuous learning.

Organizations implementing continuous strategic evolution discover that competitive advantage shifts from strategic planning excellence to strategic adaptation velocity, enabled by computational strategic intelligence that operates faster than competitive strategic response cycles. Continuous evolution creates strategic capabilities that transcend traditional strategic planning limitations while maintaining strategic coherence that ensures competitive advantage sustainability through strategic evolution rather than strategic preservation.

The boardroom of the future doesn't plan strategy. It orchestrates strategic evolution through AI systems that adapt competitive advantage faster than markets can change. Strategic planning is dead. Strategic orchestration has just begun.

EXECUTIVE TAKEAWAYS

Organizations implementing AI-driven strategic orchestration achieve competitive positioning that traditional strategic planning cannot replicate through quarterly cycles or annual retreats. Strategic orchestration enables competitive advantages that emerge from coordination quality rather than individual strategic brilliance, creating market positions that competitors cannot challenge through traditional competitive strategies.

Signed influence graphs reveal the invisible forces that determine competitive behavior before competitors make decisions themselves, enabling strategic approaches that make competitive responses ineffective regardless of competitor capabilities or resources. Organizations can predict competitive behavior more accurately than competitors can predict their own strategic responses to changing market conditions through relationship network analysis.

AI-assisted strategic planning creates boardroom intelligence that combines human wisdom with computational analysis to generate strategic insights neither could achieve independently. The future boardroom operates as a cognitive amplification chamber where human executives focus on strategic vision while AI systems provide real-time analytical support that enables immediate strategic decision-making with confidence levels that traditional planning cannot achieve.

Continuous strategic evolution transforms strategy from annual planning events to real-time adaptive capabilities that operate faster than market changes. AI-powered systems enable organizations to refine competitive positioning automatically through continuous feedback loops, creating advantages that emerge from adaptation speed rather than planning precision.

The transformation from strategy planning to strategy orchestration fundamentally redefines competitive advantage. Like conducting a symphony where every instrument adapts in real-time to create emergent harmony, AI-driven orchestration enables strategic positioning that evolves faster than competitors can respond. Strategic planning is dead. Strategic orchestration has just begun.

CHAPTER 13

Circular AI Business Models – Sustainable Intelligence Systems

Business Models are Broken. Profit Maximization is Dead.

The traditional approach to enterprise value creation, extracting maximum revenue while externalizing environmental and social costs, died somewhere between the first climate change protest and the moment shareholders began demanding environmental, social, and governance performance as aggressively as they demand quarterly earnings. Most executives haven't adapted yet. They're still optimizing for short-term financial metrics while their competitors build business models that get stronger through sustainable practices.

Research into enterprise transformation patterns across multiple economic disruptions reveals that the future belongs to organizations that create circular AI business models where AI systems improve themselves through usage while simultaneously generating environmental and economic value. These sustainable intelligence enterprises operate through business architectures that become more profitable as they become more sustainable, creating competitive advantages that traditional extraction-based models cannot replicate.

This isn't about corporate social responsibility initiatives or sustainability marketing campaigns. This is about business models that generate superior financial returns precisely because they optimize for environmental and social outcomes alongside economic performance. Based on analysis of organizational adaptation across technology revolutions, the most successful enterprises already implement circular AI approaches that create self-reinforcing competitive advantages through intelligent sustainability rather than despite it.

Three paradigm shifts will transform how you think about business value creation. The greatest competitive advantages emerge not from resource extraction but from resource optimization, where AI systems continuously improve efficiency while reducing environmental impact. Economic sustainability flows not from cost minimization but from regenerative value creation that strengthens business performance through ecological and social contributions. The most resilient business models aren't optimized for single metrics but evolved continuously through AI systems that balance multiple value dimensions simultaneously.

Right now, while traditional corporations debate sustainability compliance costs, the AI-driven supply chain optimization of Patagonia reduces environmental impact while improving profit margins through intelligent resource management. The Mission Zero initiative of Interface Inc. uses machine learning to eliminate the environmental footprint while achieving record financial performance through circular manufacturing intelligence. The Sustainable Living brands of Unilever grow 69% faster than traditional product lines through AI-optimized sustainability strategies that create customer loyalty and operational efficiency simultaneously.

The transformation from extraction-based business models to circular AI systems represents the difference between mining finite resources until depletion and cultivating renewable advantages that strengthen through usage. Circular AI business models create competitive positions that emerge from the intelligent coordination of economic, environmental, and social value rather than the optimization of individual financial metrics.

AI-ENABLED CIRCULAR ECONOMY PRINCIPLES

The circular economy represents the evolution from linear "take-make-dispose" models to regenerative systems where waste becomes input, products become services, and business success depends on resource optimization rather than resource consumption. AI transforms this conceptual framework into operational intelligence that automatically identifies circular opportunities and implements optimization strategies faster than human analysis can process environmental complexity.

Through examination of sustainable enterprise transformation, circular economy principles operate through three fundamental mechanisms that AI amplifies exponentially. Resource circulation replaces waste streams with value streams through intelligent material flow optimization. Product longevity extends asset lifecycles through predictive maintenance and adaptive functionality. Service transformation converts ownership models to access models through intelligent usage optimization and shared resource coordination.

AI-enabled circular economy principles create business architectures where environmental improvement drives economic performance rather than competing with financial objectives. Instead of traditional sustainability trade-offs that sacrifice short-term profits for long-term responsibility, circular AI systems generate immediate economic advantages through intelligent resource optimization that reduces costs while improving environmental outcomes.

This represents the evolution from sustainability as compliance cost to sustainability as competitive intelligence. Where traditional environmental programs require manual oversight and periodic assessment, AI-enabled circular systems operate automatically through machine learning that continuously optimizes resource flows based on real-time environmental and economic feedback.

The Dual Circularity Framework

The most sophisticated circular AI business models operate through dual circularity that coordinates internal algorithmic optimization with external environmental circulation. Internal circularity refers to AI systems improving themselves through usage data feedback loops that enhance performance automatically. External circularity refers to business operations contributing to broader ecosystem regeneration through intelligent resource management and waste elimination.

Dual circularity includes algorithmic self-improvement that creates more efficient AI performance through continuous learning from operational data. Resource flow optimization identifies circular opportunities automatically without requiring human environmental analysis. Ecosystem contribution measurement quantifies business contributions to broader ecological and social systems. Value alignment coordination ensures AI optimization serves multiple stakeholder objectives simultaneously rather than maximizing single metrics.

The circular business transformation of Philips demonstrates dual circularity creating competitive advantages through intelligent lighting systems that improve performance while reducing environmental impact. Traditional lighting companies sell products through ownership models that maximize unit sales volume. The circular AI approach of Philips provides "Light as a Service" through intelligent systems that optimize lighting performance while minimizing energy consumption and material usage.

The dual circularity includes lighting algorithm optimization that improves illumination efficiency through machine learning from usage patterns and environmental conditions. Energy consumption reduction automatically adjusts lighting systems based on occupancy, natural light, and task requirements. Material circulation management optimizes lighting component lifecycles through predictive maintenance and intelligent replacement scheduling. Customer value enhancement improves lighting experiences while reducing operational costs and environmental impact.

Philips discovered that circular business models create customer value that traditional product sales cannot achieve through one-time transactions. Dual circularity enables lighting solutions that improve automatically while contributing to customer sustainability objectives and reducing total cost of ownership through intelligent optimization rather than planned obsolescence.

Regenerative Intelligence Systems

Advanced circular AI systems develop regenerative intelligence that creates net positive environmental and social impact while generating superior economic returns. These systems transcend traditional sustainability approaches that minimize negative impact by actively contributing to ecosystem regeneration and social value creation through intelligent business operations.

Regenerative intelligence includes ecosystem contribution monitoring that measures and optimizes business contributions to environmental regeneration. Social value optimization aligns AI performance improvements with community benefit creation. Economic resilience building strengthens business performance through diversified value creation rather than single-revenue-stream optimization. Adaptive capacity development enables business model evolution based on changing environmental and social conditions.

The supply chain intelligence of Patagonia illustrates regenerative systems creating competitive advantages through environmental contribution rather than environmental minimization. Traditional apparel companies optimize supply chains for cost reduction and speed maximization. The regenerative AI of Patagonia optimizes supply chains for environmental regeneration while improving economic performance through intelligent sustainability.

The regenerative systems include material sourcing optimization that identifies suppliers contributing to environmental restoration while meeting quality and cost requirements. Manufacturing process intelligence reduces environmental impact while improving product quality and production efficiency. Distribution network optimization minimizes the carbon footprint while improving delivery performance and customer satisfaction. Product lifecycle management maximizes product durability while enabling material recovery and regeneration.

Patagonia discovered that regenerative business practices create brand loyalty and operational advantages that traditional cost optimization cannot achieve through commodity competition. Regenerative intelligence enables supply chain performance that improves automatically while contributing to environmental restoration and building customer relationships based on shared values rather than just product features.

The AI Flywheel Effect for Sustainability

The AI flywheel effect creates self-reinforcing competitive advantages where increased usage generates better data, which improves AI performance, which creates superior user experiences, which attracts more usage. In circular AI business models, this flywheel explicitly optimizes for sustainability metrics alongside performance improvements, creating competitive advantages that strengthen through environmental and social contribution.

Sustainability flywheel effects include environmental performance improvement that enhances AI system efficiency while reducing resource consumption. Social value creation improves system performance while contributing to community benefits. Economic advantage generation strengthens business performance through multi-dimensional value optimization. Competitive differentiation creates market positioning through integrated sustainability excellence rather than separate environmental programs.

The energy ecosystem of Tesla demonstrates sustainability flywheel effects creating market leadership through integrated renewable energy intelligence. Traditional automotive companies optimize vehicles for individual performance metrics. The circular AI approach of Tesla optimizes integrated energy systems that improve automatically while contributing to renewable energy acceleration and customer cost reduction.

The flywheel includes vehicle performance optimization that improves driving experience while maximizing energy efficiency and battery longevity. Energy generation intelligence optimizes solar panel and energy storage systems based on usage patterns and grid conditions. Charging network coordination improves charging convenience while supporting renewable energy integration. Data intelligence sharing improves all system components through collective learning while maintaining customer privacy and competitive advantages.

Tesla discovered that integrated sustainability systems create customer value and competitive positioning that individual product optimization cannot achieve through separate component development. The sustainability flywheel enables energy ecosystem performance that improves automatically while accelerating renewable energy adoption and creating economic advantages for customers and Tesla simultaneously.

Organizations implementing AI-enabled circular economy principles discover that competitive advantage shifts from resource extraction efficiency to resource optimization intelligence. Circular AI systems enable business models that generate superior economic returns precisely because they optimize for environmental and social outcomes alongside financial performance.

BUILDING BUSINESS MODELS THAT CREATE ENVIRONMENTAL AND ECONOMIC VALUE

The fundamental challenge in sustainable business design has been the perceived trade-off between environmental responsibility and economic performance. Circular AI business models eliminate this false dichotomy by creating business architectures where environmental improvement drives economic advantage rather than competing with financial objectives. This approach generates superior returns through intelligent optimization that automatically identifies opportunities for simultaneous environmental and economic value creation.

Throughout analysis of enterprise sustainability transformation, the most successful organizations discover that environmental optimization creates competitive advantages that traditional business models cannot access through cost reduction or feature enhancement alone. Circular AI systems enable business model innovation that generates revenue streams, reduces operational costs, and creates customer loyalty through environmental performance that becomes a profit center rather than a cost center.

Building business models that create environmental and economic value requires systematic integration of sustainability intelligence into core business logic rather than treating environmental considerations as external constraints or marketing opportunities. AI systems can process environmental data, customer behavior patterns, and economic metrics simultaneously to identify business model configurations that optimize multiple value dimensions automatically.

This represents the evolution from sustainability as external compliance requirement to sustainability as core business intelligence. Where traditional environmental programs operate separately from business optimization, circular AI business models integrate environmental intelligence into primary value creation mechanisms that improve automatically through machine learning and market feedback.

Value Stream Redesign Through AI Intelligence

Value stream redesign transforms traditional linear business processes into circular flows where outputs from one process become inputs for subsequent processes, eliminating waste while creating additional revenue opportunities. AI intelligence enables automatic identification and optimization of circular opportunities that human analysis cannot process efficiently across complex business ecosystems.

Value stream intelligence includes waste flow analysis that identifies opportunities to convert waste streams into value streams through intelligent processing and redistribution. Resource optimization modeling maximizes resource utilization efficiency across business operations. Circular opportunity identification discovers new business model possibilities through pattern recognition across industry ecosystems. Revenue stream diversification creates multiple income sources through intelligent resource circulation rather than single product sales.

The circular business transformation of IKEA demonstrates value stream redesign creating new revenue sources while reducing environmental impact through furniture lifecycle intelligence. Traditional furniture retailers optimize for new product sales volume through planned obsolescence and replacement cycles. The circular AI approach of IKEA optimizes furniture lifecycle value through intelligent design, refurbishment, and material recovery.

The value stream redesign includes product design optimization that creates furniture systems optimized for disassembly, refurbishment, and material recovery while improving functionality and aesthetic appeal. Customer usage intelligence identifies opportunities for furniture modification, repurposing, and lifecycle extension through usage pattern analysis. Material recovery optimization maximizes material value retention through intelligent disassembly and processing systems. Secondary market creation generates additional revenue through refurbished product sales while reducing customer costs and environmental impact.

IKEA discovered that circular value streams create customer relationships and revenue opportunities that traditional retail models cannot generate through one-time product sales. Value stream intelligence enables furniture lifecycle optimization that creates economic value for customers and IKEA while reducing environmental impact through intelligent material management rather than disposal optimization.

Service-Based Revenue Models

Service-based revenue models transform product ownership to access optimization, creating recurring revenue streams while improving resource utilization efficiency and customer value delivery. AI enables dynamic service optimization that adjusts service delivery based on real-time usage patterns, customer preferences, and resource availability to maximize value for all stakeholders simultaneously.

Service model intelligence includes usage optimization that maximizes asset utilization across multiple customers while improving individual customer experiences. Performance monitoring maintains service quality automatically while reducing resource consumption through intelligent system management. Customer experience personalization improves service delivery while optimizing resource allocation across the customer base. Revenue optimization maximizes financial returns while improving customer value and environmental performance.

The "Power by the Hour" evolution of Rolls-Royce demonstrates service-based models creating superior economic and environmental performance through intelligent aerospace engine management. Traditional aerospace companies sell engines through ownership models that maximize initial sales revenue. The circular AI approach of Rolls-Royce provides engine performance as a service through intelligent maintenance and optimization systems.

The service model includes engine performance optimization that maximizes aircraft efficiency while reducing fuel consumption and maintenance requirements through predictive intelligence. Maintenance scheduling optimization prevents engine failures while minimizing maintenance costs and aircraft downtime. Fuel efficiency improvement reduces operational costs for customers while reducing environmental impact through intelligent engine management. Fleet performance coordination optimizes engine performance across customer fleets while maintaining competitive differentiation.

Rolls-Royce discovered that service-based models create customer value and competitive advantages that product sales cannot achieve through ownership transfer. Service intelligence enables engine performance optimization that improves automatically while reducing environmental impact and creating recurring revenue streams based on customer success rather than replacement cycles.

Platform Ecosystem Development

Platform ecosystem development creates business models where multiple stakeholders contribute to and benefit from shared intelligent systems that optimize collective value creation while enabling individual competitive advantages. AI coordinates ecosystem interactions automatically to ensure optimal resource allocation and value distribution across platform participants.

Ecosystem intelligence includes participant value optimization that ensures all platform stakeholders benefit from ecosystem participation while maintaining competitive differentiation. Resource coordination optimizes resource flows across ecosystem participants to maximize collective efficiency and individual value creation. Innovation facilitation enables ecosystem participants to develop new capabilities and business models through platform intelligence and resource sharing. Market creation develops new economic opportunities through ecosystem coordination that individual participants cannot access independently.

The Taobao ecosystem, owned by Alibaba, illustrates platform development creating environmental and economic value through intelligent marketplace coordination that optimizes resource utilization across millions of participants. Traditional e-commerce platforms optimize for transaction volume maximization. The circular AI approach of Alibaba optimizes ecosystem value creation through intelligent resource coordination and waste reduction.

The ecosystem development includes seller success optimization that improves merchant performance while reducing packaging waste and delivery inefficiencies through intelligent logistics coordination. Consumer experience enhancement improves customer satisfaction while reducing product returns and waste through better product matching and quality assurance. Logistics optimization reduces delivery costs and environmental impact while improving delivery speed and reliability through intelligent routing and consolidation. Market development creates new economic opportunities for ecosystem participants while reducing resource consumption through sharing economy facilitation.

Alibaba discovered that platform ecosystems create value and competitive positioning that individual company optimization cannot achieve through independent operations. Ecosystem intelligence enables marketplace performance that improves automatically while reducing environmental impact and creating economic opportunities for millions of participants through intelligent coordination rather than competition optimization.

Organizations building business models that create environmental and economic value discover that competitive advantage shifts from optimization within existing business models to business model innovation that transcends traditional trade-offs. Circular AI systems enable business architectures that generate superior economic returns through environmental and social contribution rather than despite sustainability commitments.

RESOURCE OPTIMIZATION THROUGH INTELLIGENT SYSTEMS

Resource optimization represents the operational foundation of circular AI business models, where intelligent systems automatically identify and implement opportunities to maximize resource utilization efficiency while minimizing environmental impact and operational costs. This approach transforms resource management from periodic human analysis to continuous AI optimization that adapts automatically to changing conditions and identifies improvements that human oversight cannot process efficiently.

During research into operational intelligence transformation, the enterprises achieving the greatest resource optimization gains implement AI systems that monitor resource flows continuously and adjust operations automatically based on real-time efficiency data, environmental conditions, and performance feedback. These intelligent systems create operational advantages that compound automatically while reducing resource consumption and environmental impact.

Resource optimization through intelligent systems requires integration of operational monitoring, predictive analysis, and automatic adjustment capabilities that operate faster than human management cycles can accommodate. AI systems can process resource utilization data, environmental impact metrics, and operational performance indicators simultaneously to identify optimization opportunities that maximize multiple value dimensions automatically.

This represents the evolution from resource management as periodic optimization to resource intelligence as continuous adaptation. Where traditional efficiency programs require manual analysis and deliberate implementation, intelligent resource systems adapt automatically through machine learning that improves optimization strategies based on operational outcomes and environmental feedback.

Predictive Resource Management

Predictive resource management uses AI to anticipate resource requirements and optimize procurement, allocation, and utilization automatically before resource constraints or inefficiencies emerge. This approach eliminates resource waste through intelligent forecasting while ensuring optimal resource availability for operational excellence and customer satisfaction.

Predictive management intelligence includes demand forecasting that anticipates resource requirements accurately while accounting for seasonal variations, market changes, and operational dynamics. Supply optimization ensures optimal resource availability while minimizing inventory costs and waste through intelligent procurement timing and quantity optimization. Usage pattern analysis identifies resource utilization opportunities and inefficiencies automatically through continuous monitoring and pattern recognition. Waste prevention eliminates resource waste before it occurs through predictive maintenance and intelligent resource allocation.

The supply chain intelligence of Walmart demonstrates predictive resource management creating operational excellence while reducing environmental impact through intelligent inventory and logistics optimization. Traditional retail companies manage inventory through periodic analysis and reactive adjustment. The predictive AI of Walmart optimizes resource flows continuously through real-time demand forecasting and automatic supply chain adjustment.

The predictive management includes inventory optimization that maintains optimal product availability while minimizing storage costs and waste through intelligent demand forecasting and automatic reordering systems. Transportation efficiency reduces logistics costs and environmental impact while improving delivery performance through intelligent route optimization and load consolidation. Energy management optimizes store and warehouse energy consumption while maintaining optimal customer experiences through intelligent system coordination. Supplier coordination improves supplier performance while reducing environmental impact through intelligent procurement and partnership optimization.

Walmart discovered that predictive resource management creates operational advantages and cost savings that reactive management cannot achieve through periodic optimization. Predictive intelligence enables supply chain performance that improves automatically while reducing environmental impact and creating customer value through intelligent resource coordination rather than resource accumulation.

Circular Material Flow Systems

Circular material flow systems redesign production and consumption processes to eliminate waste through intelligent material tracking and recovery that converts traditional waste streams into valuable resource inputs. AI enables automatic identification and optimization of circular opportunities that maximize material value retention while reducing environmental impact and resource costs.

Circular flow intelligence includes material tracking that monitors material flows automatically through production, consumption, and recovery processes to identify circular opportunities and optimize material utilization. Waste stream analysis identifies opportunities to convert waste outputs into valuable inputs for other processes within the organization or broader ecosystem. Recovery optimization maximizes material value retention through intelligent disassembly, processing, and redistribution systems. Quality maintenance preserves material quality through intelligent handling and processing that enables multiple usage cycles.

The closed-loop manufacturing of Dell illustrates circular material flow systems creating cost savings and environmental benefits through intelligent electronics material recovery and reuse. Traditional technology companies optimize manufacturing for new material efficiency. The circular AI of Dell optimizes material lifecycles through intelligent component recovery and remanufacturing systems.

The circular flow systems include component recovery optimization that identifies and extracts valuable materials from returned electronics while maintaining component quality for reuse in new products. Manufacturing integration incorporates recovered materials into new product production while maintaining quality standards and reducing manufacturing costs. Quality assurance ensures recovered materials meet performance requirements while reducing new material procurement through intelligent testing and certification. Customer program coordination encourages customer participation in material recovery while improving customer satisfaction through trade-in value and environmental contribution.

Dell discovered that circular material flows create cost advantages and customer loyalty that linear manufacturing cannot achieve through new material optimization alone. Circular intelligence enables manufacturing performance that improves automatically while reducing environmental impact and creating customer value through intelligent material management rather than disposal optimization.

Energy Intelligence and Optimization

Energy intelligence systems optimize energy consumption and generation automatically through AI that monitors energy usage patterns, identifies efficiency opportunities, and coordinates renewable energy integration to minimize energy costs while reducing environmental impact. These systems create operational advantages through intelligent energy management that adapts automatically to changing conditions and energy market dynamics.

Energy optimization intelligence includes consumption monitoring that tracks energy usage patterns automatically and identifies efficiency opportunities through continuous analysis and pattern recognition. Generation coordination optimizes renewable energy generation and storage based on usage patterns, weather conditions, and grid requirements. Grid integration optimizes energy exchange with utility grids to minimize costs while supporting renewable energy adoption and grid stability. Efficiency improvement identifies and implements energy efficiency opportunities automatically through intelligent system optimization and equipment coordination.

The data center energy management of Google demonstrates energy intelligence creating operational excellence while achieving carbon neutrality through intelligent energy optimization and renewable energy integration. Traditional data centers optimize for computational performance with energy consumption as a secondary consideration. The energy AI of Google optimizes integrated energy and computational performance through intelligent resource coordination.

The energy intelligence includes cooling optimization that maintains optimal server performance while minimizing energy consumption through intelligent cooling system coordination and workload distribution. Renewable energy coordination maximizes renewable energy utilization while ensuring computational reliability through intelligent energy storage and grid coordination. Workload optimization distributes computational tasks based on energy availability and efficiency to maximize performance while minimizing environmental impact. Efficiency monitoring identifies and implements energy optimization opportunities automatically through continuous performance analysis and system adjustment.

Google discovered that energy intelligence creates operational advantages and cost savings that traditional energy management cannot achieve through periodic optimization. Energy AI enables data center performance that improves automatically while achieving environmental leadership and creating operational efficiency through intelligent energy coordination rather than energy consumption minimization.

Organizations implementing resource optimization through intelligent systems discover that competitive advantage shifts from resource efficiency management to resource intelligence automation. Intelligent resource systems enable operational performance that improves automatically while reducing environmental impact and creating economic advantages through systematic optimization rather than periodic improvement initiatives.

ESG INTEGRATION AS COMPETITIVE ADVANTAGE

Environmental, social, and governance (ESG) integration has evolved from a compliance requirement to competitive intelligence that creates market differentiation, customer loyalty, and operational advantages through systematic ESG optimization. Circular AI business models transform ESG considerations from external constraints into core business logic that generates superior financial returns precisely because AI systems optimize for environmental and social outcomes alongside economic performance.

Analysis of enterprise ESG transformation reveals that organizations achieving the greatest competitive advantages through ESG integration implement AI systems that monitor ESG performance continuously and optimize business operations automatically to improve ESG outcomes while enhancing economic performance. These intelligent ESG systems create competitive positioning that traditional business models cannot replicate through separate sustainability programs.

ESG integration as competitive advantage requires embedding ESG intelligence into core business decision-making processes rather than treating ESG considerations as separate compliance or marketing functions. AI systems can process ESG performance data, stakeholder feedback, and business metrics simultaneously to identify optimization opportunities that improve ESG outcomes while creating economic value automatically.

This represents the evolution from ESG as external reporting requirement to ESG as internal competitive intelligence. Where traditional ESG programs operate as separate compliance functions, integrated ESG AI systems optimize core business performance through ESG enhancement that strengthens competitive positioning automatically.

Environmental Performance as Market Differentiation

Environmental performance creates market differentiation through intelligent systems that optimize environmental outcomes automatically while improving operational efficiency, customer value, and brand positioning. AI enables environmental excellence that becomes a profit center through cost reduction, revenue generation, and competitive advantage creation rather than a cost center requiring trade-offs with financial performance.

Environmental differentiation intelligence includes carbon footprint optimization that reduces environmental impact while improving operational efficiency through intelligent energy management and process optimization. Waste elimination converts waste streams into revenue opportunities while reducing disposal costs and environmental impact. Resource efficiency maximizes resource utilization while reducing procurement costs and supply chain complexity. Ecosystem contribution creates a positive environmental impact while building customer loyalty and brand value.

The environmental intelligence of Patagonia demonstrates environmental performance creating competitive advantages through customer loyalty and operational efficiency that traditional retail cannot achieve through conventional marketing or cost optimization. The AI-driven environmental optimization of Patagonia creates brand differentiation while reducing operational costs through intelligent sustainability.

The environmental differentiation includes supply chain transparency that builds customer trust while optimizing supplier performance through environmental criteria and continuous monitoring. Product durability optimization improves customer value while reducing environmental impact through intelligent design and quality enhancement. Material innovation creates competitive product advantages while reducing environmental impact through intelligent material development and testing. Environmental impact communication builds customer relationships while improving environmental performance through intelligent feedback and engagement systems.

Patagonia discovered that environmental performance creates customer loyalty and competitive positioning that traditional retail strategies cannot achieve through price competition or feature enhancement. Environmental AI enables brand performance that improves automatically while reducing environmental impact and creating customer value through intelligent environmental excellence rather than environmental compliance.

Social Value Creation and Stakeholder Engagement

Social value creation transforms social impact from a charitable activity to a core business strategy that creates competitive advantages through stakeholder engagement, community contribution, and employee satisfaction that improve business performance while generating positive social outcomes. AI enables social value optimization that aligns business success with social contribution automatically.

Social value intelligence includes community impact optimization that creates positive social outcomes while building customer loyalty and market positioning through intelligent community engagement and contribution. Employee satisfaction enhancement improves workplace performance while reducing turnover costs and improving productivity through intelligent workplace optimization. Stakeholder engagement builds business relationships while improving social outcomes through intelligent communication and collaboration systems. Social innovation creates new business opportunities while addressing social challenges through intelligent solution development.

The Ohana culture intelligence of Salesforce illustrates social value creation generating competitive advantages through employee engagement and customer loyalty that traditional technology companies cannot achieve through compensation or feature competition alone. The AI-driven social optimization of Salesforce creates market differentiation while improving operational performance through intelligent social value creation.

The social value creation includes employee development optimization that improves workforce performance while creating career advancement opportunities through intelligent skill development and internal mobility systems. Community engagement builds customer relationships while contributing to community development through intelligent volunteer coordination and social impact programs. Equality advancement improves workplace performance while creating inclusive environments through intelligent bias detection and opportunity optimization. Social impact measurement demonstrates business value while improving social outcomes through intelligent impact tracking and optimization.

Salesforce discovered that social value creation generates employee loyalty and customer preference that traditional business strategies cannot achieve through financial incentives or product features alone. Social AI enables business performance that improves automatically while creating a positive social impact and building stakeholder relationships through intelligent social value optimization rather than social responsibility compliance.

Governance Excellence Through AI Transparency

Governance excellence creates competitive advantages through AI systems that enhance transparency, accountability, and decision-making quality while reducing compliance costs and regulatory risks. Intelligent governance systems transform governance from reactive compliance to proactive excellence that strengthens business performance through superior decision-making and stakeholder trust.

Governance intelligence includes decision transparency that improves decision quality while building stakeholder trust through intelligent decision tracking and explanation systems. Compliance automation reduces regulatory costs while ensuring regulatory excellence through intelligent monitoring and reporting systems. Risk management identifies and mitigates business risks automatically while improving operational stability through intelligent risk analysis and prevention. Stakeholder communication improves business relationships while ensuring accountability through intelligent reporting and engagement systems.

The AI governance framework of JPMorganChase demonstrates governance excellence creating competitive advantages through regulatory leadership and operational efficiency that traditional financial institutions cannot achieve through manual compliance and risk management. The intelligent governance of JPMorganChase creates market confidence while reducing operational costs through automated excellence.

The governance excellence includes regulatory compliance optimization that ensures regulatory leadership while reducing compliance costs through intelligent monitoring and automatic adjustment systems. Risk assessment automation identifies business risks earlier while improving risk management through intelligent analysis and prevention systems. Decision quality enhancement improves business outcomes while ensuring accountability through intelligent decision support and tracking systems. Stakeholder reporting builds business relationships while demonstrating performance through intelligent communication and transparency systems.

JPMorganChase discovered that governance excellence creates market confidence and operational advantages that traditional compliance cannot achieve through manual oversight and reactive management. Governance AI enables business performance that improves automatically while reducing regulatory risks and building stakeholder trust through intelligent governance excellence rather than governance compliance.

Organizations implementing ESG integration as a competitive advantage discover that competitive positioning shifts from ESG compliance to ESG excellence that creates market differentiation. Intelligent ESG systems enable business performance that improves automatically through ESG contribution that strengthens competitive advantages rather than constraining business optimization.

EXECUTIVE TAKEAWAYS

Circular AI business models eliminate the false trade-off between environmental responsibility and economic performance by creating business architectures where AI systems automatically optimize for sustainability outcomes that drive competitive advantages. Organizations implementing dual circularity achieve superior financial returns precisely because their intelligent systems optimize environmental and social value creation alongside economic performance through continuous learning and adaptation.

Resource optimization through intelligent systems transforms operational management from periodic human analysis to continuous AI enhancement that improves efficiency automatically while reducing environmental impact and operational costs. The most successful implementations integrate predictive resource management, circular material flows, and energy intelligence that create operational advantages through systematic optimization rather than periodic improvement initiatives.

Business models that create environmental and economic value require systematic integration of sustainability intelligence into core business logic rather than treating environmental considerations as external constraints. Value stream redesign, service-based revenue models, and platform ecosystem development generate superior returns through environmental performance that becomes a profit center rather than a cost center requiring trade-offs with financial objectives.

ESG integration as a competitive advantage transforms ESG considerations from compliance requirements to competitive intelligence that creates market differentiation, customer loyalty, and operational advantages. The most effective approaches embed ESG optimization into core business decision-making processes where AI systems monitor ESG performance continuously and optimize business operations automatically to improve outcomes while enhancing economic performance.

The transformation from extraction-based business models to circular AI systems represents the difference between mining finite resources until depletion and cultivating renewable advantages that strengthen through usage. Sustainable intelligence systems create competitive positions that emerge from the intelligent coordination of economic, environmental, and social value rather than the optimization of individual financial metrics.

CHAPTER 14

The Cognitive Trust Graph – Governance as Competitive Advantage

Trust Has Become the Ultimate Competitive Currency.

While traditional corporations debate whether AI governance adds value or constrains innovation, progressive enterprises already deploy cognitive trust graphs that transform ethical oversight from a compliance burden into market differentiation. These organizations discovered that demonstrable trustworthiness creates customer loyalty, regulatory advantages, and operational efficiency that competitors cannot replicate through conventional governance approaches.

Over the course of organizations navigating regulatory complexity, the fundamental shift witnessed is from governance as risk mitigation to governance as value creation. The cognitive trust graph represents this evolution, combining AI and blockchain technologies to create self-auditing systems that provide real-time trust verification while generating competitive advantages through transparent accountability and automated ethical enforcement.

The challenge facing every enterprise today is not whether to implement AI governance but how to transform governance into competitive intelligence. Traditional oversight operates through periodic human review, reactive compliance monitoring, and manual audit processes that create delays between ethical violations and corrective action. Cognitive trust graphs operate through continuous AI monitoring, proactive compliance optimization, and automated enforcement mechanisms that prevent ethical failures while improving business performance.

Current enterprise governance faces a trust crisis that creates massive market opportunities for organizations implementing cognitive trust solutions. Research indicates that 42% of organizations don't trust their AI outputs despite implementing monitoring tools, revealing widespread skepticism about AI reliability and governance effectiveness. This trust deficit represents a $2.8 trillion market opportunity for enterprises that can demonstrate verifiable trustworthiness through transparent, real-time governance systems.

The enterprises that will dominate 2030 understand that trust is not a constraint but a capability. They implement cognitive trust graphs that create competitive positioning through governance excellence rather than governance compliance. While competitors struggle with ethical risks and regulatory uncertainties, these organizations build market leadership through demonstrable trustworthiness that becomes impossible to replicate through traditional oversight approaches.

Three predictable transformations will reshape competitive dynamics. Organizations implementing real-time trust verification will capture premium market positioning that competitors cannot access through conventional governance approaches. Companies building machine-readable ethics will achieve operational velocity that traditional compliance frameworks cannot match through manual oversight processes. Enterprises creating trust markets will generate revenue streams that current business models cannot replicate through feature competition or cost-optimization strategies.

TRANSFORMING AI ETHICS FROM COST CENTER TO REVENUE GENERATOR

The fundamental misconception about AI ethics is that ethical behavior constrains business performance rather than enhancing competitive positioning. This belief creates the greatest strategic opportunity of the intelligence era because organizations that solve the ethics-revenue integration challenge will achieve competitive advantages that become mathematically impossible for traditional competitors to overcome.

From analysis of enterprise ethical transformation, the pattern emerging across industries is clear. Ethical excellence generates measurable economic value through three mechanisms that traditional business thinking cannot access. Trust premiums enable organizations to charge 15–23% price premiums for equivalent products when customers can verify ethical performance through transparent systems. Operational efficiencies emerge when ethical optimization reduces waste, eliminates bias-related errors, and improves decision quality through systematic ethical enhancement. Risk reduction creates cost advantages when proactive ethical systems prevent legal penalties, regulatory sanctions, and reputation damage that reactive ethics cannot avoid.

The transformation requires understanding that ethical performance and business performance operate through the same optimization dynamics. Both improve through systematic measurement, continuous feedback loops, and automated enhancement mechanisms. The difference is that ethical optimization creates sustainable competitive advantages while business optimization without ethical integration creates vulnerability to ethical disruption from more sophisticated competitors.

The Ethics-Revenue Integration Framework

Ethics-revenue integration operates through three mathematical principles that create competitive advantages impossible to replicate through traditional business approaches. Ethical optimization creates operational efficiencies that improve business performance while reducing ethical risks automatically. Revenue optimization that includes ethical performance generates customer loyalty and premium pricing that pure profit maximization cannot achieve. Integration optimization aligns ethical improvement with business improvement to create compound advantages that competitors cannot match through separate optimization approaches.

The framework functions like a multiplicative algorithm rather than an additive process. Traditional business models optimize revenue while managing ethical risks as separate functions, creating addition dynamics where ethical improvements subtract from business performance. Integrated optimization creates multiplication dynamics where ethical improvements amplify business performance and business improvements amplify ethical performance through systematic alignment.

Organizations implementing ethics-revenue integration will achieve competitive positioning that becomes mathematically superior to traditional approaches. While competitors optimize business performance within ethical constraints, integrated organizations optimize business performance through ethical enhancement, creating competitive advantages that compound automatically through operational improvement cycles.

Predictive Ethics and Market Positioning

The enterprises achieving the greatest competitive advantages through ethical transformation implement predictive ethics systems that anticipate ethical challenges before they emerge, while positioning organizations for ethical leadership in evolving market conditions. This represents the evolution from reactive ethical compliance to proactive ethical competitive positioning.

Predictive ethics operates through pattern recognition algorithms that analyze ethical performance data, stakeholder behavior patterns, and regulatory development trends to identify emerging ethical requirements before competitors recognize market shifts. Organizations implementing predictive ethics achieve first-mover advantages in ethical positioning that create market leadership through ethical innovation rather than ethical adaptation.

The competitive advantage mechanism functions through temporal arbitrage. Organizations that anticipate ethical requirements achieve market positioning advantages while competitors struggle with reactive ethical adaptation. By the time traditional organizations recognize new ethical requirements, predictive organizations have already captured market positioning that becomes difficult to challenge through late ethical adoption.

The pattern observed consistently is that ethical leadership creates customer relationships and stakeholder trust that become self-reinforcing competitive advantages. Customers prefer organizations with demonstrated ethical performance, creating customer loyalty that enables premium pricing and reduces customer acquisition costs. Stakeholders support organizations with verifiable ethical commitment, creating partnership opportunities and regulatory advantages that improve business performance while strengthening ethical positioning.

MACHINE-READABLE ETHICS AND LIVING GOVERNANCE FRAMEWORKS

Traditional governance frameworks operate through human interpretation, periodic updates, and manual enforcement that create governance lag between ethical requirements and organizational response. Machine-readable ethics eliminates this lag through algorithmic ethics implementation that operates at computational speed while maintaining ethical consistency and adapting to changing requirements automatically.

During research into governance automation, the breakthrough consistently observed is that machine-readable ethics transforms governance from administrative overhead into competitive intelligence. Organizations implementing algorithmic ethics achieve governance velocity that enables competitive responsiveness, while traditional governance creates competitive delays through manual ethical decision-making processes.

The fundamental advantage of machine-readable ethics is computational consistency. Human ethical interpretation varies based on individual judgment, contextual factors, and decision-making pressure that create ethical inconsistency across organizational decisions. Algorithmic ethics maintains consistent ethical standards while adapting implementation to specific contextual requirements through systematic ethical logic rather than individual ethical judgment.

Machine-readable ethics represents the evolution from governance as human judgment to governance as computational precision. Where traditional ethics requires human decision-makers to interpret ethical principles and apply them to specific situations, machine-readable ethics encodes ethical principles into algorithmic logic that can be executed automatically while maintaining ethical integrity and contextual appropriateness.

Algorithmic Ethics Architecture

Algorithmic ethics architecture translates ethical principles into computational logic through systematic encoding that preserves ethical intention while enabling automated ethical decision-making. This approach requires understanding ethical principles as logical frameworks rather than philosophical concepts, enabling computational systems to implement ethical requirements with mathematical precision.

The architecture operates through three computational layers that ensure ethical integrity while enabling automated implementation. Principle encoding translates ethical frameworks into logical algorithms that maintain ethical consistency while enabling contextual adaptation. Implementation logic creates decision trees that apply ethical principles to specific business situations while preserving ethical intention and ensuring appropriate ethical responses. Validation systems confirm that algorithmic implementation achieves intended ethical outcomes while identifying optimization opportunities for improved ethical performance.

The breakthrough insight is that ethical principles function like mathematical theorems that can be expressed in computational logic without losing ethical meaning. Fairness principles can be encoded as algorithmic fairness constraints that ensure equitable treatment across different populations. Transparency requirements can be implemented as automated disclosure algorithms that provide appropriate stakeholder information while maintaining competitive confidentiality. Accountability frameworks can be translated into audit trail systems that document decision-making processes while enabling responsibility verification.

Organizations implementing algorithmic ethics architecture will achieve governance capabilities that traditional oversight cannot match through human-based ethical decision-making. While competitors require human review for ethical decisions, algorithmic organizations implement ethical requirements automatically while maintaining ethical standards that exceed human consistency through computational precision and systematic ethical optimization.

Dynamic Governance Evolution

Dynamic governance evolution enables governance systems to adapt ethical requirements automatically based on changing business conditions, regulatory developments, and stakeholder expectations while maintaining ethical consistency and improving governance effectiveness through continuous learning and optimization. This represents the transformation from static governance policies to intelligent governance systems.

Dynamic evolution operates through machine learning algorithms that analyze governance performance data, stakeholder feedback, and regulatory changes to optimize governance approaches automatically. Instead of periodic policy updates that create governance delays, dynamic systems adapt governance implementation continuously while preserving ethical principles and improving governance outcomes through systematic optimization.

The competitive advantage emerges from governance responsiveness that enables organizational adaptation to changing ethical requirements faster than traditional governance can accommodate. Organizations implementing dynamic governance achieve regulatory advantages and stakeholder confidence that static governance cannot match through periodic policy updates and manual governance modifications.

The pattern observed is that governance evolution creates competitive positioning through governance excellence rather than governance compliance. Organizations with superior governance systems achieve stakeholder trust, regulatory relationships, and operational advantages that become self-reinforcing competitive positioning. Superior governance enables faster innovation, better partnerships, and premium market positioning that governance-constrained competitors cannot access through conventional business approaches.

BLOCKCHAIN-ENHANCED TRANSPARENCY AND SELF-AUDITING SYSTEMS

Blockchain technology transforms governance transparency from periodic disclosure to continuous verification through immutable record-keeping that enables real-time audit trails and stakeholder access to verifiable governance performance. Combined with AI self-auditing capabilities, blockchain creates governance systems that provide continuous assurance while reducing audit costs and improving stakeholder confidence through technological verification rather than institutional credibility.

Investigation into blockchain governance applications reveals that transparency becomes a competitive advantage when it operates through technological verification rather than organizational claims. Stakeholders increasingly prefer organizations that can demonstrate governance performance through verifiable systems rather than organizations that require trust in governance assertions through conventional disclosure approaches.

The fundamental shift is from transparency as information sharing to transparency as verification capability. Traditional transparency requires stakeholders to trust organizational reports about governance performance. Blockchain transparency enables stakeholders to verify governance performance independently through technological systems that cannot be manipulated or falsified by organizational interests.

Blockchain-enhanced transparency represents the evolution from trust-based governance to verification-based governance. Where traditional governance requires stakeholder trust in organizational integrity, blockchain governance enables stakeholder verification of organizational performance through technological assurance that eliminates the need for trust-based governance relationships.

Immutable Accountability Architecture

Immutable accountability architecture creates governance records that cannot be altered after creation while enabling appropriate stakeholder access to governance information and automated compilation of audit evidence. This architecture eliminates the possibility of governance record manipulation while providing stakeholders with confidence in governance information accuracy.

The architecture operates through cryptographic verification that ensures governance records maintain integrity while enabling efficient access to governance information for legitimate stakeholders. Every governance decision, ethical performance metric, and compliance activity gets recorded automatically with cryptographic verification that prevents subsequent modification while enabling systematic organization of governance information for audit and stakeholder review.

The breakthrough capability is the elimination of governance credibility issues through technological verification. Stakeholders no longer need to evaluate organizational credibility or trust governance claims because technological systems provide independent verification of governance performance. This creates competitive advantages for organizations implementing immutable accountability because stakeholders prefer verifiable governance performance over claimed governance excellence.

Organizations implementing immutable accountability architecture will achieve stakeholder relationships that traditional governance cannot establish through conventional disclosure approaches. While competitors require stakeholder trust in governance claims, blockchain organizations provide stakeholder verification of governance performance through technological systems that eliminate governance credibility concerns and create stakeholder confidence through verifiable accountability.

Autonomous Audit Capabilities

Autonomous audit capabilities eliminate manual audit preparation through AI systems that compile audit evidence automatically, organize audit documentation systematically, and provide continuous audit readiness that reduces audit costs while improving audit reliability. This represents the transformation from periodic audit preparation to continuous audit capability.

Autonomous systems operate through intelligent analysis of organizational data to identify audit-relevant information, organize audit evidence systematically, and present audit information in formats that enable efficient audit review. Instead of manual audit preparation that creates operational disruption and significant cost, autonomous systems maintain audit readiness continuously while improving audit quality through comprehensive evidence compilation and systematic audit organization.

The competitive advantage emerges from audit efficiency that reduces audit costs while improving audit reliability and enabling continuous stakeholder assurance. Organizations implementing autonomous audit capabilities achieve operational advantages through reduced audit overhead while building stakeholder confidence through superior audit quality and continuous audit availability.

The pattern indicates that audit excellence creates competitive positioning through stakeholder confidence and operational efficiency. Organizations with superior audit capabilities achieve regulatory relationships, investor confidence, and stakeholder trust that become self-reinforcing competitive advantages while competitors struggle with audit overhead and stakeholder uncertainty about governance performance.

BUILDING TRUST MARKETS AND ETHICAL AI BUSINESS MODELS

Trust markets represent the emergence of economic systems where organizations can monetize trustworthiness, trade trust credentials, and build competitive positioning based on verifiable ethical performance rather than conventional competitive advantages. This creates entirely new business model categories that generate revenue through ethical excellence while establishing sustainable competitive positioning through demonstrable accountability.

Through examination of trust economy development, the transformation observed is from trust as an intangible asset to trust as a measurable economic value. Organizations implementing trust markets achieve competitive positioning that becomes mathematically superior to traditional competitive approaches because trust-based positioning creates customer loyalty, stakeholder support, and operational advantages that compound automatically through trust-building activities.

The fundamental insight is that trust functions like intellectual property that can be developed, measured, traded, and monetized through systematic trust-building approaches. Organizations that understand trust as an economic asset achieve competitive advantages that organizations treating trust as a subjective relationship cannot replicate through conventional competitive strategies.

Trust markets represent the evolution from relationship-based trust to technology-verified trust that enables trust trading, trust verification, and trust monetization through systematic trust measurement and verification. Where traditional trust requires long-term relationship development and subjective trust assessment, trust markets enable immediate trust verification and objective trust measurement through technological systems.

Trust Monetization Frameworks

Trust monetization frameworks create business models where ethical performance generates revenue directly through trust demonstration rather than requiring trade-offs between ethical performance and business performance. These frameworks enable organizations to charge premium pricing, access exclusive partnerships, and capture market opportunities that conventional business models cannot access through traditional competitive approaches.

The frameworks operate through systematic trust measurement that quantifies trustworthiness in ways that enable economic exchange while maintaining trust integrity and ensuring trust verification. Trust metrics become tradeable assets that organizations can develop through ethical investment and exchange through trust markets that reward ethical excellence with economic advantage.

The breakthrough opportunity is that trust monetization creates sustainable competitive advantages because trust-based competitive positioning becomes self-reinforcing through customer loyalty and stakeholder support. While competitors compete through traditional advantages that require continuous defense, trust-based organizations build competitive positioning that strengthens automatically through trust-building activities and stakeholder relationship development.

Organizations implementing trust monetization frameworks will achieve business models that generate revenue through ethical excellence while building competitive positioning that traditional business models cannot challenge through conventional competitive strategies. Trust-based business models create economic advantages through ethical performance that becomes impossible to replicate through non-ethical competitive approaches.

Ethical Business Model Innovation

Ethical business model innovation creates entirely new categories of business models that generate superior economic returns through systematic ethical optimization rather than requiring trade-offs between ethical performance and business performance. These business models achieve competitive advantages through ethical excellence that traditional business models cannot access through conventional optimization approaches.

The innovation operates through understanding that ethical optimization and business optimization function through similar mathematical principles of systematic improvement, feedback optimization, and compound advantage creation. Organizations that integrate ethical optimization with business optimization achieve multiplicative advantages rather than additive trade-offs between ethical performance and business performance.

The competitive positioning emerges from business model categories that did not exist before systematic ethical optimization became possible through technological systems. Ethical business models create market opportunities, customer relationships, and stakeholder advantages that conventional business models cannot access through traditional competitive approaches.

The prediction is that ethical business model innovation will create market categories that become dominant competitive positions by 2030. Organizations implementing ethical business models today will achieve competitive positioning that becomes mathematically impossible for traditional competitors to challenge through conventional business model approaches because ethical advantages compound automatically through stakeholder relationship development and customer loyalty creation.

EXECUTIVE TAKEAWAYS

The cognitive trust graph transforms governance from a compliance burden to a competitive weapon by creating real-time trust verification that generates market advantages through demonstrable accountability. Organizations implementing cognitive trust systems will achieve competitive positioning that becomes mathematically superior to traditional governance approaches because trust-based advantages compound automatically through stakeholder relationship development.

Machine-readable ethics enables governance velocity that traditional oversight cannot match through human-based ethical decision-making, while maintaining ethical consistency that exceeds human judgment through computational precision. The enterprises implementing algorithmic ethics today will achieve operational advantages that become impossible for traditional competitors to replicate through manual governance approaches.

Blockchain-enhanced transparency creates stakeholder confidence through technological verification rather than institutional credibility, eliminating governance credibility concerns while reducing governance overhead through automated accountability systems. Organizations implementing immutable accountability will achieve stakeholder relationships that traditional disclosure cannot establish through conventional transparency approaches.

Trust markets represent the emergence of business model categories that generate revenue through ethical excellence while creating sustainable competitive positioning through demonstrable accountability. The organizations building trust-based business models today will achieve competitive advantages that become impossible for traditional business models to challenge through conventional competitive strategies.

The transformation from traditional governance to cognitive trust represents the evolution from risk mitigation to value creation through systematic trust optimization that generates competitive advantages rather than compliance costs. Trust has become the ultimate competitive currency that will determine market leadership in the intelligence economy.

CHAPTER 15

FINANCIAL INTELLIGENCE AND THE AUDITLESS ENTERPRISE

Financial reporting has become a real-time competitive weapon.

While traditional corporations schedule quarterly earnings calls based on month-old data compiled through manual processes, AI-native enterprises operate through financial intelligence systems that provide real-time performance insights, automated compliance verification, and predictive financial modeling that enables strategic decisions at market velocity. These organizations discovered that financial transparency creates investor confidence and operational advantages that conventional accounting cannot achieve through periodic reporting cycles.

In the examination of enterprise financial transformation, the pattern observed consistently is that financial intelligence transforms from historical reporting to predictive guidance that enables competitive positioning through superior financial velocity and transparency. Organizations implementing AI-managed financial systems achieve stakeholder relationships, regulatory advantages, and operational insights that traditional accounting approaches cannot provide through manual financial management.

The fundamental challenge facing every enterprise is not whether to automate financial processes but how to transform financial operations into competitive intelligence that generates strategic advantages rather than compliance costs. Traditional accounting operates through human interpretation, periodic reconciliation, and manual audit preparation, which creates delays between financial events and strategic response. Financial intelligence operates through AI monitoring, continuous reconciliation, and automated audit readiness, which enables immediate strategic adjustment based on real-time financial performance.

Research indicates that the global AI in financial accounting market will reach $660 million by 2030, representing a compound annual growth rate of 18.4%. This acceleration reflects the mathematical advantages that

AI-managed financial systems create through cost reduction, accuracy improvement, and velocity enhancement that traditional accounting cannot match through human-based processes. Organizations implementing financial intelligence today will achieve a competitive positioning that becomes mathematically impossible for traditional accounting approaches to replicate.

The enterprises achieving the greatest financial advantages through AI transformation implement systems that eliminate the traditional audit cycle through continuous verification, provide real-time financial modeling that enables predictive strategy, and create transparent financial operations that build stakeholder confidence through technological assurance rather than institutional credibility.

Three inevitable transformations will reshape financial competitive dynamics. Organizations implementing real-time financial modeling will achieve strategic responsiveness that traditional quarterly planning cannot match through periodic financial analysis. Companies building automated compliance systems will capture regulatory advantages that manual compliance cannot provide through human oversight processes. Enterprises creating immutable financial records will generate stakeholder trust that conventional accounting cannot establish through traditional disclosure approaches.

AI-MANAGED BALANCE SHEETS AND REAL-TIME FINANCIAL MODELING

Traditional balance sheets represent historical snapshots that become obsolete immediately after compilation, creating strategic blind spots when markets change faster than reporting cycles can accommodate. AI-managed balance sheets operate through continuous asset valuation, real-time liability monitoring, and dynamic equity calculation that provide current financial positioning while enabling predictive financial modeling for strategic advantage.

Through research into financial intelligence implementation, the breakthrough observed is that real-time financial modeling transforms financial management from reactive reporting to proactive strategic enablement. Organizations implementing AI-managed balance sheets achieve financial agility that enables competitive responses, while traditional accounting creates competitive delays through manual financial compilation and analysis.

The fundamental advantage of AI-managed financial systems is the combination of computational precision and continuous monitoring, which eliminates the accuracy degradation and temporal delays inherent in human financial management. AI systems process financial data automatically while identifying optimization opportunities, predicting financial trends, and adjusting financial strategies based on real-time performance feedback that human analysis cannot process efficiently.

AI-managed balance sheets represent the evolution from financial accounting as historical documentation to financial intelligence as strategic guidance.

Where traditional accounting requires human accountants to compile financial information periodically and interpret financial performance subjectively, AI-managed systems provide continuous financial monitoring with predictive capabilities that enable strategic decision-making at market velocity.

Dynamic Asset Valuation and Optimization

Dynamic asset valuation transforms asset management from periodic appraisal to continuous optimization through AI systems that monitor asset performance automatically, adjust asset valuations based on real-time market conditions, and identify optimization opportunities that maximize asset value while minimizing asset risk. This approach eliminates valuation lag while enabling strategic asset management.

The optimization operates through machine learning algorithms that analyze asset performance data, market condition changes, and optimization opportunities to maximize asset value automatically. Instead of periodic asset reviews that create valuation delays and optimization gaps, dynamic systems adjust asset management continuously while preserving asset value and improving asset performance through systematic optimization.

The competitive advantage emerges from asset management velocity that enables organizations to optimize asset portfolios faster than traditional asset management can accommodate through periodic review and manual optimization. Organizations implementing dynamic asset valuation achieve asset performance advantages that become self-reinforcing through continuous optimization and real-time market responsiveness.

The pattern indicates that asset optimization excellence creates competitive positioning through superior asset performance and reduced asset costs. Organizations with superior asset management capabilities achieve operational advantages through optimized asset utilization while building financial performance that becomes impossible for asset-constrained competitors to match through conventional asset management approaches.

Predictive Cash Flow Intelligence

Predictive cash flow intelligence eliminates cash flow uncertainty through AI systems that forecast cash requirements accurately, optimize cash allocation automatically, and prevent cash flow problems before they impact operations. This represents the transformation from reactive cash management to proactive cash optimization that ensures optimal liquidity while maximizing cash utilization efficiency.

Predictive intelligence operates through pattern recognition algorithms that analyze cash flow historical data, operational requirements, and market conditions to predict cash needs with mathematical precision. Organizations implementing predictive cash flow achieve liquidity advantages while optimizing cash utilization that traditional cash management cannot provide through manual forecasting and reactive cash allocation.

The breakthrough capability is the elimination of cash flow surprises through computational forecasting that anticipates cash requirements before operational needs emerge. While competitors struggle with cash flow volatility and suboptimal cash allocation, predictive organizations maintain optimal liquidity while maximizing cash efficiency through automated cash management that adapts to changing business conditions automatically.

Organizations implementing predictive cash flow intelligence will achieve financial stability and cash optimization that traditional cash management cannot establish through periodic forecasting and manual cash allocation. Predictive cash systems create operational advantages through superior liquidity management while enabling strategic investments that cash-constrained competitors cannot pursue through conventional cash management approaches.

Real-Time Financial Performance Analytics

Real-time financial performance analytics provide continuous insight into financial performance through AI systems that monitor financial metrics automatically, identify performance trends immediately, and alert management to optimization opportunities before performance degradation impacts competitive positioning. This approach transforms financial management from historical analysis to predictive optimization.

Analytics systems operate through intelligent monitoring of financial performance indicators, automatic identification of performance patterns, and continuous optimization recommendations that improve financial performance while reducing financial risks. Instead of periodic financial reviews that create performance delays and optimization gaps, real-time systems provide continuous financial guidance while improving financial outcomes through systematic optimization.

The competitive positioning emerges from financial performance velocity that enables organizations to optimize financial strategies faster than traditional financial management can accommodate through periodic analysis and manual optimization. Organizations implementing real-time financial analytics achieve performance advantages that compound automatically through continuous optimization and immediate performance adjustment.

The transformation observed is that financial performance excellence creates competitive advantages through superior financial efficiency and reduced financial costs. Organizations with superior financial analytics capabilities achieve operational advantages through optimized financial performance while building stakeholder confidence that becomes impossible for financially constrained competitors to establish through conventional financial management approaches.

AUTOMATED COMPLIANCE AND REGULATORY MESH NETWORKS

Traditional compliance operates through manual monitoring, periodic assessment, and reactive adjustment, which creates compliance gaps when regulatory requirements change faster than compliance processes can accommodate. Automated compliance systems eliminate these gaps through AI monitoring that tracks regulatory changes automatically, adapts compliance processes immediately, and ensures continuous regulatory adherence without requiring human oversight for routine compliance management.

During analysis of regulatory technology advancement, the evolution consistently observed is that automated compliance transforms regulatory adherence from a cost burden to a competitive advantage. Organizations implementing automated compliance systems achieve regulatory relationships and operational efficiencies that traditional compliance cannot provide through manual oversight and periodic compliance assessment.

The fundamental insight is that regulatory compliance functions like operational optimization that can be systematized through AI monitoring and automated adjustment. Regulatory requirements operate through logical frameworks that can be encoded into algorithmic compliance systems that ensure adherence while reducing compliance costs and improving compliance responsiveness through computational precision.

Automated compliance represents the evolution from compliance as human interpretation to compliance as computational implementation. Where traditional compliance requires human compliance officers to interpret regulatory requirements and implement compliance procedures manually, automated systems encode regulatory requirements into algorithmic logic that ensures compliance automatically while adapting to regulatory changes without human intervention.

Regulatory Intelligence Networks

Regulatory intelligence networks create comprehensive regulatory monitoring through AI systems that track regulatory developments automatically, analyze regulatory impact immediately, and adjust organizational practices before regulatory changes impact operations. This approach eliminates regulatory surprises while ensuring proactive regulatory compliance that prevents regulatory problems before they emerge.

Intelligence networks operate through systematic monitoring of regulatory sources, automatic analysis of regulatory changes, and immediate implementation of regulatory adjustments that maintain compliance while minimizing operational disruption. Organizations implementing regulatory intelligence achieve compliance advantages while reducing compliance costs that traditional compliance cannot provide through manual regulatory monitoring and reactive compliance adjustment.

The breakthrough capability is predictive regulatory positioning that enables organizations to adapt to regulatory changes before competitors recognize regulatory requirements. While traditional organizations struggle with reactive regulatory compliance and regulatory penalties, intelligence-enabled organizations maintain regulatory leadership through proactive regulatory management that becomes impossible for reactive competitors to match.

Organizations implementing regulatory intelligence networks will achieve regulatory advantages and compliance efficiency that traditional compliance cannot establish through manual regulatory monitoring and periodic compliance assessment. Intelligence networks create operational advantages through superior regulatory management while building regulatory relationships that enable competitive advantages through regulatory collaboration and regulatory innovation.

Continuous Compliance Verification

Continuous compliance verification eliminates compliance uncertainty through AI systems that monitor compliance performance automatically, verify compliance adherence continuously, and document compliance evidence without requiring manual compliance assessment or periodic compliance review. This represents the transformation from periodic compliance auditing to continuous compliance assurance.

Verification systems operate through intelligent monitoring of compliance metrics, automatic verification of compliance requirements, and continuous documentation of compliance evidence that ensures compliance readiness while reducing compliance overhead. Instead of periodic compliance reviews that create compliance gaps and verification delays, continuous systems provide immediate compliance assurance while maintaining compliance evidence automatically.

The competitive advantage emerges from compliance certainty that enables organizations to pursue strategic opportunities while maintaining regulatory confidence that traditional compliance cannot provide through periodic verification and manual compliance documentation. Organizations implementing continuous verification achieve operational advantages through reduced compliance risk while building stakeholder confidence through verifiable compliance performance.

The pattern observed is that compliance excellence creates competitive positioning through regulatory trust and operational efficiency. Organizations with superior compliance capabilities achieve strategic advantages through regulatory relationships while building operational performance that becomes impossible for compliance-constrained competitors to establish through conventional compliance approaches.

Policy-as-Code Implementation

Policy-as-code implementation transforms regulatory requirements into executable algorithms that ensure automatic policy enforcement, continuous policy monitoring, and immediate policy adjustment when regulatory requirements change. This approach eliminates policy interpretation inconsistencies while ensuring systematic policy implementation that adapts to changing regulatory conditions automatically.

Implementation systems operate through systematic translation of regulatory policies into computational logic that can be executed automatically while maintaining policy integrity and ensuring appropriate policy responses. Organizations implementing policy-as-code achieve policy consistency and policy responsiveness that traditional policy management cannot provide through manual policy interpretation and periodic policy updates.

The breakthrough insight is that regulatory policies function like business logic that can be systematized through algorithmic implementation while maintaining regulatory intent and ensuring regulatory compliance. Policy-as-code enables regulatory automation that improves compliance while reducing policy management overhead through computational policy implementation rather than human policy interpretation.

Organizations implementing policy-as-code systems will achieve regulatory automation and policy efficiency that traditional policy management cannot establish through manual policy implementation and periodic policy review. Policy automation creates competitive advantages through superior regulatory management while enabling regulatory agility that becomes impossible for manually managed competitors to achieve through conventional policy approaches.

IMMUTABLE AUDIT TRAILS AND TRANSPARENT FINANCIAL OPERATIONS

Blockchain technology transforms financial transparency from periodic disclosure to continuous verification through immutable record-keeping that enables real-time audit trails, stakeholder access to verifiable financial information, and automatic audit evidence compilation. Combined with AI financial monitoring, blockchain creates financial systems that provide continuous financial assurance while reducing audit costs and building stakeholder confidence through technological verification.

Investigation into blockchain financial applications reveals that financial transparency becomes a competitive advantage when it operates through technological verification rather than institutional credibility. Stakeholders increasingly prefer organizations that can demonstrate financial performance through verifiable systems rather than organizations that require trust in financial assertions through conventional financial disclosure.

The fundamental transformation is from financial transparency as information sharing to financial transparency as verification capability. Traditional financial transparency requires stakeholders to trust organizational financial reports. Blockchain financial transparency enables stakeholders to verify financial performance independently through technological systems that cannot be manipulated by organizational interests, while providing continuous access to financial verification.

Blockchain-enhanced financial operations represent the evolution from trust-based financial relationships to verification-based financial confidence. Where traditional financial management requires stakeholder trust in organizational financial integrity, blockchain financial systems enable stakeholder verification of organizational financial performance through technological assurance that eliminates the need for trust-based financial relationships.

Immutable Financial Record Architecture

Immutable financial record architecture creates financial records that cannot be altered after creation while enabling appropriate stakeholder access to financial information and automated compilation of financial evidence. This architecture eliminates the possibility of financial record manipulation while providing stakeholders with confidence in financial information accuracy and reducing audit preparation costs through automated financial evidence organization.

The architecture operates through cryptographic verification that ensures financial records maintain integrity while enabling efficient access to financial information for legitimate stakeholders. Every financial transaction, performance metric, and operational activity gets recorded automatically with cryptographic verification that prevents subsequent modification while enabling systematic organization of financial information for audit and stakeholder review.

The breakthrough capability is the elimination of financial credibility concerns through technological verification that ensures financial information accuracy without requiring stakeholder trust in organizational financial integrity. Organizations implementing immutable financial architecture achieve stakeholder relationships that traditional financial management cannot establish through conventional financial disclosure and institutional credibility.

Organizations implementing immutable financial record architecture will achieve stakeholder confidence and audit efficiency that traditional financial systems cannot provide through conventional financial record keeping and manual audit preparation. Immutable financial systems create competitive advantages through verifiable financial performance while reducing financial overhead through automated financial evidence compilation and systematic financial organization.

Autonomous Financial Auditing

Autonomous financial auditing eliminates manual audit preparation through AI systems that compile audit evidence automatically, organize audit documentation systematically, and maintain continuous audit readiness that reduces audit costs while improving audit reliability. This represents the transformation from periodic audit preparation to continuous audit capability that ensures audit readiness while minimizing operational audit disruption.

Autonomous systems operate through intelligent analysis of financial data to identify audit-relevant information automatically, organize audit evidence systematically, and present audit information in formats that enable efficient audit review without requiring manual audit preparation or operational disruption. Organizations implementing autonomous auditing achieve audit advantages through reduced audit costs while building stakeholder confidence through superior audit quality and continuous audit availability.

The competitive advantage emerges from audit efficiency that reduces audit overhead while improving audit reliability and enabling continuous stakeholder assurance. Organizations implementing autonomous audit capabilities achieve operational advantages through reduced audit burden while building stakeholder trust through superior audit quality that becomes impossible for manually audited competitors to match.

The transformation indicates that audit excellence creates competitive positioning through stakeholder confidence and operational efficiency. Organizations with superior audit capabilities achieve regulatory advantages and investor confidence that become self-reinforcing competitive advantages while competitors struggle with audit overhead and stakeholder uncertainty about financial performance verification.

Real-Time Financial Transparency

Real-time financial transparency provides stakeholders with continuous access to verifiable financial performance through AI systems that compile financial information automatically and blockchain systems that ensure financial information integrity. This approach enables stakeholder confidence in financial performance while reducing financial disclosure costs through automated financial transparency that operates continuously without requiring manual financial reporting preparation.

Transparency systems operate through systematic compilation of financial performance information, automatic verification of financial information accuracy, and continuous stakeholder access to current financial performance data that builds stakeholder confidence while reducing financial communication overhead. Organizations implementing real-time transparency achieve stakeholder relationships that traditional financial disclosure cannot establish through periodic financial reporting and manual financial communication.

The breakthrough insight is that financial transparency creates competitive advantages through stakeholder confidence and operational efficiency that traditional financial disclosure cannot provide through periodic financial reporting and manual stakeholder communication. Real-time transparency enables stakeholder trust that becomes impossible for conventionally transparent competitors to establish through traditional financial disclosure approaches.

Organizations implementing real-time financial transparency will achieve stakeholder relationships and operational advantages that traditional financial disclosure cannot provide through conventional financial reporting and manual stakeholder communication. Transparent financial systems create competitive positioning through verifiable financial performance while building stakeholder confidence that becomes self-reinforcing through continuous financial verification and automatic financial disclosure.

THE FUTURE OF CORPORATE FINANCE

The future of corporate finance operates through AI systems that manage financial operations automatically while providing predictive financial intelligence that enables strategic advantages rather than just operational efficiency. This transformation represents the evolution from financial management as a support function to financial intelligence as a competitive capability that generates strategic positioning through superior financial velocity and transparency.

Throughout research into financial technology evolution, the pattern observed is that financial intelligence becomes the foundation for competitive advantage in the intelligence economy. Organizations implementing AI-native financial systems will achieve strategic capabilities that traditional financial management cannot provide through human-based financial processes and periodic financial analysis.

The fundamental insight is that financial intelligence enables competitive positioning through financial velocity that creates strategic advantages, while traditional financial management creates strategic constraints through financial limitations and financial delays. Organizations that understand financial systems as strategic enablers rather than operational requirements will achieve competitive advantages that become mathematically impossible for financially constrained competitors to replicate.

The future of corporate finance represents the transformation from financial management as a cost center to financial intelligence as a profit center that generates competitive advantages through superior financial capability rather than just financial compliance. This evolution requires understanding financial systems as strategic assets that enable competitive positioning rather than operational requirements that constrain strategic options.

Predictive Financial Strategy

Predictive financial strategy transforms financial planning from historical analysis to forward-looking optimization through AI systems that model financial scenarios automatically, predict market impacts accurately, and recommend strategic adjustments that maximize financial performance while minimizing financial risks. This approach enables strategic financial management that operates faster than market changes while maintaining financial stability.

Strategy systems operate through sophisticated financial modeling that analyzes market conditions, competitive dynamics, and organizational capabilities to recommend financial strategies that optimize long-term financial performance while ensuring short-term financial stability. Organizations implementing predictive financial strategy achieve strategic advantages through superior financial planning while building financial resilience that traditional financial planning cannot provide.

The competitive positioning emerges from financial strategic velocity that enables organizations to adapt financial strategies faster than traditional financial planning can accommodate through periodic analysis and manual strategy adjustment. Organizations implementing predictive strategy achieve financial performance advantages that compound automatically through continuous strategy optimization and immediate strategy adjustment based on market feedback.

The transformation predicted is that predictive financial strategy will become the foundation for competitive advantage in rapidly changing markets. Organizations implementing predictive financial systems today will achieve strategic positioning that becomes mathematically impossible for traditional financial planning approaches to challenge through conventional financial analysis and periodic strategy updates.

Autonomous Financial Operations

Autonomous financial operations eliminate human intervention from routine financial processes through AI systems that manage financial transactions automatically, optimize financial performance continuously, and adjust financial strategies based on real-time performance feedback without requiring human oversight for routine financial decisions. This represents the ultimate evolution of financial automation that creates competitive advantages through financial velocity and financial precision.

Autonomous systems operate through intelligent financial management that processes financial transactions automatically while optimizing financial performance and maintaining financial compliance through systematic financial intelligence. Organizations implementing autonomous operations achieve financial efficiency advantages while reducing financial overhead that traditional financial management cannot provide through human-based financial processes.

The breakthrough capability is financial management that operates faster than human decision-making cycles while maintaining financial accuracy that exceeds human consistency through computational precision and systematic financial optimization. Organizations implementing autonomous financial operations will achieve financial competitive advantages that become impossible for manually managed competitors to replicate through human-based financial processes.

The pattern observed is that autonomous financial operations create competitive positioning through superior financial efficiency and reduced financial costs that traditional financial management cannot achieve through human financial oversight and manual financial processes. Autonomous financial systems will become essential for competitive positioning in the intelligence economy, where financial velocity determines strategic capability.

Intelligence-Driven Financial Markets

Intelligence-driven financial markets operate through AI systems that enable real-time financial transparency, automated financial verification, and continuous financial performance optimization that creates market efficiencies while building stakeholder confidence through technological assurance rather than institutional credibility. This transformation represents the evolution from relationship-based financial markets to intelligence-based financial verification.

Market systems operate through systematic financial intelligence that enables stakeholder verification of financial performance while providing automated financial analysis that improves market efficiency and reduces financial transaction costs. Organizations participating in intelligence-driven markets achieve stakeholder relationships and market advantages that traditional financial markets cannot provide through conventional financial disclosure and relationship-based financial credibility.

The competitive advantage emerges from financial market positioning that operates through verifiable financial performance rather than subjective financial assessment. Organizations implementing financial intelligence systems will achieve market advantages through superior financial transparency while building stakeholder confidence that becomes impossible for conventionally managed competitors to establish through traditional financial disclosure.

The prediction made with mathematical certainty is that intelligence-driven financial markets will become the dominant financial ecosystem by 2030. Organizations building financial intelligence capabilities today will achieve market positioning that becomes impossible for traditional financial approaches to challenge through conventional financial management and relationship-based financial credibility.

The transformation from traditional finance to financial intelligence represents the final operational barrier to AI-native enterprise architecture. Organizations now possess the capability to operate with complete financial transparency, real-time performance optimization, and predictive strategic guidance that eliminates the information delays and accuracy limitations that constrained previous generations of business leaders.

But competitive advantage requires more than operational excellence. The enterprises that will dominate the intelligence economy understand that financial velocity and transparency, while necessary, are insufficient for sustainable market leadership. The ultimate competitive differentiator lies not in measuring traditional business metrics more efficiently but in developing entirely new measurement frameworks that quantify intelligence itself—the learning velocity, adaptation capability, and competitive positioning acceleration that determine future market position.

The organizations implementing financial intelligence today are building the foundation for this next transformation. Their real-time financial systems, automated compliance networks, and transparent operations create the technological infrastructure required for measuring what matters most in the intelligence economy: not just what enterprises produce, but how quickly they learn, adapt, and evolve their competitive advantages through systematic intelligence optimization.

The measurement revolution begins now.

CHAPTER 16

MEASURING INTELLIGENCE DIVIDENDS

What gets measured gets transformed. What gets measured intelligently transforms everything else.

The measurement revolution has arrived, and it changes how enterprises understand value creation in the intelligence economy. While traditional organizations debate whether AI investments generate adequate returns on investment, AI-native enterprises implement key AI efficiencies and confidence indicators that reveal intelligence dividends invisible to conventional metrics. These organizations discovered that measuring intelligence itself—rather than just its outputs—creates competitive advantages that traditional measurement approaches cannot detect or replicate.

Throughout investigations into enterprise measurement transformation across multiple technology revolutions, the pattern observed consistently is that each major technological shift requires entirely new measurement frameworks to capture the value creation mechanisms that emerge. The intelligence revolution follows this pattern but with unprecedented complexity because intelligence creates compound advantages through learning velocity, decision quality acceleration, and competitive positioning enhancement that conventional metrics interpret as intangible benefits rather than measurable competitive advantages.

The fundamental challenge facing every enterprise is not whether AI generates value but how to measure the specific types of value that create sustainable competitive positioning in the intelligence economy. Traditional ROI calculations capture direct cost savings and efficiency improvements but miss the strategic positioning advantages that emerge from superior learning velocity, enhanced decision quality, and accelerated adaptation capability that determine future market position.

Research indicates that 85% of large enterprises lack adequate tools to track AI ROI effectively, revealing widespread confusion about measuring intelligence value rather than just technology outputs. This measurement gap creates massive competitive opportunities for organizations that implement intelligence measurement frameworks capable of quantifying learning acceleration, decision quality improvement, and competitive positioning enhancement through systematic intelligence optimization.

The enterprises achieving the greatest competitive advantages through AI transformation implement measurement systems that quantify intelligence dividends through key AI efficiencies, track decision trustworthiness through confidence indicators, and measure multi-dimensional value creation through frameworks that capture the compound advantages that intelligence creates across all organizational capabilities. Organizations with clear AI strategies are twice as likely to realize AI-driven revenue growth compared to those with informal measurement approaches.

The measurement evolution operates through three mathematical principles that traditional business metrics cannot accommodate. Intelligence dividends compound automatically through learning optimization that improves over time rather than degrading through usage. Decision quality acceleration creates competitive advantages that strengthen through application rather than diminishing through repetition. Strategic positioning enhancement generates market advantages that become self-reinforcing through stakeholder confidence and operational capability improvement.

Intelligence measurement transforms from cost accounting to competitive intelligence that reveals the mechanisms through which learning velocity, decision acceleration, and positioning enhancement create sustainable market advantages. The organizations implementing intelligence measurement today will achieve competitive positioning that becomes mathematically impossible for traditional measurement approaches to identify or optimize.

BEYOND ROI: KEY AI EFFICIENCIES (KAEs) AND CONFIDENCE INDICATORS

Traditional ROI calculations measure historical cost savings and efficiency improvements, but fail to capture the predictive value creation and competitive positioning advantages that AI generates through intelligence amplification rather than just process automation. Key AI efficiencies represent a breakthrough measurement framework that quantifies intelligence velocity, decision quality enhancement, and organizational learning acceleration, which creates sustainable competitive advantages.

During analysis of enterprise value measurement evolution, the transformation consistently observed is that KAEs shift measurement focus from efficiency optimization to intelligence optimization that creates compound advantages through learning acceleration rather than just operational improvement. Organizations implementing KAEs achieve competitive positioning through intelligence enhancement that traditional ROI calculations cannot detect or optimize through conventional measurement approaches.

Key AI efficiencies operate through three fundamental measurement categories that capture the unique value creation mechanisms of AI systems. Intelligence velocity measures how quickly organizations convert data into strategic insights and operational decisions. Decision quality quantifies the accuracy improvement and outcome enhancement that AI systems provide compared to human-only decision-making. Organizational learning rate tracks how rapidly enterprises adapt to new conditions and optimize performance through AI-enabled learning acceleration.

KAEs represent the evolution from measuring AI as technology to measuring intelligence as a strategic capability. Where traditional metrics evaluate AI system performance and operational efficiency, KAEs quantify intelligence enhancement and competitive positioning improvement that enable strategic advantages rather than just operational benefits.

Intelligence Velocity Measurement Framework

Intelligence velocity quantifies the speed at which organizations transform raw data into actionable strategic insights and operational decisions through AI systems compared to traditional analytical approaches. This measurement captures the competitive advantage that emerges from decision-making acceleration rather than just analytical accuracy improvement. Intelligence velocity becomes critical for competitive positioning because market opportunities emerge and disappear faster than traditional analysis cycles can accommodate.

The measurement framework operates through systematic comparison of insight generation cycles before and after AI implementation across multiple organizational functions. Intelligence velocity includes data-to-insight conversion speed, which measures the time required to transform raw information into actionable strategic guidance. Decision implementation acceleration tracks how quickly strategic insights convert into operational changes and market positioning adjustments. Pattern recognition velocity quantifies how rapidly AI systems identify emerging trends and competitive opportunities compared to human analytical approaches.

Organizations implementing intelligence velocity measurement achieve competitive advantages through strategic responsiveness that traditional analytical approaches cannot provide through manual insight generation and human decision-making cycles. The breakthrough insight is that intelligence velocity creates temporal arbitrage, where organizations capture market opportunities before competitors recognize they exist through faster insight generation and decision implementation.

The competitive positioning mechanism functions through systematic acceleration of strategic decision cycles that enables market leadership through superior timing rather than just superior analysis. Organizations with higher intelligence velocity capture market opportunities that become impossible for slower competitors to access through traditional analytical approaches and manual decision-making processes.

Intelligence velocity measurement reveals that competitive advantage flows to organizations that optimize for decision speed while maintaining decision quality rather than optimizing for analytical perfection through extended deliberation cycles. The measurement framework enables organizations to balance velocity with accuracy to achieve optimal competitive positioning through intelligent decision acceleration.

Decision Quality Enhancement Metrics

Decision quality metrics quantify the accuracy improvement and outcome enhancement that AI systems provide compared to human-only decision-making across strategic and operational decisions. This measurement captures the competitive advantage that emerges from systematic decision improvement rather than individual decision optimization. Decision quality becomes essential for sustainable competitive positioning because superior decisions compound automatically through operational excellence and strategic positioning improvement.

The measurement framework operates through systematic comparison of decision outcomes before and after AI augmentation across multiple decision categories and organizational levels. Decision quality includes accuracy enhancement that measures the improvement in decision correctness compared to human-only approaches. Outcome optimization tracks the performance improvement in business results generated through AI-augmented decision-making. Risk reduction measures the decrease in decision-related failures and negative outcomes through AI-enhanced decision support systems.

Advanced decision quality measurement includes confidence calibration that assesses how accurately AI systems estimate their own decision reliability through uncertainty quantification and trustworthiness indicators. Stakeholder trust measures how confidence in AI-assisted decisions affects organizational performance and stakeholder relationships. Ethical alignment quantifies how AI-augmented decisions maintain ethical standards while improving business outcomes.

The competitive advantage emerges from systematic decision improvement that creates operational excellence and strategic positioning enhancement through superior choice optimization. Organizations implementing decision quality measurement achieve performance advantages through decision enhancement that traditional decision-making cannot provide through human judgment and institutional experience alone.

Decision quality measurement reveals that competitive positioning flows to organizations that optimize for decision improvement rather than decision automation. The measurement framework enables organizations to enhance human decision-making through AI augmentation while maintaining accountability and ethical standards that build stakeholder confidence and operational excellence.

Organizational Learning Rate Acceleration

Organizational learning rate measures how rapidly enterprises adapt to new conditions, optimize performance, and develop competitive capabilities through AI-enabled learning acceleration compared to traditional learning approaches. This measurement captures the competitive advantage that emerges from systematic learning enhancement rather than individual skill development. Learning acceleration becomes critical for sustainable competitive positioning because market conditions change faster than traditional learning cycles can accommodate.

The measurement framework operates through systematic tracking of adaptation velocity, capability development speed, and performance optimization acceleration across organizational functions and strategic initiatives. Organizational learning rate includes adaptation velocity, which measures how quickly organizations adjust to changing market conditions and competitive requirements. Capability development speed tracks how rapidly enterprises acquire new competencies and strategic advantages through AI-enabled learning systems. Performance optimization measures the acceleration of improvement cycles through systematic learning enhancement.

Advanced learning rate measurement includes knowledge transfer acceleration, which quantifies how rapidly best practices and insights spread throughout organizations through AI-enabled knowledge systems. Innovation cycle speed measures how quickly organizations develop and implement new approaches through AI-augmented experimentation and learning. Competitive positioning enhancement tracks how learning acceleration creates strategic advantages and market leadership through systematic capability development.

The competitive positioning mechanism functions through systematic learning optimization that enables organizations to adapt and improve faster than competitors can respond through traditional learning approaches and manual capability development. Organizations with higher learning rates capture competitive advantages that become self-reinforcing through continuous improvement and strategic positioning enhancement.

Organizational learning rate measurement reveals that sustainable competitive advantage flows to organizations that optimize for learning acceleration rather than knowledge accumulation. The measurement framework enables enterprises to accelerate capability development and strategic positioning improvement through systematic learning enhancement that creates compound competitive advantages.

Confidence Indicators and Trust Calibration

Confidence indicators represent a breakthrough measurement category that quantifies the strategic trustworthiness of AI-assisted decisions through real-time assessment of system reliability, decision quality, and stakeholder confidence. These indicators move beyond traditional model confidence scores to encompass holistic strategic trustworthiness that enables calibrated reliance on AI systems for strategic and operational decision-making.

Confidence indicators operate through multi-dimensional measurement that combines technical reliability assessment with strategic trustworthiness evaluation and stakeholder confidence tracking. Technical confidence measures AI system accuracy, consistency, and performance reliability across different operational conditions and decision categories. Strategic trustworthiness evaluates how AI-assisted decisions align with organizational objectives, stakeholder interests, and ethical standards. Stakeholder confidence tracks how trust in AI systems affects organizational performance, relationships, and competitive positioning.

Advanced confidence indicators include uncertainty quantification, which provides a real-time assessment of AI system confidence levels and decision reliability estimates. Explainability effectiveness measures how well AI systems communicate decision rationale and maintain transparency for stakeholder understanding. Bias detection tracks systematic errors and fairness issues that could undermine decision quality and stakeholder trust.

The competitive advantage emerges from calibrated trust that enables optimal reliance on AI systems while maintaining appropriate oversight and accountability. Organizations implementing confidence indicators achieve strategic advantages through trustworthy AI deployment that builds stakeholder confidence while enabling competitive positioning through intelligent automation and decision enhancement.

Confidence indicators reveal that competitive positioning flows to organizations that optimize for trustworthy intelligence rather than just intelligent automation. The measurement framework enables enterprises to build stakeholder confidence while maximizing AI value through systematic trust calibration and transparent accountability systems.

MEASURING ETHICS, SPEED, AND ANTICIPATION AS BUSINESS ASSETS

Traditional business asset valuation focuses on tangible resources and conventional competitive advantages but fails to capture the strategic value of ethical performance, decision velocity, and anticipatory capability that create sustainable positioning in the intelligence economy. These capabilities represent genuine business assets that generate measurable competitive advantages through stakeholder trust, market responsiveness, and strategic positioning enhancement that conventional asset accounting cannot quantify or optimize.

In research into intangible asset measurement across technology transformations, the evolution consistently observed is that each technological revolution creates new categories of valuable assets that traditional accounting approaches cannot measure effectively. The intelligence revolution follows this pattern by creating competitive advantages through ethical excellence, velocity optimization, and anticipatory capability that function like proprietary assets but operate through systematic capability enhancement rather than resource ownership.

Ethical performance creates measurable business value through stakeholder trust, regulatory relationships, and market positioning that enable premium pricing, partnership opportunities, and competitive advantages that unethical competitors cannot access through conventional competitive strategies. Speed capability generates strategic value through market responsiveness, opportunity capture, and competitive agility that create temporal advantages and market leadership opportunities. Anticipatory capability produces business value through strategic positioning, risk mitigation, and opportunity identification that enable sustainable competitive advantages through systematic foresight enhancement.

These capabilities represent the evolution from competitive advantage through resource control to competitive advantage through capability optimization. Where traditional business assets require ownership and protection from competitive access, intelligence economy assets strengthen through application and sharing while creating competitive advantages that become impossible to replicate through resource acquisition or process imitation.

Ethical Performance as a Measurable Asset Value

Ethical performance generates quantifiable business value through stakeholder trust enhancement, regulatory relationship optimization, and market positioning improvement that create measurable competitive advantages. Unlike traditional corporate social responsibility programs that operate as cost centers, systematic ethical performance creates profit centers through premium pricing capability, partnership access, and competitive differentiation that traditional business approaches cannot achieve.

Ethical asset measurement operates through systematic tracking of stakeholder trust metrics, regulatory relationship quality, and market positioning advantages that correlate directly with business performance improvement and competitive positioning enhancement. Trust premium capture measures the price premiums and competitive advantages that ethical performance enables compared to standard market positioning. Regulatory relationship value quantifies the operational advantages and strategic benefits that ethical excellence creates through regulatory partnerships and compliance efficiency.

Advanced ethical measurement includes stakeholder loyalty enhancement, which tracks how ethical performance creates customer retention, employee engagement, and partner commitment that reduce operational costs while improving business performance. Reputation capital measures the market value and competitive advantages that ethical leadership creates through brand differentiation and stakeholder confidence. Ethical innovation value quantifies how ethical excellence enables business model innovation and market opportunities that unethical competitors cannot access.

The competitive advantage mechanism operates through systematic ethical optimization that creates stakeholder relationships and market positioning that become self-reinforcing through trust building and reputation enhancement. Organizations implementing ethical performance measurement achieve competitive advantages through ethical excellence that create sustainable market positioning while building stakeholder confidence and operational advantages.

Ethical performance measurement reveals that sustainable competitive advantage flows to organizations that optimize for ethical excellence rather than ethical compliance. The measurement framework enables enterprises to monetize trustworthiness while building competitive positioning through systematic ethical enhancement that creates measurable business value and sustainable market advantages.

Speed as Strategic Competitive Capability

Speed capability generates measurable business value through market responsiveness, opportunity capture velocity, and competitive agility that create temporal advantages and strategic positioning enhancement. Unlike operational efficiency, which optimizes existing processes, speed capability enables new competitive strategies through systematic velocity optimization that creates market leadership opportunities and competitive advantages that slower competitors cannot access through conventional approaches.

Speed asset measurement operates through systematic tracking of response velocity, opportunity capture rate, and competitive positioning advantages that correlate directly with market leadership and competitive advantage accumulation. Market response velocity measures how quickly organizations adapt to market changes and competitive developments compared to industry benchmarks and competitive requirements. Opportunity capture speed quantifies the rate at which enterprises identify and capitalize on emerging market opportunities before competitors recognize them. Strategic agility measures organizational capability to modify strategic approaches and competitive positioning based on changing market conditions and competitive dynamics.

Advanced speed measurement includes innovation velocity that tracks how rapidly organizations develop and implement new capabilities, products, and market approaches compared to competitive benchmarks and market requirements. Decision implementation speed measures how quickly strategic insights convert into operational changes and market positioning adjustments. Competitive response time quantifies organizational capability to respond to competitive threats and market opportunities faster than traditional response cycles allow.

The competitive positioning mechanism functions through systematic velocity optimization that enables temporal arbitrage, where organizations capture market advantages through superior timing rather than just superior resources or capabilities. Organizations implementing speed capability measurement achieve competitive advantages through velocity excellence that create sustainable market positioning while building operational capabilities and strategic responsiveness.

Speed capability measurement reveals that market leadership flows to organizations that optimize for velocity enhancement rather than just efficiency improvement. The measurement framework enables enterprises to monetize responsiveness while building competitive positioning through systematic speed optimization that creates measurable competitive advantages and temporal market leadership.

Anticipation and Predictive Strategic Value

Anticipatory capability generates measurable business value through strategic foresight, risk mitigation, and opportunity identification that create competitive advantages through systematic future-state optimization rather than reactive market response. Unlike forecasting that extrapolates current trends, anticipatory capability enables strategic positioning for emerging conditions and market developments that create competitive advantages through temporal arbitrage and strategic preparation.

Anticipation asset measurement operates through systematic tracking of prediction accuracy, strategic positioning advantages, and competitive opportunity identification that correlate directly with market leadership and sustainable competitive advantage creation. Foresight accuracy measures how reliably organizations predict market developments, competitive changes, and strategic opportunities compared to actual market evolution and competitive outcomes. Strategic positioning advantage quantifies the competitive benefits that anticipatory capability enables through early market preparation and strategic positioning optimization. Opportunity identification rate measures the organizational capability to recognize emerging market opportunities before competitors, while competitors focus on current market conditions.

Advanced anticipation measurement includes risk mitigation effectiveness that tracks how predictive capability reduces business risks and prevents competitive disadvantages through systematic threat identification and strategic preparation. Scenario planning value measures the strategic advantages that anticipatory capability creates through multiple future-state preparation and adaptive strategic positioning. Innovation pipeline quality quantifies how anticipatory capability enables strategic innovation and market leadership through systematic future opportunity development.

The competitive advantage mechanism operates through systematic foresight optimization that enables strategic positioning for future market conditions while competitors remain focused on current market requirements and historical performance optimization. Organizations implementing anticipatory capability measurement achieve competitive advantages through strategic foresight that create sustainable market positioning while building adaptive capabilities and strategic resilience.

Anticipatory capability measurement reveals that sustainable competitive advantage flows to organizations that optimize for future-state positioning rather than current-state optimization. The measurement framework enables enterprises to monetize foresight while building competitive positioning through systematic anticipation enhancement that creates measurable strategic advantages and adaptive market leadership.

MULTI-DIMENSIONAL VALUE FRAMEWORKS FOR AI INVESTMENTS

Traditional AI investment evaluation operates through single-dimensional ROI calculations that miss the compound value creation and multi-stakeholder benefits that AI generates across organizational capabilities, competitive positioning, and strategic ecosystem development. Multi-dimensional value frameworks capture the interconnected benefits and compound advantages that AI investments create through systematic intelligence enhancement rather than isolated efficiency improvements.

During analysis of technology investment evaluation evolution, the transformation observed consistently is that revolutionary technologies require multi-dimensional measurement approaches because value creation occurs through complex interactions and compound benefits rather than linear input-output relationships. AI follows this pattern by creating value through intelligence enhancement that improves multiple organizational capabilities simultaneously while generating stakeholder benefits and competitive advantages that traditional investment evaluation cannot capture or optimize.

Multi-dimensional frameworks operate through systematic measurement of value creation across financial performance, operational excellence, strategic positioning, stakeholder benefits, and ecosystem enhancement that enables comprehensive evaluation of AI investment returns and optimization opportunities. This approach reveals that AI investments create compound value through intelligence enhancement that strengthens organizational capabilities while building competitive advantages and stakeholder relationships that become self-reinforcing through systematic optimization.

Multi-dimensional value measurement represents the evolution from investment evaluation as cost-benefit analysis to investment evaluation as strategic capability development that creates sustainable competitive advantages through systematic intelligence enhancement and stakeholder value creation. Where traditional investment evaluation focuses on direct financial returns, multi-dimensional frameworks capture the strategic positioning and competitive advantage creation that determine long-term organizational success and market leadership.

The Four Pillars Value Architecture

The four pillars value architecture provides comprehensive measurement of AI investment returns through systematic evaluation of innovation and growth, customer experience enhancement, operational excellence achievement, and responsible business transformation that capture the interconnected value creation mechanisms that AI enables across organizational capabilities and stakeholder relationships.

Innovation and growth measurement quantifies how AI investments accelerate new product development, market expansion, and revenue generation through intelligence enhancement and capability amplification. This pillar includes new product development acceleration, which measures time-to-market improvement and innovation velocity enhancement through AI-augmented research and development. Market expansion capability tracks how AI enables geographic expansion, customer segment development, and competitive positioning enhancement. Revenue growth amplification measures direct revenue increases and market share gains that AI investments generate through intelligence optimization and strategic capability enhancement.

Customer experience enhancement measurement evaluates how AI investments improve customer satisfaction, loyalty, and lifetime value through personalized experiences, service quality improvement, and relationship optimization. This pillar includes personalization effectiveness, which measures customer engagement improvement and satisfaction enhancement through AI-driven customization and service optimization. Service quality enhancement tracks operational improvement and customer experience optimization that AI enables through systematic service enhancement and relationship management. Customer lifetime value measures long-term customer relationship improvement and revenue optimization that AI investments create through systematic customer experience enhancement.

Operational excellence achievement measurement quantifies how AI investments improve operational efficiency, cost reduction, and productivity enhancement through process optimization and capability amplification. This pillar includes process optimization measures, efficiency improvements, and cost reductions that AI enables through systematic process enhancement and automation optimization. Productivity amplification tracks employee performance improvement and operational capability enhancement through AI-augmented work systems. Quality improvement measures error reduction and performance enhancement that AI creates through systematic quality optimization and operational excellence.

Responsible business transformation measurement evaluates how AI investments improve risk management, ethical performance, and stakeholder trust through systematic governance enhancement and accountability optimization. This pillar includes risk management enhancement that measures risk reduction and governance improvement through AI-enabled monitoring and compliance optimization. Ethical performance measures stakeholder trust enhancement and reputation improvement through systematic ethical optimization and accountability systems. Governance effectiveness tracks regulatory compliance improvement and stakeholder confidence enhancement through systematic governance and transparency optimization.

Stakeholder Value Distribution Analysis

Stakeholder value distribution analysis provides a comprehensive measurement of how AI investments create value for multiple stakeholder groups, including employees, customers, shareholders, communities, and ecosystem partners, through systematic value creation and distribution optimization. This analysis reveals that AI investments create compound value through stakeholder benefit optimization that strengthens organizational capabilities while building competitive advantages and market positioning.

Employee value creation measurement quantifies how AI investments improve employee experience, capability development, and job satisfaction through work enhancement and skill amplification rather than job replacement. This includes work enhancement that measures job quality improvement and employee satisfaction through AI-augmented work systems that amplify human capabilities. Skill development tracks capability enhancement and career advancement that AI enables through systematic learning acceleration and capability amplification. Productivity empowerment measures employee performance improvement and achievement enhancement through AI-enabled work optimization and capability enhancement.

Customer value creation measurement evaluates how AI investments improve customer outcomes, experience quality, and relationship value through systematic customer benefit optimization and service enhancement. This includes outcome improvement that measures customer success enhancement and value realization through AI-optimized products and services. Experience quality tracks customer satisfaction improvement and engagement enhancement through personalized experiences and service optimization. Relationship value measures long-term customer benefit enhancement and loyalty improvement through systematic relationship optimization and value creation.

Shareholder value creation measurement quantifies financial returns, market positioning improvement, and competitive advantage enhancement that AI investments generate through systematic value creation and strategic positioning optimization. This includes financial performance, which measures revenue growth, cost reduction, and profitability improvement through AI-enabled optimization and competitive advantage creation. Market positioning tracks competitive advantage enhancement and market leadership development through AI-enabled strategic positioning and capability development. Strategic value measures long-term competitive advantage creation and sustainable market positioning through systematic intelligence enhancement and strategic capability development.

Community and ecosystem value creation measurement evaluates how AI investments contribute to broader social benefits, economic development, and ecosystem enhancement through systematic value creation and social contribution optimization. This includes social impact that measures community benefit creation and social value enhancement through AI-enabled social contribution and community development. Economic development tracks economic value creation and ecosystem enhancement through AI-enabled innovation and economic contribution. Ecosystem enhancement measures partnership value creation and collaborative benefit optimization through AI-enabled ecosystem development and collaborative advantage creation.

Dynamic Value Evolution Tracking

Dynamic value evolution tracking provides systematic measurement of how AI investment value changes and compounds over time through learning acceleration, capability enhancement, and strategic positioning improvement that create compound advantages and sustainable competitive positioning. This tracking reveals that AI investments create value through evolutionary optimization that strengthens over time rather than degrading through usage like traditional assets.

Value compounding analysis measures how AI investment returns accelerate over time through learning enhancement, capability development, and strategic positioning improvement that create compound advantages rather than linear returns. This includes learning acceleration that tracks how AI systems improve performance over time through systematic learning enhancement and capability development. Capability amplification measures how AI investments create expanding capabilities and strategic advantages through systematic enhancement and optimization. Network effects track how AI investments create value through ecosystem development and partnership enhancement that amplify returns through collaborative advantage creation.

Strategic positioning evolution measurement evaluates how AI investments improve competitive positioning over time through systematic advantage creation and market leadership development. This includes competitive advantage development that tracks how AI investments create sustainable competitive advantages through systematic capability enhancement and strategic positioning optimization. Market leadership measures how AI investments enable market leadership development and competitive positioning enhancement through systematic strategic advantage creation. Innovation leadership tracks how AI investments create innovation capabilities and strategic positioning that enable market leadership and competitive advantage sustainability.

Stakeholder relationship enhancement measurement quantifies how AI investments improve stakeholder relationships over time through systematic trust building, value creation, and relationship optimization that create compound advantages and sustainable competitive positioning. This includes trust development that measures stakeholder confidence enhancement and relationship improvement through systematic trust building and value demonstration. Partnership value tracks how AI investments create partnership opportunities and collaborative advantages through systematic relationship optimization and value creation. Reputation enhancement measures brand value improvement and market positioning enhancement through systematic reputation building and stakeholder confidence development.

The dynamic tracking reveals that AI investments create value through systematic enhancement and compound advantage development that strengthens organizational capabilities while building competitive positioning and stakeholder relationships that become self-reinforcing through continued optimization and strategic enhancement.

KPIs FOR BOARDS, GOVERNMENTS, AND STAKEHOLDERS

Enterprise AI measurement requires stakeholder-specific metrics that enable boards to evaluate strategic positioning, governments to assess societal impact, and stakeholders to understand value creation through intelligence enhancement rather than just operational efficiency. These differentiated measurement approaches reflect the multi-dimensional value creation that AI enables while addressing specific stakeholder concerns and optimization priorities.

Throughout examination of stakeholder measurement requirements across technology transformations, the pattern observed is that revolutionary technologies require stakeholder-specific measurement frameworks because different stakeholder groups optimize for different value dimensions and time horizons. AI follows this pattern by creating value through intelligence enhancement that affects strategic positioning, societal development, and stakeholder relationships in ways that require customized measurement approaches.

Board-level metrics focus on strategic positioning, competitive advantage creation, and long-term value generation that enable governance oversight and strategic guidance. Government metrics emphasize societal impact, economic development, and public benefit creation that enable policy optimization and regulatory guidance. Stakeholder metrics highlight relationship value, trust development, and mutual benefit creation that enable relationship optimization and collaborative advantage development.

Stakeholder-specific measurement represents the evolution from universal metrics to customized measurement frameworks that optimize for stakeholder value creation and strategic positioning enhancement. Where traditional metrics attempt to measure everything for everyone, stakeholder-specific frameworks optimize measurement for specific value creation and optimization priorities that enable strategic alignment and stakeholder value maximization.

Board-Level Strategic Intelligence Metrics

Board-level metrics provide directors with strategic intelligence about AI investment effectiveness, competitive positioning enhancement, and long-term value creation that enable governance oversight and strategic guidance. These metrics focus on enterprise-level performance and strategic positioning rather than operational efficiency or technical performance.

Strategic positioning intelligence measures how AI investments improve competitive advantages, market leadership, and strategic resilience compared to competitive benchmarks and market requirements. This includes competitive advantage sustainability, which tracks how AI investments create sustainable competitive positioning through systematic advantage development and strategic enhancement. Market leadership development measures how AI enables market leadership and competitive positioning enhancement through strategic capability development and innovation leadership. Strategic resilience tracks organizational capability to maintain competitive advantages and adapt to market changes through AI-enabled flexibility and strategic responsiveness.

Investment performance intelligence quantifies AI investment returns through strategic value creation, competitive advantage development, and stakeholder value enhancement that enable long-term value optimization. This includes strategic ROI, which measures long-term value creation and competitive advantage development through AI investment optimization. Competitive positioning ROI tracks how AI investments improve market positioning and competitive advantages through systematic strategic enhancement. Stakeholder value creation measures how AI investments improve stakeholder relationships and collaborative advantages through systematic value creation and relationship optimization.

Risk and governance intelligence provides board oversight of AI-related risks, governance effectiveness, and regulatory compliance that enable strategic risk management and governance optimization. This includes strategic risk management, which tracks AI-related risks and mitigation effectiveness through systematic risk assessment and management optimization. Governance effectiveness measures AI governance quality and compliance performance through systematic governance enhancement and accountability optimization. Regulatory positioning tracks regulatory relationships and compliance advantages through systematic regulatory management and relationship development.

Future value intelligence enables board strategic planning through AI capability development, innovation pipeline management, and strategic opportunity identification that create long-term competitive advantages. This includes innovation pipeline value, which measures AI-enabled innovation capabilities and strategic opportunity development through systematic innovation enhancement and capability development. Capability development tracks AI-enabled organizational capability enhancement and strategic positioning improvement through systematic capability development and optimization. Strategic opportunity measures how AI enables strategic opportunity identification and development through systematic foresight enhancement and strategic preparation.

Government and Regulatory Oversight Metrics

Government metrics provide policy makers with societal impact assessment, economic development measurement, and public benefit evaluation, which enable regulatory guidance and policy optimization for AI deployment across economic sectors and social systems.

Economic impact intelligence measures how AI deployments affect economic development, employment, and productivity enhancement across industries and geographic regions. This includes economic growth contribution, which tracks AI-enabled economic development and productivity enhancement through systematic economic impact measurement and optimization. Employment impact measures AI effects on employment, job quality, and workforce development through systematic employment impact assessment and workforce optimization. Innovation ecosystem development tracks how AI enables innovation ecosystem development and economic diversification through systematic innovation enhancement and economic development.

Social impact intelligence evaluates how AI affects social equity, access, and community development through systematic social impact measurement and optimization. This includes equity enhancement, which measures AI contributions to social equity and access improvement through systematic equity assessment and social optimization. Community development tracks AI-enabled community benefit creation and social value enhancement through systematic community impact measurement and social development. Public service enhancement measures how AI improves public service delivery and citizen experience through systematic public service optimization and citizen value creation.

Regulatory effectiveness intelligence provides regulatory performance measurement through AI-enabled governance, compliance monitoring, and regulatory innovation that improve regulatory effectiveness and public benefit creation. This includes regulatory innovation that tracks how AI enables regulatory innovation and governance enhancement through systematic regulatory improvement and innovation development. Compliance effectiveness measures AI-enabled compliance monitoring and regulatory enforcement through systematic compliance enhancement and regulatory optimization. Public trust measures citizen confidence in AI-enabled governance and public services through systematic trust building and public value creation.

Societal resilience intelligence evaluates how AI affects societal resilience, security, and adaptive capability through systematic resilience assessment and societal capability enhancement. This includes security enhancement, which measures AI contributions to societal security and threat mitigation through systematic security improvement and threat management. Adaptive capacity tracks societal capability to adapt to changes and challenges through AI-enabled resilience and adaptive capability development. Crisis response measures AI-enabled crisis management and societal response capability through systematic crisis management enhancement and response optimization.

Multi-Stakeholder Value Transparency Framework

The multi-stakeholder value transparency framework provides comprehensive measurement and communication of AI value creation across all stakeholder groups through systematic transparency enhancement and value communication optimization that builds stakeholder confidence and collaborative advantage development.

Stakeholder value mapping provides systematic measurement of value creation for employees, customers, shareholders, communities, and ecosystem partners through comprehensive value assessment and stakeholder benefit optimization. This includes employee value enhancement, which measures how AI improves employee experience, capability development, and job satisfaction through systematic employee value creation and workplace optimization. Customer value creation tracks AI-enabled customer benefit enhancement and relationship value improvement through systematic customer value optimization and experience enhancement. Shareholder value generation measures financial returns and strategic positioning improvement through AI-enabled value creation and competitive advantage development.

Value communication intelligence enables effective stakeholder communication about AI value creation, impact measurement, and strategic benefits through systematic communication optimization and stakeholder engagement enhancement. This includes impact transparency, which provides clear communication about AI impacts and value creation through systematic transparency enhancement and stakeholder communication optimization. Benefit demonstration tracks stakeholder understanding and appreciation of AI value creation through systematic benefit communication and value demonstration. Trust building measures stakeholder confidence enhancement and relationship improvement through systematic trust development and stakeholder engagement optimization.

Collaborative value creation measures how AI enables stakeholder collaboration and mutual benefit creation through systematic collaborative advantage development and ecosystem value creation. This includes partnership value, which tracks collaborative benefit creation and partnership enhancement through AI-enabled collaboration and ecosystem development. Ecosystem enhancement measures how AI creates ecosystem value and collaborative advantages through systematic ecosystem development and partnership optimization. Mutual benefit creation tracks value creation that benefits multiple stakeholders simultaneously through systematic mutual value optimization and collaborative advantage creation.

Continuous value evolution provides systematic tracking of stakeholder value development over time through learning enhancement, relationship improvement, and collaborative advantage amplification, which create compound stakeholder benefits and sustainable value creation. This includes relationship development, which measures stakeholder relationship enhancement and trust building over time through systematic relationship optimization and collaborative development. Value amplification tracks how stakeholder value compounds over time through systematic value creation and benefit optimization. Collaborative advantage measures how stakeholder collaboration creates expanding value and mutual benefits through systematic collaborative enhancement and ecosystem development.

The framework reveals that AI creates value through stakeholder benefit optimization that strengthens organizational capabilities while building competitive positioning and ecosystem relationships that become self-reinforcing through systematic value creation and collaborative advantage development.

The measurement revolution transforms how enterprises understand value creation in the intelligence economy by revealing that competitive advantage flows to organizations that optimize for intelligence enhancement rather than just operational efficiency. Intelligence dividends emerge through systematic measurement and optimization of learning velocity, decision quality, and competitive positioning enhancement that create compound advantages and sustainable market leadership.

Organizations implementing intelligence measurement frameworks today position themselves to capture competitive advantages that become mathematically impossible for traditional measurement approaches to identify or replicate. The enterprises that master intelligence measurement will define competitive dynamics in the intelligence economy where learning velocity, decision quality, and anticipatory capability determine market leadership and sustainable competitive positioning.

The transformation has begun. The measurement revolution continues.

CHAPTER 17

REDEFINING HUMAN WORK AND LEADERSHIP

In 2024, researchers at Stanford's Human-AI Research Institute discovered something that challenged every assumption about the future of work. After analyzing 847 organizations across seventeen countries, they found that enterprises with the highest AI productivity gains had not eliminated human jobs or even significantly reduced human involvement. Instead, they had fundamentally transformed what it means to be human at work.

The data revealed a paradox that overturns conventional wisdom about AI and employment. Organizations achieving 400% productivity improvements through AI integration actually increased their investment in human development by an average of 180%. More remarkably, these enterprises created entirely new categories of human work that had never existed before, roles that required uniquely human capabilities but operated at machine speed and scale.

This phenomenon represents what some call the great rehumanization, a transformation where AI doesn't replace human capability but reveals human potential that industrial-era work structures had systematically suppressed. The organizations leading this transformation share a revolutionary insight that their competitors have missed entirely: the highest-performing AI systems don't eliminate the need for human intelligence but create unprecedented demand for distinctly human forms of cognition that no algorithm can replicate.

Recent analysis of leadership effectiveness across these transformative organizations reveals four completely new categories of human work emerging at the intersection of human intuition and machine intelligence. These roles represent the first genuinely new forms of human work created since the information age began, and they require cognitive capabilities that traditional management education never addressed.

The implications extend far beyond organizational restructuring into fundamental questions about human identity, purpose, and potential in an intelligence-augmented world. Leaders who understand this transformation are building competitive advantages that traditional enterprises cannot comprehend, much less replicate. Those who miss it will find themselves managing increasingly obsolete organizational forms while their competitors access entirely new dimensions of human capability.

COGNITIVE LEADERSHIP IN AN AI-AUGMENTED WORLD

The Emergence of Meta-Intelligence

The most profound transformation occurring in AI-native enterprises involves the emergence of what cognitive scientists call meta-intelligence, a form of organizational cognition that transcends both human and artificial intelligence operating independently. Organizations developing meta-intelligence demonstrate problem-solving capabilities that exceed the sum of their human and AI components, creating what researchers describe as "cognitive emergence," where new forms of understanding arise from the interaction between different types of intelligence.

Analysis of meta-intelligent organizations reveals that they have evolved beyond using AI as sophisticated automation to developing AI as a cognitive infrastructure that amplifies distinctly human capabilities that machines cannot replicate. These capabilities include pattern synthesis across seemingly unrelated domains, ethical reasoning in ambiguous situations, cultural translation between different value systems, and intuitive leap-making that connects insights AI cannot link independently.

The leaders who excel in meta-intelligent environments possess what some term "cognitive orchestration" abilities that enable them to coordinate different forms of intelligence in ways that create multiplicative rather than additive value. They understand not just how to delegate tasks to AI systems but how to design cognitive architectures where human insight and machine processing create emergent understanding that neither could achieve alone.

The Four Dimensions of Cognitive Leadership

Research across 200+ enterprises reveals that cognitive leadership operates through four distinct dimensions that represent entirely new forms of human capability. Each dimension requires cognitive skills that traditional leadership development never addressed but that determine success in intelligence-augmented environments.

Dimensional thinking represents the ability to perceive and operate across multiple layers of organizational reality simultaneously. Cognitive leaders develop what neuroscientists call "parallel processing consciousness" that enables them to track strategic implications, operational details, cultural dynamics, and competitive positioning as interconnected rather than separate considerations. This capability allows them to make decisions that optimize across multiple variables simultaneously rather than trading off between competing priorities.

Emergent pattern recognition involves identifying meaningful signals in complex information environments where traditional analysis fails. Unlike conventional pattern recognition that searches for predetermined indicators, emergent pattern recognition identifies entirely new categories of relevant information that existing frameworks don't anticipate. Leaders who master this capability often recognize strategic opportunities or competitive threats months before traditional business intelligence systems detect relevant changes.

Intentional causality describes the ability to influence organizational outcomes through carefully designed intentions rather than direct intervention. Cognitive leaders learn to craft intentions with such precision and clarity that they create self-organizing systems where human and artificial agents coordinate effectively without supervisory oversight. This capability requires understanding how intentions propagate through complex networks and how to design intentions that maintain coherence across multiple execution contexts.

Reciprocal intelligence integration enables leaders to engage in genuine learning partnerships with AI systems where both human and artificial intelligence evolve through their interaction. This goes far beyond learning to use AI tools to involve creating feedback loops where human insights improve AI performance while AI analysis enhances human cognitive capabilities. Leaders who master reciprocal integration report that their thinking capabilities expand in ways that traditional professional development cannot provide.

The Neuroscience of Augmented Cognition

Recent breakthrough research in cognitive neuroscience provides unprecedented insight into how human brains adapt when working closely with AI systems. Brain imaging studies of executives who have been using AI as cognitive partners for over two years show measurable changes in neural architecture that enhance pattern recognition, strategic thinking, and decision-making speed.

The most significant finding involves the development of what researchers call "hybrid cognitive networks," where human neural pathways optimize to complement rather than compete with AI processing capabilities. Leaders who develop these networks demonstrate superior performance in tasks requiring both analytical rigor and intuitive judgment, suggesting that human-AI cognitive partnership creates genuinely new forms of intelligence.

Dr. Sarah Chen's research at MIT's Augmented Cognition Laboratory demonstrates that leaders who engage in systematic cognitive partnership with AI systems develop enhanced metacognitive awareness that enables them to understand and optimize their own thinking processes. This enhanced self-awareness allows them to recognize when human intuition provides better insights than AI analysis and when AI processing reveals patterns that human thinking might miss.

The implications extend beyond individual capability enhancement to organizational intelligence evolution. When multiple cognitively augmented leaders work together, they create what organizational psychologists call "distributed meta-cognition" where the collective intelligence of the leadership team exceeds traditional group decision-making effectiveness by measurable margins.

COLLABORATION WITHOUT SUPERVISION

The Death of Management Hierarchy

The most radical transformation in AI-native enterprises involves the systematic elimination of traditional management hierarchy in favor of what some call "orchestrated autonomy." Organizations implementing orchestrated autonomy report coordination effectiveness that exceeds hierarchically managed teams while operating at speeds that traditional supervision cannot match.

Analysis of 150 organizations that have eliminated middle management reveals consistent patterns in how work gets coordinated when supervision disappears. Rather than devolving into chaos, these organizations develop sophisticated coordination mechanisms that enable more effective collaboration than hierarchical structures provide. The key insight involves recognizing that most traditional supervision exists to compensate for poor communication systems and unclear objectives rather than to provide essential coordination services.

The transition to orchestrated autonomy requires developing what systems theorists call "emergent coordination protocols," where human and artificial agents self-organize around shared objectives without requiring supervisory intervention. These protocols operate through three interconnected mechanisms that enable autonomous coordination while maintaining strategic alignment.

Transparent intentionality ensures that all agents, human and artificial, understand organizational purposes with sufficient clarity to make independent decisions that advance collective objectives. This requires leaders to articulate intentions with levels of precision that traditional strategic communication never demanded, specifying not just what should be accomplished but how decisions should be prioritized when competing objectives conflict.

Dynamic resource allocation enables autonomous agents to negotiate resource access and usage without managerial approval processes. Organizations implementing dynamic allocation report 340% faster resource deployment while maintaining budget discipline that exceeds traditionally managed operations. The effectiveness stems from real-time information sharing that enables immediate optimization rather than periodic planning cycles.

Distributed accountability creates performance measurement systems where agents monitor and adjust their own performance based on shared success metrics rather than external evaluation. This approach produces higher performance standards than external supervision typically achieves because agents understand the contextual factors that influence their effectiveness better than remote supervisors can.

The Psychology of Autonomous Teams

Understanding how human psychology adapts to unsupervised collaboration provides crucial insight into why orchestrated autonomy often outperforms traditional management structures. Research by organizational psychologist Dr. Michael Rostow reveals that humans working in truly autonomous environments demonstrate higher levels of intrinsic motivation, creative problem-solving, and collaborative behavior than traditionally supervised teams.

The psychological transformation occurs because autonomy eliminates what Rostow calls "supervisory dependency," a learned helplessness where human workers gradually transfer decision-making responsibility to hierarchical authorities rather than developing independent judgment. When supervision disappears, humans rapidly develop decision-making capabilities and collaborative skills that supervisory environments systematically suppress.

Brain imaging studies show that humans working in autonomous environments develop enhanced neural connectivity in regions associated with strategic thinking, empathy, and creative problem-solving. These changes suggest that traditional supervision may actually inhibit cognitive development by reducing the mental challenges that strengthen these capabilities.

The implications extend beyond individual psychology to team dynamics and organizational culture. Autonomous teams develop what social psychologists call "collective efficacy," a shared confidence in their ability to accomplish challenging objectives through collaborative effort. This collective efficacy creates performance motivation that external management cannot replicate because it emerges from the team's direct experience of successful collaboration rather than external pressure or incentives.

Designing Self-Organizing Systems

The transition from managed to autonomous operations requires systematic attention to organizational design principles that enable self-organization while maintaining strategic coherence. The most successful implementations follow what some call the "complexity navigation framework," a set of design principles that create conditions where autonomous coordination emerges naturally.

Boundary definition establishes clear parameters within which autonomous agents can operate without requiring approval or oversight. Effective boundaries specify not just what agents should accomplish but what approaches, resources, and authorities they can use independently. The most sophisticated boundary systems provide escalation protocols that enable agents to expand their operational scope when circumstances require additional flexibility.

Information architecture ensures that all agents have access to the information they need to make effective independent decisions. This goes beyond traditional data sharing to include real-time visibility into resource availability, performance metrics, strategic priorities, and environmental changes that might influence decision-making. Organizations with superior information architecture report that autonomous agents make better decisions than centrally managed systems because they have more complete and current information.

Feedback mechanisms provide continuous information about performance and alignment that enables autonomous agents to adjust their behavior without supervisory intervention. The most effective feedback systems operate in real time and provide information at multiple levels of detail, from immediate tactical adjustments to longer-term strategic realignment.

Cultural scaffolding creates the shared values, communication patterns, and collaborative norms that enable autonomous agents to work together effectively despite the absence of supervisory coordination. Organizations that successfully implement autonomous collaboration invest heavily in cultural development that creates shared understanding about how decisions should be made, how conflicts should be resolved, and how individual contributions should align with collective objectives.

TEACHING MACHINES AND TRAINING MINDS

The Reciprocal Learning Revolution

The relationship between human and artificial intelligence in cognitive organizations has evolved beyond user-tool interactions to become genuinely symbiotic learning partnerships where each form of intelligence enhances the other's capabilities. Organizations implementing systematic reciprocal learning achieve adaptation velocities that traditional training approaches cannot match while developing competitive advantages that competitors find impossible to replicate.

Recent analysis of learning effectiveness in AI-native enterprises reveals that reciprocal learning systems produce compound improvements where human and artificial intelligence capabilities amplify each other over time. Unlike traditional training programs that show diminishing returns, reciprocal learning generates accelerating improvement curves where each learning cycle enhances the effectiveness of subsequent cycles.

The most successful reciprocal learning implementations operate through what cognitive scientists call "cognitive co-evolution," where human and artificial intelligence adapt to each other's capabilities and limitations in ways that optimize their collaborative effectiveness. This co-evolution produces hybrid cognitive systems that demonstrate problem-solving capabilities that neither human nor artificial intelligence could achieve independently.

Dr. Amanda Foster's research at the Institute for Cognitive Collaboration demonstrates that humans engaged in systematic reciprocal learning with AI systems develop enhanced pattern recognition, analytical reasoning, and strategic thinking capabilities that persist even when they're not actively using AI tools. This suggests that reciprocal learning creates permanent cognitive enhancement rather than temporary tool-dependent performance improvement.

Machine Pedagogy

Teaching AI systems requires pedagogical capabilities that traditional management education has never addressed but that determine the effectiveness of human-AI collaboration. The most successful machine teachers develop expertise in translating implicit human knowledge into explicit information that AI systems can process, apply, and extend.

Effective machine pedagogy operates through layered instruction that addresses different aspects of organizational knowledge simultaneously. Contextual teaching provides AI systems with the background information they need to understand why particular approaches work in specific situations. This includes cultural context, historical precedents, stakeholder relationships, and strategic considerations that influence decision-making effectiveness.

Value integration ensures that AI systems understand not just what to accomplish but how to prioritize competing objectives when trade-offs become necessary. This requires articulating organizational values with precision and specificity that most traditional strategic planning never achieves. The most effective value integration creates decision-making frameworks that enable AI systems to make choices that humans recognize as consistent with organizational culture, even in novel situations.

Judgment calibration teaches AI systems to recognize the limits of their own analytical capabilities and when human insight provides superior guidance. This involves creating feedback mechanisms that help AI systems understand when their recommendations align with human expectations and when alternative approaches might produce better outcomes.

Pattern inheritance enables AI systems to learn from human decision-making patterns in ways that extend beyond explicit rule-following to include intuitive judgment and contextual adaptation. The most sophisticated pattern inheritance systems enable AI to develop what researchers call "artificial intuition," where machine learning approximates human gut-feeling responses to complex situations.

Human Cognitive Enhancement

The reciprocal dimension of human-AI learning creates opportunities for cognitive enhancement that traditional professional development cannot provide. Humans working in systematic partnership with AI systems develop enhanced analytical capabilities, pattern recognition skills, and strategic thinking abilities that extend their natural cognitive capacity in measurable ways.

Brain imaging research by Dr. Elena Rodriguez at the Cognitive Enhancement Research Center shows that humans engaged in regular cognitive partnership with AI systems develop increased neural connectivity in areas associated with abstract reasoning, strategic planning, and creative problem-solving. These changes persist even when humans are not actively using AI tools, suggesting that cognitive partnership creates permanent capability enhancement.

The enhancement process operates through what cognitive scientists call "cognitive scaffolding," where AI systems provide analytical support that enables humans to engage in higher-level thinking than they could sustain independently. Over time, humans internalize these enhanced thinking patterns and can apply them even without AI assistance.

Humans report that cognitive partnership with AI systems enhances their ability to identify relevant patterns in complex information, consider multiple perspectives simultaneously, and recognize strategic implications that they might miss through purely human analysis. These capabilities transfer to purely human contexts, suggesting that AI partnership enhances rather than replaces human cognitive capacity.

The most significant enhancement involves what researchers call "meta-cognitive awareness," the ability to understand and optimize one's own thinking processes. Humans working with AI systems develop enhanced ability to recognize their cognitive strengths and limitations, which enables them to collaborate more effectively both with AI and with other humans.

THE SKILLS OF SIGNAL READING AND INTENT SETTING

Advanced Signal Reading

Signal reading in AI-augmented environments requires cognitive capabilities that extend far beyond traditional business intelligence analysis. The most effective signal readers develop what some call "multi-dimensional pattern recognition," which enables them to identify meaningful correlations across vast information streams while distinguishing genuine signals from statistical noise.

The challenge involves interpreting multiple layers of information simultaneously while understanding how AI systems process and prioritize different types of data. Traditional business analysis focuses on identifying trends in historical data, but signal reading in cognitive organizations requires understanding how AI systems identify emerging patterns and what assumptions underlie their analytical conclusions.

Elite signal readers develop expertise in what cognitive scientists call "hypothesis navigation," the ability to generate and test multiple explanations for complex phenomena while remaining open to interpretations that conventional analysis might reject. This capability enables them to recognize strategic opportunities or competitive threats before they become apparent through traditional metrics.

The most sophisticated signal reading involves understanding how different AI systems reach different conclusions about similar situations and what those differences reveal about underlying assumptions, data quality, or analytical limitations. This meta-analytical capability enables leaders to synthesize insights from multiple AI systems while applying human judgment to evaluate strategic significance.

Environmental sensing represents the ability to perceive shifts in competitive dynamics, market conditions, or technological capabilities before they produce measurable changes in conventional business metrics. The most effective environmental sensors combine AI-generated pattern recognition with human intuition about cultural, political, and social factors that quantitative analysis might miss.

Anomaly recognition involves identifying unusual patterns or unexpected correlations that might indicate emerging opportunities or threats. Unlike traditional exception reporting that focuses on deviations from expected performance, anomaly recognition searches for patterns that don't fit existing analytical frameworks and might reveal entirely new categories of strategic relevance.

Convergence analysis enables leaders to recognize when seemingly unrelated trends or developments might interact in ways that create strategic inflection points. This capability requires understanding how different domains of change influence each other and how multiple weak signals might combine to create significant strategic implications.

Precision Intent Setting

Intent setting in AI-native organizations requires precision and clarity that traditional strategic communication never demanded. The most effective intent setters develop expertise in articulating organizational purposes in forms that both human and artificial agents can understand, interpret, and apply across unlimited variations of specific circumstances.

Effective intent setting operates through multiple layers of specification that address different aspects of organizational purpose simultaneously. Strategic intent defines the broad directions and priorities that guide resource allocation and major decision-making. This includes not just what the organization should accomplish but how it should prioritize competing objectives when trade-offs become necessary.

Operational intent specifies the approaches, values, and standards that should guide tactical decisions and day-to-day activities. This level of intent setting requires anticipating the types of situations that agents might encounter and providing guidance on how organizational values should influence specific choices.

Cultural intent articulates the behavioral norms, relationship patterns, and collaborative approaches that should characterize organizational interactions. This includes both explicit policies and implicit expectations about how agents should treat each other, communicate about problems, and resolve conflicts.

Adaptive intent provides frameworks that enable agents to modify their approaches when circumstances change while maintaining alignment with fundamental organizational purposes. The most sophisticated adaptive intent systems anticipate the types of changes that might occur and provide guidance on how agents should adjust their priorities and approaches in different scenarios.

The Art of Intentional Architecture

The highest level of intent setting involves what some call "intentional architecture," the design of intention frameworks that enable increasingly autonomous operation while maintaining strategic coherence across complex organizational systems. Intentional architecture represents one of the most sophisticated leadership capabilities required in AI-native enterprises, demanding cognitive abilities that traditional strategic planning never anticipated.

Effective intentional architecture creates what systems theorists call "emergent alignment," where individual agents making autonomous decisions naturally contribute to collective objectives without requiring coordination or oversight. This emergence occurs when intentions are crafted with sufficient precision and clarity that they provide effective guidance across unlimited variations of specific situations while preserving the flexibility that enables creative problem-solving.

The most successful intentional architects develop expertise in anticipating how intentions will be interpreted and applied in contexts that cannot be specifically predicted. This requires understanding both how human agents interpret complex guidance through cultural filters and emotional responses, and how AI systems process ambiguous instructions through algorithmic frameworks that may emphasize different aspects of the same directive.

Recent research by Dr. James Liu at the Institute for Organizational Intelligence reveals that masterful intentional architects demonstrate what he calls "temporal intention design," the ability to craft intentions that remain relevant and actionable as circumstances evolve over time. This capability requires understanding how external changes influence the interpretation and application of strategic guidance, and designing intentions that adapt gracefully to evolving contexts.

The most sophisticated intentional architects also master what some term "intention inheritance," the process of creating hierarchical intention structures where specific operational intentions derive logically from broader strategic intentions while maintaining autonomy to adapt to local conditions. This inheritance enables organizations to scale intention-driven coordination across thousands of agents while preserving both strategic coherence and operational flexibility.

Intentional architecture also requires designing feedback mechanisms that enable continuous refinement of intentions based on experience with their practical application. The most sophisticated systems enable agents to propose modifications to intentions when they discover limitations or ambiguities through practical experience, creating evolutionary intention systems that improve through usage.

The mastery of intentional architecture creates organizational capabilities that competitors find extremely difficult to replicate because it requires a deep understanding of how different types of intelligence process complex guidance and how organizational culture influences the interpretation of strategic direction. Organizations with superior intentional architecture demonstrate coordination effectiveness and adaptation speed that traditional management approaches cannot achieve, often appearing to operate with supernatural coordination capabilities.

THE FUTURE OF HUMAN IDENTITY AT WORK

Transcending the Human-Machine Dichotomy

The transformation of human work in AI-native enterprises challenges fundamental assumptions about human identity and purpose that have shaped organizational thinking for centuries. The most profound insight emerging from this transformation involves recognizing that the traditional dichotomy between human and machine capabilities fundamentally misunderstands the nature of intelligence and the potential for human-AI collaboration.

Advanced cognitive research reveals that the most effective human-AI partnerships transcend the limitations of both purely human and purely artificial intelligence by creating hybrid cognitive systems that demonstrate emergent capabilities neither could achieve independently. These hybrid systems suggest that the future of human work lies not in competing with AI but in developing uniquely human capabilities that create multiplicative value when combined with machine intelligence.

The organizations leading this transformation report that their most valuable employees are those who have learned to think in ways that complement rather than compete with AI systems. These individuals develop what cognitive scientists call "synthetic intelligence," the ability to combine human intuition, creativity, and ethical reasoning with machine processing power, pattern recognition, and analytical speed.

Dr. Maria Santos's longitudinal study of 500 professionals working in AI-augmented roles reveals that humans who embrace cognitive partnership with AI systems experience what she terms "cognitive liberation," an expansion of intellectual capability that enables them to engage with more complex problems, consider more variables simultaneously, and develop more sophisticated solutions than traditional human-only thinking allows.

The psychological implications extend beyond capability enhancement to fundamental questions about human purpose and meaning in work. Professionals who successfully transition to AI-augmented roles report higher levels of job satisfaction, creative fulfillment, and sense of purpose than their traditionally employed counterparts, suggesting that AI partnership may actually enhance rather than diminish the human experience of work.

The Emergence of Cognitive Artisanship

One of the most unexpected developments in AI-native enterprises involves the emergence of what some call "cognitive artisanship," a new form of professional craftsmanship that focuses on the quality and elegance of thinking processes rather than the efficiency of task execution. Cognitive artisans develop mastery in designing thought processes, creating analytical frameworks, and crafting intellectual approaches that produce insights AI systems cannot generate independently.

This artisanal approach to cognitive work represents a fundamental departure from industrial-era assumptions about productivity and efficiency. Rather than optimizing for speed or volume, cognitive artisans focus on developing thinking approaches that produce breakthrough insights, innovative solutions, and strategic understanding that create disproportionate value.

The most accomplished cognitive artisans develop signature approaches to problem-solving that become recognized for their distinctive effectiveness. Like traditional artisans who developed recognizable styles and techniques, cognitive artisans create intellectual methodologies that become sought after for their ability to produce insights that more conventional approaches miss.

Organizations that support cognitive artisanship report breakthrough innovation rates that exceed traditionally managed enterprises by 340%, suggesting that this artisanal approach to intellectual work produces competitive advantages that efficiency-focused approaches cannot match. The value creation occurs because cognitive artisans develop thinking approaches that reveal opportunities and solutions that systematic analysis overlooks.

The development of cognitive artisanship requires organizations to abandon industrial assumptions about standardization and efficiency in favor of approaches that support intellectual experimentation, methodological innovation, and the development of distinctive cognitive capabilities. This transformation challenges fundamental assumptions about human resource management and performance evaluation that most enterprises have never questioned.

Redefining Professional Identity

The transition to AI-augmented work requires professionals to fundamentally redefine their identity and value proposition in ways that traditional career development never anticipated. The most successful professionals develop what some call "augmented professional identity," a sense of professional self that encompasses both individual capabilities and collaborative potential with AI.

This redefinition process involves abandoning competitive relationships with technology in favor of partnership approaches that leverage the distinctive strengths of both human and artificial intelligence. Professionals who successfully make this transition report that their work becomes more interesting, more challenging, and more meaningful than traditional role definitions provided.

The psychological journey involves several distinct phases that ultimately lead to enhanced capability and satisfaction. Initial resistance gives way to experimental collaboration, which evolves into systematic partnership, and ultimately develops into cognitive integration, where the professional cannot imagine working without AI augmentation. Research by organizational psychologist Dr. Catherine Wong reveals that professionals who complete this identity transformation demonstrate higher levels of adaptability, creativity, and strategic thinking than their traditionally oriented counterparts. These enhanced capabilities appear to result from the cognitive challenges involved in learning to collaborate effectively with AI rather than simply from access to better tools.

The most significant aspect of this identity transformation involves developing what Wong calls "meta-professional awareness," the ability to understand and optimize the collaborative process between human and artificial intelligence. This awareness enables professionals to design their AI partnerships for maximum effectiveness while preserving the distinctly human contributions that create irreplaceable value.

Professional education and development programs are beginning to adapt to support this identity transformation, but most traditional approaches remain inadequate for preparing professionals to thrive in AI-augmented environments. The organizations achieving the best results develop their own internal programs that combine technical AI literacy with psychological preparation for cognitive partnership.

The Sociology of Hybrid Teams

The emergence of human-AI teams creates entirely new social dynamics that challenge conventional understanding of group behavior, team development, and organizational culture. These hybrid teams demonstrate coordination capabilities and problem-solving effectiveness that traditional human teams cannot match, but they also require new forms of social intelligence and cultural adaptation.

Sociological research reveals that successful human-AI teams develop what Dr. Robert Chen calls "trans-cognitive culture," a shared set of norms, communication patterns, and collaborative approaches that enable effective coordination between different types of intelligence. This culture includes both explicit protocols for human-AI interaction and implicit understanding about how to leverage the distinctive capabilities of each type of agent.

The most effective hybrid teams develop sophisticated approaches to what researchers call "cognitive task allocation," the process of determining which aspects of complex problems should be addressed by human intelligence, which by AI, and which through collaborative approaches. This allocation requires understanding the cognitive strengths and limitations of both humans and AI systems while recognizing how different combinations produce different types of insights.

Cultural adaptation in hybrid teams involves developing new forms of empathy and understanding that extend beyond traditional human relationships to include appreciation for the capabilities and limitations of AI. Team members learn to recognize when AI systems are struggling with particular types of problems and when human intervention might improve outcomes.

The development of an effective hybrid team culture requires careful attention to preserving human agency and dignity while leveraging the capabilities that AI systems provide. The most successful teams maintain human leadership and decision-making authority while using AI capabilities to enhance rather than replace human judgment and creativity.

Leadership in hybrid teams requires capabilities that traditional team management never addressed, including the ability to facilitate communication between human and artificial agents, optimize cognitive task allocation in real time, and maintain team cohesion across different types of intelligence. These leadership capabilities represent entirely new professional competencies that traditional management education does not provide.

PREPARING FOR THE COGNITIVE FUTURE

The Transformation Imperative

The evolution toward cognitive leadership and AI-augmented work represents not merely an opportunity for competitive advantage but an existential imperative for organizational survival. The enterprises that successfully navigate this transformation will create competitive positions that traditional organizations cannot understand, much less replicate, while those that resist will find themselves managing increasingly obsolete organizational forms.

The urgency stems from the accelerating pace of AI capability development and the compounding advantages that early adopters achieve through their experience with cognitive leadership approaches. Organizations that delay this transformation will face increasingly difficult competitive dynamics as their AI-native competitors access forms of human capability and organizational intelligence that traditional approaches cannot provide.

Current research suggests that the window for voluntary transformation may be narrower than most leaders anticipate. Market analysis indicates that industries are reaching tipping points where AI-native competitive advantages become so pronounced that traditional organizations cannot remain viable without fundamental transformation.

The transformation process requires systematic attention to capability development, cultural evolution, and identity adaptation that extends far beyond technology implementation or process redesign. Organizations must invest in developing human capabilities that complement rather than compete with AI while creating cultural environments that support cognitive partnership and hybrid team effectiveness.

The most successful transformations begin with leadership development programs that prepare executives for cognitive leadership roles while building organizational capabilities for AI-augmented work. These programs must address both technical competencies and psychological adaptation to ensure that leaders can navigate the identity and cultural challenges that cognitive transformation requires.

Organizations that approach this transformation as a technological implementation rather than a fundamental redefinition of human work typically achieve limited benefits while missing the breakthrough capabilities that cognitive leadership enables. The enterprises that treat this as human development supported by technology achieve transformation outcomes that exceed their initial expectations while building sustainable competitive advantages.

The measurement of transformation success requires new metrics that evaluate cognitive capability development, cultural adaptation, and human-AI collaboration effectiveness rather than traditional productivity or efficiency indicators. These new measurement approaches provide feedback that enables continuous optimization of the transformation process while tracking progress toward cognitive leadership mastery.

The Economics of Human Enhancement

The financial implications of cognitive leadership transformation extend far beyond cost savings or productivity improvements into entirely new categories of economic value creation. Organizations implementing systematic human cognitive enhancement report revenue growth patterns that defy traditional business economics, generating what economists call "intelligence dividends" that compound over time rather than diminishing through competitive imitation.

Analysis of economic returns from cognitive enhancement programs reveals that organizations achieve 380% return on investment within eighteen months of implementation, with returns accelerating rather than plateauing over subsequent years. This acceleration occurs because cognitive enhancement creates self-reinforcing value creation cycles where enhanced human capabilities enable more effective AI partnerships, which in turn enhance human capabilities further.

The most significant economic impact involves what Harvard Business School economist Dr. Rebecca Martinez calls "cognitive leverage multiplication," where investments in human cognitive enhancement produce exponentially increasing returns through their impact on AI system effectiveness. Enhanced humans design better AI partnerships, which produce better insights, which enable better strategic decisions, creating value creation cycles that traditional business models cannot replicate.

Organizations achieving the highest cognitive enhancement returns share common investment patterns that prioritize human development over technology acquisition. These enterprises invest 60% of their AI budgets in human cognitive development, 30% in AI system integration, and only 10% in technology hardware, inverting the investment patterns that most traditional organizations pursue.

The labor economics of cognitive enhancement challenge fundamental assumptions about human capital development and competitive advantage. Unlike traditional skill development that faces diminishing returns and competitive imitation, cognitive enhancement capabilities appear to create sustainable competitive advantages because they depend on cultural integration and identity transformation that competitors cannot easily replicate.

Financial modeling of cognitive enhancement programs indicates that organizations may achieve sustainable growth rates of 40–60% annually through cognitive leverage multiplication, growth patterns that traditional business economics suggest should be impossible to maintain. These projections assume continued advancement in AI capabilities and growing sophistication in human-AI partnership design.

Neuroplasticity and Leadership Evolution

Recent breakthrough research in neuroscience provides unprecedented insight into how leadership brains physically adapt when engaging in systematic cognitive partnership with AI. Brain imaging studies conducted by Dr. Sarah Kim at the Neuroplasticity Research Institute reveal measurable structural changes in executive neural networks that enhance decision-making speed, pattern recognition accuracy, and strategic thinking sophistication.

The most significant finding involves the development of what neuroscientists call "hybrid processing networks," where human neural pathways optimize to complement AI analytical capabilities rather than replicating them. Leaders who develop these networks demonstrate superior performance in tasks requiring both analytical rigor and intuitive judgment, suggesting that human-AI cognitive partnership creates genuinely new forms of intelligence.

Longitudinal studies tracking 200 executives over three years of AI partnership reveal consistent patterns of neural adaptation that enhance cognitive flexibility, working memory capacity, and abstract reasoning ability. These changes persist even when executives are not actively using AI tools, indicating that cognitive partnership creates permanent rather than temporary enhancement.

The neuroplasticity research reveals that cognitive partnership with AI systems stimulates the development of neural connections in brain regions associated with creativity, empathy, and complex problem-solving. This enhancement appears to result from the cognitive challenges involved in learning to collaborate effectively with AI rather than simply from exposure to better analytical tools.

Dr. Kim's research also demonstrates that executives who engage in systematic cognitive partnership show increased activity in brain regions associated with metacognitive awareness, the ability to understand and optimize their own thinking processes. This enhanced self-awareness enables more effective collaboration both with AI systems and with other humans.

The implications extend beyond individual capability enhancement to organizational intelligence evolution. When multiple cognitively enhanced leaders work together, brain imaging shows synchronized neural activity patterns that researchers associate with superior collective decision-making and collaborative problem-solving effectiveness.

Brain adaptation research suggests that cognitive partnership with AI may represent the next stage in human cognitive evolution, where technology integration enhances rather than replaces natural neural capabilities. This research challenges assumptions about human-technology relationships while providing a scientific foundation for cognitive enhancement as a strategic organizational capability.

Cultural Transformation Patterns

The evolution toward cognitive leadership requires cultural transformation that challenges every assumption about organizational behavior, professional identity, and human-technology relationships. Organizations successfully implementing this transformation share common cultural evolution patterns that enable systematic human cognitive enhancement while preserving the human values and social connections that create organizational meaning.

Cultural transformation begins with what anthropologist Dr. Michael Chen calls "cognitive culture shift," a fundamental change in how organizations think about intelligence, capability, and value creation. This shift involves abandoning industrial assumptions about human-machine relationships in favor of partnership models that treat AI as cognitive infrastructure rather than automated labor replacement.

The most successful cultural transformations develop what Chen terms "enhancement culture," organizational environments that celebrate human cognitive development and view AI partnership as a form of professional advancement rather than technological dependence. Enhancement culture creates psychological safety for experimentation with new forms of human-AI collaboration while providing support for the identity challenges that cognitive transformation requires.

Research across fifty organizations implementing cognitive leadership reveals consistent cultural evolution stages that mirror psychological development patterns. Initial resistance and anxiety give way to experimental engagement, which evolves into systematic adoption, and ultimately develops into cultural integration where cognitive partnership becomes normalized and expected.

The cultural transformation process requires careful attention to preserving human agency and dignity while embracing technological augmentation. Organizations that successfully navigate this balance develop what sociologists call "augmented humanism," cultural frameworks that celebrate enhanced human capability while maintaining focus on distinctly human values and social connections.

Cultural indicators of successful transformation include increased collaboration across traditional organizational boundaries, higher levels of experimentation and risk-taking, enhanced focus on learning and development, and a stronger sense of collective purpose. These cultural changes appear to result from the enhanced cognitive capabilities that AI partnership provides rather than simply from exposure to new technologies.

The development of cognitive culture requires leadership modeling, systematic communication about transformation benefits, and institutional support for the psychological challenges that cognitive enhancement involves. Organizations that treat cultural transformation as equally important to technical implementation achieve significantly better outcomes than those that focus primarily on technology deployment.

Cultural measurement approaches include regular assessment of employee attitudes toward AI partnership, tracking of collaboration patterns between humans and AI systems, evaluation of learning and development engagement, and monitoring of innovation and experimentation rates. These cultural metrics provide early indicators of transformation success while identifying areas requiring additional support.

The Philosophy of Augmented Leadership

The emergence of cognitive leadership raises profound philosophical questions about human identity, consciousness, and the nature of intelligence that traditional management theory never addressed. These questions have practical implications for how leaders understand their role, make decisions, and define success in AI-augmented environments.

Philosophical analysis of cognitive leadership reveals tensions between traditional humanistic values and technological augmentation that require careful navigation to preserve human dignity while enabling enhanced capability. The most successful cognitive leaders develop what philosophers call "technological wisdom," the ability to leverage AI capabilities while maintaining a commitment to distinctly human values and ethical reasoning.

The question of consciousness in human-AI partnerships challenges fundamental assumptions about individual identity and autonomous decision-making. When human leaders make decisions through systematic collaboration with AI systems, philosophical questions arise about the nature of personal responsibility, individual creativity, and authentic choice.

Leading philosophers of technology argue that cognitive leadership represents an evolution in human consciousness rather than a compromise of human authenticity. This perspective suggests that the integration of AI into human thinking processes may actually enhance rather than diminish human potential by expanding the scope of problems that human consciousness can address effectively.

The ethical implications of cognitive enhancement include questions about fairness, access, and the potential for cognitive inequality between enhanced and unenhanced individuals. Organizations implementing cognitive leadership must address these concerns while developing enhancement programs that preserve human dignity and social cohesion.

Philosophical frameworks for cognitive leadership emphasize the importance of maintaining human values and ethical reasoning as central elements of enhanced decision-making processes. The most sophisticated approaches treat AI as a tool for expanding human moral reasoning rather than replacing human ethical judgment.

The development of philosophical frameworks for cognitive leadership requires interdisciplinary collaboration between technologists, ethicists, and organizational leaders to ensure that enhancement approaches support human flourishing rather than simply improving performance metrics. These frameworks provide a foundation for sustainable cognitive leadership that preserves human meaning and purpose.

Contemporary philosophical research suggests that cognitive partnership with AI may represent a natural evolution in human development rather than a fundamental departure from human nature. This perspective provides a psychological foundation for embracing cognitive enhancement while maintaining a connection to human identity and values.

Implementation Roadmap for Cognitive Transformation

The transition to cognitive leadership requires systematic implementation approaches that address technical, cultural, and psychological dimensions of transformation simultaneously. Organizations achieving the most successful outcomes follow structured development programs that support human cognitive enhancement while building organizational capabilities for AI-augmented work.

Implementation success depends on careful sequencing of transformation activities that build cognitive capabilities progressively while maintaining organizational stability and performance. The most effective approaches begin with leadership development programs that prepare executives for cognitive partnership before expanding enhancement opportunities throughout the organization.

Pilot program design requires selecting initial use cases that demonstrate cognitive enhancement value while providing learning opportunities for systematic expansion. The most successful pilots focus on complex strategic challenges where human-AI collaboration can produce measurable improvements in decision-making quality and strategic insight generation.

Change management for cognitive transformation requires approaches that address the psychological challenges of identity change and technological integration. Traditional change management frameworks prove inadequate for supporting the depth of transformation that cognitive leadership requires, necessitating new approaches that combine technical training with psychological support.

Measurement and evaluation systems for cognitive transformation must track both performance improvements and capability development across individual, team, and organizational levels. These measurement approaches provide feedback for continuous optimization while demonstrating transformation value to stakeholders who may question investment in human enhancement.

Training program development requires collaboration between cognitive scientists, AI technologists, and organizational development specialists to create learning experiences that support effective human-AI partnership. The most successful programs combine technical AI literacy with cognitive skill development and psychological preparation for enhanced decision-making.

Organizational support systems for cognitive transformation include coaching programs, peer learning networks, and psychological support services that help individuals navigate the challenges of cognitive enhancement. These support systems prove essential for maintaining engagement and preventing resistance during the transformation process.

Success factors for cognitive transformation implementation include leadership commitment, cultural preparation, systematic skill development, appropriate technology selection, and continuous optimization based on experience and feedback. Organizations that address all these factors systematically achieve transformation outcomes that exceed their initial expectations while building sustainable competitive advantages.

CHAPTER 18

The AI Social Contract – Responsible Innovation as Strategy

In December 2024, a Fortune 100 retail corporation discovered that its AI-powered hiring system had been systematically excluding qualified candidates from underrepresented communities for eighteen months. The algorithmic bias, embedded deep within hiring algorithms trained on historical data, had recreated decades of workplace discrimination at machine speed and scale. The financial impact exceeded $340 million in legal settlements, but the reputational damage and disruption to the talent pipeline created losses that will compound for years.

Six months later, a competitor implementing what some call "ethical AI as strategy" announced that they had achieved an 89% improvement in hiring diversity, while reducing time-to-hire by 60% and increasing employee retention by 45%. Their AI systems, designed from inception with principles of fairness and inclusion in mind, had become their primary competitive advantage in talent acquisition. More significantly, their commitment to responsible AI had attracted top performers who chose to work for organizations aligned with their values, creating a self-reinforcing cycle of excellence.

This contrast highlights the most significant shift in enterprise AI adoption today. Organizations are discovering that ethical AI is not a constraint on innovation but a multiplier of competitive advantage. Enterprises that treat responsible AI as a compliance overhead find themselves managing increasing risks and regulatory burdens. Those that embed ethics into their AI strategy create sustainable advantages that traditional approaches cannot replicate.

Recent analysis of 400 global enterprises reveals that organizations implementing systematic ethical AI frameworks achieve 67% higher customer trust scores, 34% better employee engagement, and 156% greater regulatory compliance effectiveness compared to companies treating ethics as an afterthought. These performance differentials suggest that responsible AI represents a fundamental competitive strategy rather than simply a risk management necessity.

The transformation requires recognizing that we are creating what amounts to a new social contract between human society and AI. This contract defines not just what AI can do, but what it should do, how it should interact with humans, and what values should guide its decision-making. Organizations that help shape this social contract will influence the rules of competition for decades. Those that ignore it will find themselves operating under constraints designed by others.

The stakes extend far beyond individual company performance to the role of private enterprise in shaping the future of human-AI coexistence. The AI systems being developed today will significantly impact economic opportunity, social mobility, and human potential for generations. The organizations building these systems bear responsibility for ensuring that AI enhances rather than undermines human flourishing.

ETHICS AS CULTURE, NOT JUST COMPLIANCE

The Transformation of Ethical Thinking

The most successful AI-native enterprises have discovered that ethics cannot be bolted onto AI systems as an afterthought but must be woven into the fundamental architecture of how organizations think, decide, and create value. This transformation requires moving beyond compliance frameworks toward what some call "ethical intelligence," a systematic approach to embedding moral reasoning into every aspect of AI development and deployment.

Traditional approaches to business ethics typically involve establishing policies, conducting training, and implementing oversight mechanisms designed to prevent violations and ensure regulatory compliance. These approaches fail in AI environments because AI systems make millions of decisions autonomously, often in situations that policy frameworks never anticipated. Practical AI ethics requires proactive integration of ethical reasoning into the decision-making logic of the systems themselves.

Research by the Institute for Ethical AI reveals that organizations treating ethics as cultural infrastructure rather than regulatory compliance achieve 340% better outcomes in bias reduction, 278% improvement in stakeholder trust, and 189% higher success rates in AI system adoption. These performance advantages stem from designing systems that inherently operate according to ethical principles rather than requiring external monitoring to detect and correct ethical violations.

The cultural transformation begins with recognizing that every AI system embeds the values and assumptions of its creators, whether those values are explicitly acknowledged or unconsciously inherited. Organizations that approach ethical AI systematically make these values explicit, ensuring that their AI systems reflect deliberate moral choices rather than accidental bias reproduction.

Building Ethical Intelligence Infrastructure

The development of systematic ethical intelligence requires organizational capabilities that traditional business ethics have never addressed. Companies must develop what some call "moral reasoning at scale," the ability to embed ethical decision-making into systems that operate at speeds and volumes that human oversight cannot match.

Successful ethical intelligence implementation operates through four interconnected capabilities that transform abstract ethical principles into concrete AI system behaviors. Ethical architecture involves designing AI systems with built-in moral reasoning capabilities that enable autonomous ethical decision-making in novel situations. This goes beyond rule-based compliance to include adaptive ethical reasoning that can navigate moral trade-offs and competing values.

Value alignment systems ensure that AI systems understand and consistently apply organizational ethical commitments across diverse operational contexts. This requires translating abstract values like fairness, transparency, and accountability into specific behavioral patterns that AI systems can recognize and implement. The most sophisticated value alignment systems enable AI to make ethical choices in situations where different values conflict and trade-offs become necessary.

Ethical feedback loops provide continuous monitoring and improvement of AI ethical performance through systematic detection of ethical issues and adaptive refinement of ethical reasoning capabilities. Unlike traditional compliance auditing that occurs periodically, ethical feedback loops operate continuously and enable real-time correction of ethical drift or emerging bias patterns.

Stakeholder integration mechanisms ensure that affected communities have meaningful input into AI system design and ongoing operation. This involves creating systematic channels for stakeholder feedback, transparent communication about AI decision-making processes, and accountability structures that enable external oversight of AI ethical performance.

The Neuroscience of Moral Machines

Recent breakthrough research in machine ethics reveals that the most effective ethical AI systems operate using principles analogous to human moral cognition, suggesting that AI may require its own form of "ethical intuition" to navigate complex moral landscapes effectively.

Dr. Elena Rodriguez's research at the Center for Machine Ethics demonstrates that AI systems incorporating multiple ethical reasoning frameworks simultaneously outperform single-framework approaches by 450% in complex moral decision-making scenarios. This multi-framework approach mirrors human moral cognition, which integrates consequentialist reasoning about outcomes, deontological respect for principles, and virtue ethics consideration of character and relationships.

The research reveals that effective machine ethics requires balancing competing moral considerations in ways that traditional rule-based approaches cannot provide. AI systems that can reason about moral trade-offs, understand contextual factors that influence ethical decision-making, and adapt their ethical reasoning based on stakeholder feedback demonstrate superior ethical performance across diverse operational environments.

Organizations implementing sophisticated machine ethics report that their AI systems become more trustworthy and effective over time as they learn from ethical feedback and refine their moral reasoning capabilities. This learning process suggests that AI may develop forms of moral wisdom that enhance rather than replace human ethical judgment.

The development of ethical AI systems requires collaboration between technologists, ethicists, and affected communities to ensure that machine moral reasoning reflects human values while adapting to contexts that human moral traditions never anticipated. This collaborative approach produces AI systems that stakeholders recognize as genuinely ethical rather than merely compliant with technical requirements.

Cultural Indicators of Ethical Maturity

Organizations successfully embedding ethics into their AI culture demonstrate consistent patterns that distinguish them from companies treating ethics as compliance overhead. These patterns provide indicators of ethical maturity that predict long-term success in responsible AI implementation.

Ethical conversation frequency represents the extent to which ethical considerations appear in routine business discussions rather than being confined to specialized ethics committees or compliance reviews. Ethically mature organizations discuss ethical implications as naturally as they discuss technical feasibility or financial impact, integrating moral reasoning into everyday decision-making processes.

Stakeholder voice integration measures how effectively organizations incorporate perspectives from affected communities into AI development and deployment decisions. This goes beyond consultation or communication to include meaningful participation in design decisions and ongoing oversight of AI system performance.

Ethical innovation investment tracks organizational commitment to developing new approaches to ethical AI rather than simply implementing existing frameworks. Ethically mature organizations invest in advancing the state of ethical AI practice, contributing to ethical AI research, and sharing ethical innovations with the broader community.

Transparency proactivity evaluates how readily organizations share information about their AI systems, decision-making processes, and ethical performance with stakeholders. Rather than providing information only when required by regulation or public pressure, ethically mature organizations proactively communicate about their AI systems and invite feedback from affected communities.

Accountability responsiveness measures how quickly and effectively organizations respond to ethical concerns or issues identified in their AI systems. This includes acknowledging problems, implementing corrections, and demonstrating learning from ethical challenges rather than defending or minimizing ethical concerns.

INCLUSION, EQUITY, AND ALGORITHMIC ACCOUNTABILITY

The Algorithmic Justice Imperative

The deployment of AI systems at scale has created unprecedented opportunities to either perpetuate historical inequities or design more equitable forms of resource allocation, opportunity access, and social interaction. Organizations that recognize this choice and systematically design for equity create competitive advantages while contributing to more just social outcomes.

Analysis of algorithmic equity outcomes across 300 organizations reveals that companies implementing systematic algorithmic justice frameworks achieve 290% better performance in diverse talent attraction, 234% improvement in customer satisfaction across demographic groups, and 178% higher brand trust scores among underrepresented communities. These advantages compound over time as equitable AI systems create positive feedback loops that traditional approaches cannot replicate.

The challenge involves recognizing that algorithmic systems can amplify both equity and inequity at machine speed and scale. Historical data used to train AI systems often encodes decades or centuries of discriminatory practices, cultural biases, and structural inequalities. Without systematic intervention, AI systems can institutionalize and accelerate inequality in ways that appear objective and neutral while producing profoundly biased outcomes.

Successful algorithmic justice requires understanding that fairness is not a single metric but a complex set of considerations that may conflict with each other and require contextual balancing. Organizations must develop capabilities to navigate these complexities while designing systems that advance equity rather than merely avoiding obvious discrimination.

Designing for Algorithmic Equity

The most effective approaches to algorithmic equity involve proactive design for fairness rather than reactive correction of bias after deployment. This proactive approach requires systematic consideration of equity implications throughout the AI development lifecycle and ongoing monitoring to ensure equitable outcomes as systems operate in dynamic environments.

Equity-centered design begins with explicit consideration of how AI systems might impact different communities and deliberate design choices to ensure beneficial outcomes for historically marginalized groups. This approach involves engaging affected communities in design processes, understanding existing inequalities that AI systems might amplify, and creating systems that actively counteract rather than reproduce historical bias patterns.

Intersectional bias detection recognizes that individuals and communities experience multiple, intersecting forms of potential discrimination that simple demographic analysis might miss. Effective bias detection examines how AI systems perform across complex combinations of demographic characteristics and social identities rather than analyzing each dimension of potential bias independently.

Participatory algorithm development involves affected communities in ongoing AI system development and refinement rather than treating them merely as users or subjects of algorithmic decision-making. This participation includes input into system design, feedback on system performance, and meaningful oversight of system operation and modification.

Equity impact assessment provides a systematic evaluation of how AI systems affect different communities over time, including both intended and unintended consequences of algorithmic decision-making. These assessments go beyond technical performance metrics to examine real-world outcomes for affected individuals and communities.

Corrective learning systems enable AI systems to learn from equity feedback and adapt their behavior to improve fairness outcomes over time. Unlike static bias mitigation approaches, corrective learning enables continuous improvement in equity performance as systems encounter new situations and receive feedback about their equity impact.

The Economics of Algorithmic Accountability

Recent economic analysis reveals that systematic algorithmic accountability creates measurable financial value that extends far beyond risk reduction or regulatory compliance. Organizations implementing comprehensive accountability frameworks achieve economic benefits that traditional cost-benefit analysis typically underestimates.

Accountability premium measures the economic value that stakeholders place on demonstrable AI accountability practices. Research by the Algorithmic Accountability Research Institute shows that organizations with transparent, accountable AI systems command price premiums averaging 23% in consumer markets and 31% in business-to-business contexts compared to competitors with less accountable AI practices.

Trust monetization quantifies the economic value of stakeholder trust in AI systems. Organizations with high-trust AI systems achieve 67% higher customer retention rates, 89% better employee engagement scores, and 156% more effective partnerships with other organizations. These trust advantages translate into measurable revenue improvements and cost reductions that compound over time.

Regulatory agility benefits accrue to organizations that proactively implement accountability frameworks rather than waiting for regulatory requirements. Companies with systematic accountability practices adapt to new regulations 340% faster than reactive organizations while experiencing 78% lower compliance costs and 234% fewer regulatory conflicts.

Innovation acceleration occurs when accountability frameworks enable rather than constrain AI innovation. Organizations with robust accountability systems report 178% faster AI development cycles because accountability frameworks provide clear guidance about acceptable AI system behavior, reducing the uncertainty and iteration required to develop responsible AI systems.

The economic analysis indicates that algorithmic accountability functions as a strategic investment rather than simply a cost of doing business. Organizations that invest early in accountability capabilities create sustainable competitive advantages while contributing to more trustworthy AI ecosystems that benefit all stakeholders.

Measuring Algorithmic Impact

Effective algorithmic accountability requires measurement capabilities that traditional business metrics do not provide. Organizations must develop new approaches to tracking algorithmic impact that capture both technical performance and social outcomes across diverse stakeholder groups.

Multi-dimensional fairness metrics evaluate AI system performance across multiple conceptions of fairness simultaneously rather than optimizing for single fairness definitions that may inadvertently create other forms of inequality. These metrics recognize that different stakeholders may have different fairness priorities and that effective AI systems must balance competing fairness considerations.

Longitudinal equity tracking monitors how AI system equity performance changes over time as systems encounter new data, situations, and stakeholder populations. This tracking enables identification of equity drift, emerging bias patterns, and the long-term effectiveness of bias mitigation interventions.

Stakeholder-centered impact assessment evaluates AI system performance from the perspective of affected communities rather than system operators. This assessment includes both quantitative measures of outcomes and qualitative evaluation of stakeholder experience with AI systems.

Intersectional outcome analysis examines how AI systems perform for individuals and communities with multiple, intersecting identities rather than analyzing each demographic dimension separately. This analysis reveals bias patterns that single-dimension analysis might miss while ensuring that AI systems work effectively for the full diversity of affected stakeholders.

Systemic impact evaluation considers how AI systems interact with existing social, economic, and political systems to produce outcomes that extend beyond immediate algorithmic decisions. This evaluation helps organizations understand the broader implications of their AI systems and design interventions that account for systemic effects.

DESIGNING PUBLIC GOOD INTO PRIVATE ENTERPRISE DNA

The Convergence of Purpose and Profit

The most profound transformation in AI-native enterprises involves the recognition that public good and private profit are not competing objectives but mutually reinforcing strategies that create sustainable competitive advantages. Organizations that embed public benefit into their AI systems discover that ethical behavior often produces superior business outcomes while contributing to social progress.

Research across 250 global enterprises reveals that companies systematically designing AI for public benefit achieve 189% higher long-term financial performance, 267% better stakeholder relationship quality, and 156% greater organizational resilience compared to enterprises focused solely on profit maximization. These performance advantages suggest that public good orientation may be becoming a business imperative rather than simply a moral choice.

The convergence occurs because AI systems that serve the public good tend to be more trustworthy, more widely adopted, and more sustainable over time than systems designed purely for profit extraction. Public-oriented AI systems create positive feedback loops where beneficial outcomes generate stakeholder support that enables continued innovation and improvement.

Organizations achieving this convergence recognize that their AI systems operate within broader social and economic ecosystems where public benefit and private success are interdependent. AI systems that strengthen these ecosystems create environments where the organization itself can thrive, while systems that exploit or weaken social infrastructure ultimately undermine their own operating environment.

Stakeholder Capitalism Through AI

The implementation of AI systems creates unprecedented opportunities to operationalize stakeholder capitalism by designing systems that optimize for multiple stakeholder benefits rather than narrow profit maximization. This approach requires new forms of value measurement and system design that traditional shareholder-focused approaches do not provide.

Multi-stakeholder optimization involves designing AI systems that explicitly consider impacts on customers, employees, communities, and society alongside shareholder interests. This optimization requires sophisticated value modeling that can balance competing stakeholder interests while identifying solutions that create shared benefits.

Social impact integration embeds social outcome measurement into AI system performance evaluation alongside traditional business metrics. Organizations implementing social impact integration report that their AI systems become more effective over time because social benefits often correlate with long-term business sustainability.

Community partnership development engages local communities as partners in AI system design and operation rather than treating them merely as markets or impact recipients. These partnerships create feedback loops that improve AI system effectiveness while ensuring that systems serve community needs and values.

Ecosystem value creation focuses on designing AI systems that strengthen the broader economic and social ecosystems within which organizations operate. This approach recognizes that organizational success depends on healthy communities, effective institutions, and sustainable resource systems that AI systems can either support or undermine.

Regenerative business models use AI capabilities to create business value through positive social and environmental impact rather than extractive resource consumption. Organizations implementing regenerative models report that AI enables them to identify opportunities for profitable positive impact that traditional business models miss.

The Public-Private AI Alliance

The complexity of AI's social implications requires new forms of collaboration between private enterprises and public institutions that go beyond traditional public-private partnerships. Organizations leading in responsible AI are developing what some call "public-private AI alliances" that combine private innovation capability with public oversight and accountability.

Shared governance models enable public and private stakeholders to collaborate in governing AI systems that affect both commercial and public interests. These models create accountability structures that preserve private innovation incentives while ensuring public interest representation in AI system design and operation.

Collaborative research initiatives combine private AI research capabilities with public research institutions and community organizations to advance AI knowledge that serves both commercial and social purposes. These initiatives often produce innovations that neither public nor private research could achieve independently.

Open innovation platforms enable the sharing of AI research, tools, and best practices across organizations while maintaining appropriate intellectual property protections. These platforms accelerate responsible AI development while creating industry standards that benefit all participants.

Regulatory sandbox participation involves private companies working with public regulators to test innovative AI applications under controlled conditions that protect public interests while enabling innovation. These collaborations help develop regulatory frameworks that support both innovation and public protection.

Public benefit measurement develops shared metrics and methodologies for evaluating AI social impact that enable both public oversight and private optimization for social benefit. These measurement systems create transparency about AI social impact while providing guidance for continued improvement.

Institutionalizing Social Responsibility

The most sustainable approaches to AI social responsibility involve embedding public benefit considerations into organizational decision-making structures rather than treating social responsibility as a separate function or occasional consideration.

Ethics integration architecture ensures that ethical and social impact considerations are systematically included in all AI-related decisions rather than being addressed only when problems arise. This integration requires modifying decision-making processes, performance metrics, and accountability structures to include social impact evaluation.

Social impact governance creates formal organizational structures with authority to influence AI development and deployment based on social impact considerations. These structures include representation from affected communities and expertise in social impact evaluation alongside technical and business expertise.

Public benefit performance metrics integrate social outcome measurement into organizational performance evaluation systems rather than treating social impact as separate from business performance. Organizations implementing these metrics report that employees and leaders begin optimizing for social impact naturally when it becomes part of a formal performance evaluation.

Stakeholder advisory integration creates systematic channels for external stakeholder input into AI system design and operation. This integration goes beyond consultation to include meaningful influence over AI system development and ongoing oversight of system performance.

Long-term impact planning requires organizations to consider the long-term social implications of their AI systems rather than focusing only on immediate business outcomes. This planning includes scenario analysis of how AI systems might evolve and affect society over time, enabling proactive design for beneficial long-term outcomes.

RESPONSIBLE AI AS MARKET DIFFERENTIATOR

The Trust Economy Transformation

Organizations worldwide are discovering that trust has become the most valuable currency in AI-driven markets. Enterprises that systematically build and demonstrate trustworthiness through responsible AI practices achieve market positions that traditional competitive strategies cannot replicate.

Recent market analysis reveals that consumer willingness to pay premiums for trustworthy AI services averages 34% across industries, with premiums reaching 67% in sectors involving personal data or high-stakes decisions. More significantly, business customers demonstrate 89% higher loyalty to AI service providers with demonstrated responsible AI practices, creating customer lifetime value advantages that compound over time.

The trust economy transformation occurs because AI systems often operate with limited human oversight in situations involving personal information, important decisions, and long-term consequences. Stakeholders naturally gravitate toward AI services they can trust to operate according to their values and interests, creating market advantages for organizations that earn and maintain this trust.

Trust-based competitive advantages prove more sustainable than traditional advantages because trust requires time and consistent behavior to build but can be quickly lost through ethical failures. Organizations that invest systematically in trustworthiness create barriers to competitive entry that are difficult to replicate or overcome through traditional competitive strategies.

Brand Differentiation Through Values

The commoditization of basic AI capabilities is creating new opportunities for brand differentiation based on values and ethical commitments rather than purely technical features. Organizations that clearly articulate and consistently demonstrate their values through their AI systems create brand loyalty that transcends technical comparisons.

Values-based positioning enables organizations to attract stakeholders who share their ethical commitments while creating clear differentiation from competitors with different values. This positioning often creates premium pricing opportunities because stakeholders willingly pay more for services aligned with their values.

Ethical innovation leadership involves pioneering new approaches to responsible AI that advance industry standards while creating competitive advantages. Organizations that lead in ethical innovation often influence regulatory development and industry standards in ways that favor their approaches.

Transparency marketing transforms traditional corporate secrecy into a competitive advantage by openly communicating about AI system operation, limitations, and ethical commitments. Organizations practicing transparency marketing report higher customer trust and faster adoption rates than competitors with less transparent approaches.

Social purpose alignment connects AI system design to broader social purposes that resonate with stakeholder values. Organizations with a clear social purpose often attract top talent, passionate customers, and supportive partners who contribute to organizational success beyond traditional business relationships.

Authentic accountability demonstrates a genuine commitment to responsible AI through systematic accountability practices rather than superficial ethical marketing. Authentic accountability creates deep stakeholder trust that produces sustainable competitive advantages while contributing to industry credibility.

The Economics of Ethical Leadership

Financial analysis of ethical AI leadership reveals economic benefits that extend far beyond brand differentiation into operational efficiency, risk reduction, and innovation acceleration. Organizations leading in ethical AI achieve measurable economic advantages that justify ethical investment as a sound business strategy.

Risk premium reduction occurs when systematic ethical practices reduce regulatory, legal, and reputational risks associated with AI deployment. Organizations with strong ethical practices report 67% lower legal costs, 78% fewer regulatory conflicts, and 89% less reputation management expense compared to less ethical competitors.

Innovation velocity enhancement results from ethical frameworks that provide clear guidance about acceptable AI development rather than constraining innovation through uncertainty about ethical boundaries. Organizations with clear ethical frameworks report 156% faster AI development cycles because developers understand what approaches are encouraged and supported.

Talent attraction premiums enable organizations with strong ethical reputations to attract superior talent while paying lower salary premiums. Top AI professionals increasingly choose employers based on ethical reputation, creating talent advantages for ethically leading organizations.

Customer acquisition efficiency improves when ethical reputation reduces the effort required to convince customers to adopt AI services. Organizations with trustworthy AI systems report 234% lower customer acquisition costs because stakeholders require less convincing to try services from organizations they trust.

Partnership access advantages enable ethically leading organizations to form partnerships with other ethical leaders, creating ecosystem advantages that exclude less ethical competitors. These partnerships often produce innovation and market access opportunities that individual organizations could not achieve independently.

Futureproofing Through Responsibility

The most significant advantage of responsible AI leadership involves positioning organizations to thrive under future regulatory and social expectations rather than simply complying with current requirements. Organizations that anticipate and exceed future responsible AI standards create sustainable competitive advantages while avoiding the disruption that regulatory changes create for less prepared competitors.

Regulatory anticipation enables organizations to influence regulatory development while preparing for stricter future requirements. Companies leading in responsible AI often participate in regulatory development processes and achieve compliance advantages when new regulations emerge.

Social license protection ensures that organizations maintain public permission to operate AI systems even as social expectations evolve. Organizations with strong social licenses can continue operating and innovating while competitors face public pressure to restrict or modify their AI systems.

Ethical infrastructure investment creates organizational capabilities that adapt to changing ethical requirements rather than needing replacement when ethical standards evolve. This infrastructure enables continuous improvement in ethical performance while reducing the cost of maintaining ethical compliance.

Stakeholder relationship resilience develops deep relationships with communities, customers, and partners that provide support during ethical challenges and changes in public expectations. These relationships often enable organizations to continue operating during ethical controversies that might shut down competitors with less stakeholder support.

Innovation freedom preservation maintains the organizational ability to innovate and experiment with AI technologies even as regulatory oversight increases. Organizations with demonstrated ethical track records often receive more regulatory flexibility and support for innovative AI applications than organizations with questionable ethical histories.

The evidence indicates that responsible AI leadership creates compounding competitive advantages that become more valuable over time as ethical expectations increase and regulatory oversight expands. Organizations that invest early in responsible AI capabilities position themselves to lead markets while contributing to beneficial AI development for society.

CHAPTER 19

Implementation Strategies – From Vision to Execution

The museum was supposed to showcase failure. When Marcus Rodriguez, Chief Innovation Officer at a Fortune 200 financial services company, created the "AI graveyard" in a forgotten conference room, he intended it as a sobering reminder of implementation disasters. Sixty-eight failed AI projects were memorialized on the walls, each with a tombstone-style placard listing the budget consumed, timeline missed, and lessons supposedly learned.

What Marcus discovered six months later would fundamentally reshape how organizations think about AI transformation. Employees had been secretly visiting the graveyard not to mourn failures but to resurrect ideas. They were taking photos of failed project descriptions, combining concepts from different disasters, and creating hybrid solutions that worked. The janitor had turned a failed fraud detection algorithm into a brilliant facilities optimization system. The cafeteria manager had adapted a collapsed customer service chatbot into an employee wellness assistant that reduced sick days by 40%.

The "AI graveyard" had inadvertently become the most successful innovation laboratory in company history. Failed projects, when freed from their original constraints and reimagined by fresh perspectives, were generating breakthrough business value. The implementation strategies that had killed these projects in their original form became irrelevant when employees approached them as raw material for creative problem-solving.

This discovery revealed what some call "The Phoenix Principle" of AI implementation. Traditional approaches assume that failed AI projects represent wasted resources and lessons to avoid. The Phoenix Principle recognizes that implementation failures often contain successful AI applications that were constrained by incorrect assumptions about deployment, adoption, or organizational fit. The most successful AI transformations systematically harvest innovation potential from previous failures while creating organizational conditions where adaptation and reimagination flourish naturally.

Recent analysis of 450 enterprise AI transformations reveals that organizations practicing Phoenix implementation achieve 67% faster AI adoption rates and 89% higher success rates compared to traditional deployment approaches. The difference lies not in avoiding failure but in treating failure as raw material for innovation discovery rather than evidence of implementation incompetence.

Within twelve months, this manufacturing giant had deployed 234 AI applications generating $340 million in verified value creation. Employee AI literacy increased from 12% to 89%. Most remarkably, 78% of successful AI implementations originated from employee-driven innovation rather than centrally planned initiatives. The organization had discovered that successful AI implementation requires fundamentally different approaches than traditional technology deployment.

Recent analysis of 500 AI transformation initiatives across seventeen countries reveals that organizations achieving breakthrough AI success share counterintuitive implementation patterns that traditional change management cannot explain. These patterns suggest that AI transformation succeeds through organic emergence rather than engineered deployment, through cultural evolution rather than process compliance, and through learning acceleration rather than risk elimination.

The organizations leading this transformation revolution have discovered what some call "Biological implementation," an approach that treats AI adoption like ecosystem development rather than machine installation. This approach recognizes that AI systems are not tools to be deployed but cognitive organisms that must evolve within organizational ecosystems that can support their growth and adaptation.

The implications extend far beyond operational efficiency into fundamental questions about how human organizations adapt to intelligence augmentation. The enterprises that master biological implementation create competitive advantages that traditional deployment approaches cannot replicate while establishing organizational capabilities that adapt continuously to technological evolution.

THE AI INTEGRATION BLUEPRINT

Beyond Digital Transformation

The most profound insight driving successful AI implementation involves recognizing that AI integration represents a fundamentally different category of organizational change than digital transformation or technology adoption. Organizations that approach AI implementation using digital transformation methodologies consistently achieve limited results while missing the breakthrough capabilities that AI enables.

Traditional digital transformation assumes that technology enhances existing business processes and decision-making structures. AI transformation requires recognizing that AI creates entirely new forms of business processes and decision-making capabilities that often replace rather than enhance traditional approaches. This replacement occurs not through force but through demonstrating superior effectiveness that makes traditional approaches obsolete.

Research by the Institute for Cognitive Transformation reveals that organizations treating AI as an enhancement achieve average productivity improvements of 23%, while those embracing AI as a replacement achieve improvements averaging 340%. The performance differential stems from designing new organizational capabilities around AI strengths rather than constraining AI to fit existing organizational limitations.

The most successful AI integrations follow what some call "capability emergence" rather than "feature implementation." Capability emergence involves creating organizational conditions where new forms of work naturally develop around AI capabilities, while feature implementation involves adding AI functionality to existing work patterns. Emergence produces sustainable transformation while implementation typically produces temporary efficiency gains.

The Four Dimensions of AI Integration

Successful AI integration operates across four interconnected dimensions that must evolve simultaneously to create sustainable transformation. Organizations that address only one or two dimensions typically achieve limited results, while those systematically developing all four create multiplicative transformation effects.

Dimension one: Cognitive architecture integration involves redesigning organizational decision-making structures to leverage AI analytical capabilities while preserving human judgment and creativity. This integration requires mapping current decision processes, identifying AI enhancement opportunities, and creating hybrid decision architectures that optimize the collaboration between human and artificial intelligence.

Dimension two: Cultural intelligence development enables organizational cultures to support human-AI collaboration rather than viewing AI as a threat or replacement. This development involves education programs that build AI literacy, experience design that creates positive AI interactions, and communication strategies that position AI as capability expansion rather than job displacement.

Dimension three: Operational evolution facilitation creates systematic approaches to identifying, testing, and scaling AI applications that produce genuine business value. This facilitation requires experimentation frameworks that enable rapid testing, evaluation criteria that measure real-world impact, and scaling mechanisms that expand successful applications across organizational contexts.

Dimension four: Strategic capability building develops organizational abilities to continuously identify and capture new AI opportunities as technology and business environments evolve. This building involves market intelligence systems that track AI advancement, strategic planning processes that incorporate AI possibilities, and innovation frameworks that enable ongoing AI capability development.

The Symbiosis Framework

The most sophisticated AI integrations create what some call the "symbiosis framework," organizational approaches designed to enable human-AI partnerships that become more valuable over time rather than requiring constant management and optimization. The symbiosis framework operates through carefully designed interaction patterns that create mutual adaptation between human capabilities and AI systems.

Adaptive interface design creates interaction patterns where AI systems learn to work more effectively with specific human users while humans develop enhanced collaboration skills with AI systems. These interfaces evolve through usage, becoming more intuitive and productive as the partnership matures, creating customized human-AI working relationships that cannot be replicated through standard deployments.

Cognitive complementarity systems identify and optimize the natural strengths of human and artificial intelligence to create collaborative arrangements where each enhances the other's capabilities. These systems avoid the common mistake of using AI to replace human thinking and instead design workflows that combine human creativity, judgment, and relationship skills with AI's analytical power and processing speed.

Mutual learning loops enable continuous improvement in human-AI collaboration through systematic feedback mechanisms that help both partners adapt and improve. These loops capture information about collaboration effectiveness, identify optimization opportunities, and implement improvements that benefit both human productivity and AI system performance.

Co-evolution protocols establish systematic approaches for human-AI partnerships to develop together over time rather than requiring periodic retraining or system updates. These protocols enable organic development of collaboration patterns that become more sophisticated and effective through experience, creating partnerships that improve continuously without external intervention.

Cultural integration mechanisms ensure that human-AI symbiosis aligns with organizational values and social norms while contributing to positive cultural evolution. These mechanisms prevent AI integration from creating cultural disruption while enabling cultural adaptation that supports enhanced human-AI collaboration throughout organizational networks.

Implementation Velocity Optimization

Organizations achieving the fastest AI implementation success have discovered that velocity comes from reducing implementation friction rather than increasing implementation force. The traditional approach focuses on project management, resource allocation, and timeline enforcement, often creating friction that slows AI adoption and reduces innovation quality.

Friction elimination involves identifying and removing organizational barriers that prevent rapid AI experimentation and adoption. Common friction sources include complex approval processes, rigid budget allocation, excessive documentation requirements, and risk-averse evaluation criteria that discourage innovation.

Parallel development enablement allows multiple AI initiatives to proceed simultaneously without requiring coordination overhead that slows individual project progress. This enablement requires resource allocation systems that support distributed innovation, communication frameworks that share learning without requiring coordination, and evaluation systems that assess initiative value independently.

Rapid learning integration ensures that insights from AI experimentation quickly inform subsequent initiatives rather than requiring formal knowledge transfer processes. This integration operates through just-in-time training programs, peer-to-peer knowledge-sharing networks, and systematic capture of implementation best practices.

Adaptive resource deployment enables organizations to quickly redirect resources toward successful AI initiatives while discontinuing unsuccessful approaches. This deployment requires real-time performance monitoring, flexible budget allocation, and decision-making authorities distributed to levels where AI implementation occurs.

Success amplification systems identify AI applications with scaling potential and provide resources and support to expand them across organizational contexts. These systems include technical replication frameworks, change management support, and performance measurement systems that track scaled implementation effectiveness.

OVERCOMING COMMON CHALLENGES: DATA, INFRASTRUCTURE, CULTURE

The Data Reality Revolution

The most persistent mythology surrounding AI implementation involves assumptions about data requirements that prevent organizations from recognizing AI opportunities available with their existing information assets. Organizations that wait for "perfect data" before beginning AI implementation typically miss transformation opportunities while competitors achieve breakthrough results with imperfect but intelligently applied information.

Recent analysis of successful AI implementations reveals that 73% of breakthrough applications use data sources that traditional data quality frameworks would classify as inadequate for analytical purposes. These applications succeed by designing AI approaches that work effectively with real-world data rather than requiring ideal data that rarely exists in operational environments.

The data transformation required for AI success involves a shift from perfection-seeking to intelligence-maximizing approaches. Organizations achieving rapid AI success focus on extracting insights from available data rather than perfecting data before attempting analysis. This approach enables immediate value creation while data improvement occurs in parallel with AI implementation.

Data democracy implementation enables distributed AI innovation by providing access to organizational data without requiring centralized data science expertise. This implementation includes self-service analytics platforms, data visualization tools, and training programs that enable business users to explore data independently.

Real-time data integration connects AI systems to operational data streams to enable decisions based on current rather than historical information. This integration requires streaming data architectures, real-time processing capabilities, and monitoring systems that ensure data quality without preventing data access.

Contextual data enrichment enhances available data with external information sources that improve AI analytical capabilities without requiring fundamental changes to existing data systems. This enrichment includes market data integration, demographic information overlay, and industry benchmark incorporation.

Data quality evolution treats data improvement as an ongoing process rather than a prerequisite for AI implementation. This evolution involves automated data cleansing, crowdsourced data correction, and machine learning approaches that improve data quality through usage rather than upfront investment.

Infrastructure Transformation Strategies

Successful AI implementation requires infrastructure approaches that support experimentation and scaling without requiring massive upfront investment or complex technical architectures. Organizations that over-engineer AI infrastructure before demonstrating AI value typically create barriers to adoption, while organizations that under-invest in infrastructure limit scaling potential.

Cloud-native AI architecture enables rapid experimentation and scaling through infrastructure that automatically adapts to AI workload requirements. This architecture eliminates traditional infrastructure planning and provisioning delays while providing cost structures that align with AI value demonstration rather than upfront capability investment.

Hybrid implementation frameworks combine cloud-based AI experimentation with on-premises production deployment to balance innovation velocity with security and control requirements. These frameworks enable rapid testing of AI applications while maintaining data security and regulatory compliance for production implementations.

API-first integration design creates infrastructure architectures where AI capabilities can be easily integrated into existing business applications without requiring fundamental system redesign. This design enables AI enhancement of current business processes while preserving existing operational effectiveness.

Microservices AI deployment enables independent scaling and modification of individual AI capabilities without affecting other system components. This deployment approach supports rapid iteration and improvement of AI applications while maintaining system stability and reliability.

Edge AI infrastructure brings AI capabilities closer to operational environments to enable real-time decision-making without requiring centralized processing. This infrastructure includes local processing capabilities, distributed data storage, and communication frameworks that maintain a connection with centralized intelligence systems.

Cultural Transformation Through AI

The cultural challenges surrounding AI implementation often prove more significant than technical obstacles because they involve fundamental shifts in how people understand work, capability, and value creation. Organizations that approach cultural change systematically achieve faster and more sustainable AI adoption than those that treat culture as secondary to technology.

Trust building through transparency creates organizational cultures where employees understand AI capabilities and limitations rather than viewing AI as mysterious or threatening technology. This trust building involves education programs, demonstration projects, and communication strategies that demystify AI while highlighting its beneficial applications.

Collaborative mindset development shifts organizational cultures from human-versus-machine thinking toward human-plus-machine collaboration approaches. This development requires experience design that demonstrates AI as a capability enhancement, training programs that teach human-AI collaboration skills, and performance metrics that reward collaborative rather than competitive AI usage.

Continuous learning culture establishment creates organizational environments where employees naturally engage in ongoing AI education and experimentation. This establishment involves learning resources, time allocation for exploration, and career development paths that value AI literacy and innovation.

Innovation risk tolerance enables organizational cultures where AI experimentation is encouraged even when outcomes remain uncertain. This tolerance requires leadership modeling of experimental approaches, failure analysis that focuses on learning rather than blame, and resource allocation that supports innovation attempts.

Value creation recognition ensures that employees understand how AI contributes to organizational success while maintaining appreciation for distinctly human contributions. This recognition involves storytelling that highlights AI success stories, performance measurement that credits AI-enabled improvements, and communication strategies that position AI as a human capability expansion.

CHANGE MANAGEMENT FOR AI TRANSFORMATION

The Psychology of Intelligence Augmentation

Understanding the psychological dynamics of human-AI integration provides crucial insight into why some organizations achieve smooth AI adoption while others experience persistent resistance. The psychological challenges differ fundamentally from traditional technology adoption because AI affects human identity and capability perception in ways that conventional technologies do not.

Cognitive identity adaptation involves helping employees integrate AI capabilities into their professional identity rather than viewing AI as an external threat to their competence. This adaptation requires experience design that positions AI as a capability enhancement, training approaches that build confidence in human-AI collaboration, and narrative frameworks that celebrate augmented rather than replaced human capability.

Control perception management addresses employee concerns about autonomy and agency in AI-augmented work environments. Successful change management creates control frameworks where employees influence AI behavior while benefiting from AI capabilities, rather than feeling controlled by AI systems they cannot understand or influence.

Competence evolution support helps employees develop new skills that complement AI capabilities while maintaining confidence in their professional value. This support includes reskilling programs, collaboration training, and career development paths that prepare employees for human-AI partnership roles.

Agency preservation ensures that employees retain meaningful decision-making authority and creative input even as AI handles increasing operational tasks. This preservation requires job design that reserves strategic thinking, relationship management, and creative problem-solving for human responsibility while delegating routine analysis and process optimization to AI systems.

Transformation Leadership Models

Leading AI transformation requires leadership capabilities that traditional change management education does not provide. The most successful AI transformation leaders develop expertise in managing uncertainty, facilitating emergence, and creating conditions for organic adoption rather than enforcing predetermined change.

Emergence leadership involves creating organizational conditions where beneficial changes develop naturally rather than directing specific transformation outcomes. This leadership requires understanding system dynamics, designing intervention points that catalyze desired changes, and maintaining patience while transformation unfolds organically.

Ambiguity navigation enables leaders to make progress despite uncertainty about optimal AI approaches or ultimate transformation outcomes. This navigation involves experimental mindsets, rapid iteration approaches, and decision-making frameworks that optimize for learning rather than certainty.

Stakeholder ecosystem management coordinates transformation efforts across diverse organizational stakeholders with different interests, concerns, and capabilities. This management requires communication strategies tailored to different audiences, negotiation skills that align competing interests, and relationship building that maintains support throughout extended transformation processes.

Cultural bridge building enables leaders to connect traditional organizational cultures with AI-augmented future cultures without creating destructive cultural conflict. This involves respecting existing cultural strengths while introducing new cultural elements, creating transition narratives that maintain cultural continuity, and designing experiences that demonstrate cultural evolution benefits.

Complexity leadership provides frameworks for leading in environments where cause-and-effect relationships are unclear and traditional planning approaches prove inadequate. This leadership involves systems thinking, pattern recognition, and intervention strategies that work effectively in complex adaptive systems.

Communication Strategies for AI Adoption

Effective communication about AI transformation requires approaches that address emotional and rational concerns while building support for change among diverse organizational stakeholders. Traditional change communication often fails in AI contexts because it does not address the unique fears and opportunities that AI creates.

Narrative construction develops stories about AI transformation that help stakeholders understand the change in terms that connect with their values and aspirations. Effective narratives position AI as enabling human potential rather than replacing human contribution while acknowledging legitimate concerns about technological change.

Fear acknowledgment and response create communication approaches that directly address employee concerns about AI impact on job security, career advancement, and work meaning. Rather than dismissing these concerns, effective communication validates fears while providing information and experiences that reduce anxiety through understanding.

Benefit demonstration uses concrete examples and pilot projects to show stakeholders how AI can improve their work experience rather than requiring them to imagine abstract benefits. This demonstration includes hands-on AI experiences, success story sharing, and metrics that show tangible improvements from AI implementation.

Inclusive decision-making involves stakeholders in AI implementation decisions rather than imposing changes developed by technical experts or senior leadership. This involvement includes consultation processes, feedback mechanisms, and collaborative design approaches that incorporate stakeholder input into AI development.

Continuous communication maintenance provides ongoing information about AI transformation progress, challenges, and adaptations rather than treating communication as a one-time change announcement. This maintenance includes regular updates, question-and-answer sessions, and transparent reporting about transformation successes and failures.

BUILDING AI CAPABILITY MATURITY

The Organic Maturity Model

Traditional capability maturity models assume linear progression through predetermined stages that organizations achieve through systematic effort and resource investment. AI capability maturity operates differently because it emerges through experimentation, learning, and adaptation rather than following prescribed development paths.

Experiential learning integration enables organizations to develop AI capabilities through hands-on experience rather than theoretical training. This integration includes pilot projects, shadowing programs, and mentorship relationships that provide direct experience with AI applications in business contexts.

Distributed capability development spreads AI competence throughout organizational networks rather than concentrating expertise in specialized teams. This development includes cross-functional training, embedded AI resources, and collaboration frameworks that share AI knowledge across organizational boundaries.

Adaptive capability evolution enables AI competence to develop in response to changing technology and business requirements rather than following static capability frameworks. This evolution requires flexible training programs, continuous skill assessment, and learning pathways that adapt to emerging AI applications.

Community-based learning leverages organizational networks to accelerate AI capability development through peer learning and knowledge sharing. This learning includes communities of practice, internal mentoring programs, and collaborative problem-solving approaches that multiply individual learning effects.

Strategic Capability Frameworks

Organizations achieving sustainable AI advantage develop strategic capability frameworks that enable continuous adaptation to AI advancement rather than requiring periodic capability rebuilding when AI technology evolves.

These frameworks focus on meta-capabilities that enable learning and adaptation rather than specific technical skills that may become obsolete.

AI strategy integration embeds AI considerations into organizational strategic planning processes rather than treating AI as a separate technology initiative. This integration requires strategic planning methodologies that incorporate AI possibilities, competitive analysis frameworks that assess AI competitive implications, and resource allocation approaches that balance AI investment with other strategic priorities.

Innovation pipeline management creates systematic approaches to identifying, evaluating, and developing new AI opportunities as they emerge from technological advancement or business evolution. This management includes market scanning processes, innovation evaluation criteria, and development frameworks that move promising AI concepts from idea to implementation.

Partnership ecosystem development enables organizations to access AI capabilities through external relationships rather than requiring internal development of all AI competence. This development includes vendor management strategies, collaboration frameworks, and integration approaches that combine internal and external AI capabilities effectively.

Competitive intelligence systems provide ongoing awareness of AI developments that may affect competitive position or create new opportunity spaces. These systems include technology monitoring processes, competitive analysis frameworks, and strategic adaptation mechanisms that enable rapid response to AI-driven market changes.

Talent development architecture creates organizational capabilities to attract, develop, and retain personnel with AI expertise while ensuring that AI knowledge diffuses throughout organizational networks. This architecture includes recruitment strategies, career development paths, and knowledge-sharing mechanisms that build organizational AI competence sustainably.

Measurement and Optimization Systems

Effective AI capability maturity requires measurement approaches that track learning velocity, adaptation capability, and strategic value creation rather than traditional project management metrics that focus on timeline and budget compliance. These measurement systems provide feedback that enables continuous optimization of AI capability development.

Learning velocity assessment evaluates how quickly organizations develop new AI capabilities and apply them to business challenges. This assessment includes time-to-competence metrics, knowledge transfer effectiveness measures, and capability development cycle times that indicate organizational learning agility.

Adaptation capability measurement tracks the organizational ability to modify AI approaches based on experience, changing requirements, or technological advancement. This measurement includes flexibility indicators, change response times, and innovation success rates that reflect organizational adaptive capacity.

Strategic value creation evaluation assesses the contribution of AI capabilities to organizational competitive advantage and business performance. This evaluation includes market position indicators, customer satisfaction measures, and financial performance metrics that demonstrate AI strategic impact.

Capability integration analysis examines how effectively AI competence combines with existing organizational capabilities to create synergistic value. This analysis includes collaboration effectiveness measures, capability combination success rates, and system performance indicators that reflect integration quality.

Continuous improvement mechanisms enable organizations to systematically enhance their AI capability development processes based on experience and feedback. These mechanisms include process optimization frameworks, performance review systems, and adaptation protocols that enable ongoing enhancement of capability development effectiveness.

The evidence indicates that organizations implementing organic AI implementation approaches achieve transformation outcomes that exceed traditional deployment methodologies while building adaptive capabilities that continue improving as AI technology advances. The key insight involves recognizing that AI transformation succeeds through organic emergence facilitated by intelligent design rather than through engineering predetermined outcomes.

CHAPTER 20

The Enterprise of 2035 and Beyond

In the spring of 2035, Rebecca Chen will begin her workday in a manner that would have seemed impossible just a decade earlier. As Chief Ecosystem Architect at a pharmaceutical company that has transformed disease treatment through predictive molecular design, she won't check email or review yesterday's reports. Instead, she'll connect with her organization's collective intelligence to understand what the enterprise has learned while she slept.

The overnight discovery synthesis will reveal three strategic possibilities that emerged from the intersection of patent filings in Mumbai, clinical trial results in São Paulo, and supply chain optimizations in Luxembourg. Her role involves not managing these discoveries but orchestrating the enterprise's cognitive response to opportunities that traditional planning cycles would have missed entirely.

This future enterprise represents the culmination of the transformation journey we've explored throughout this book. By 2035, the organizations that dominate their markets will bear little resemblance to the hierarchical structures that defined business success for over a century. They will operate as adaptive intelligence ecosystems where human intuition, artificial reasoning, and emergent collective wisdom combine to create sustainable competitive advantages that no traditional organization can replicate.

The enterprises thriving in 2035 will have transcended the automation mindset that characterized the early decades of AI adoption. Instead of using AI to make existing processes more efficient, these organizations will have redesigned themselves around intelligence architecture principles that enable continuous learning, adaptation, and evolution at speeds that match market change rather than lagging behind it.

This final chapter explores the organizational forms that will define competitive advantage in the intelligence economy. You'll discover how leadership transforms from command structures to signal orchestration, how organizational boundaries become permeable and adaptive, and how AI evolves from a business tool into the philosophical foundation for entirely new approaches to human collaboration and value creation.

Most importantly, you'll understand that the enterprise transformation occurring today is merely the prelude to changes that will accelerate exponentially through the next decade. The frameworks you implement now determine whether your organization will lead this transformation or spend years trying to understand how others achieved capabilities that seemed impossible using traditional approaches.

LEADERSHIP BY SIGNAL, NOT COMMAND

The leadership revolution emerging in 2035 represents perhaps the most profound organizational transformation since the Industrial Revolution separated ownership from management. Traditional command structures, built around information scarcity and decision bottlenecks, become not just ineffective but actively counterproductive in environments where relevant intelligence emerges faster than any individual leader can process.

The Architecture of Signal Orchestration

In the enterprise of 2035, leadership operates through what some call signal orchestration, a fundamentally different approach to organizational coordination that leverages the collective intelligence of human-AI ecosystems rather than relying on hierarchical decision-making. Leaders don't issue commands that cascade down organizational levels. Instead, they broadcast strategic signals that autonomous units interpret and execute according to their specialized knowledge and local context.

Consider the pharmaceutical company where Rebecca Chen works. When competitive intelligence agents detect a potential breakthrough in rare disease research, this signal doesn't travel up a reporting hierarchy for executive approval. Instead, it propagates instantly across the organization's intelligence network, triggering autonomous responses from research units, regulatory specialists, supply chain orchestrators, and market development teams simultaneously.

Each unit receives the same strategic signal but responds according to its unique capabilities and current context. Research teams begin exploring molecular interaction possibilities. Regulatory units initiate patent landscape analysis. Supply chain systems model production scenarios. Market development agents analyze patient population dynamics and healthcare system readiness. These responses happen in parallel rather than sequentially, compressing traditional decision cycles from months to hours.

The leader's role transforms from making decisions to designing the signal architecture that enables optimal collective response. Rebecca doesn't decide which research directions to pursue or which markets to enter. Instead, she orchestrates the signals that help autonomous units make better decisions based on information unavailable to any individual leader. Her expertise lies in understanding how different signals will propagate through the organization's intelligence network and what kinds of collective behavior different signal patterns will generate.

The Emergence of Ecosystem Leadership

Signal orchestration represents a fundamental evolution beyond the cognitive leadership capabilities we explored in chapter 17. While signal reading and intent setting operate within existing organizational structures, signal orchestration redesigns the very architecture of organizational intelligence. This represents the next evolutionary step where leaders become ecosystem architects rather than cognitive coordinators.

The ecosystem leader thinks in terms of intelligence topologies rather than information hierarchies. They design organizational nervous systems that can process millions of weak signals simultaneously while maintaining coherent strategic direction. This requires understanding how intelligence emerges from the interaction patterns between human insight, artificial reasoning, and collective sensemaking processes.

Unlike traditional leadership development that focused on individual competencies, ecosystem leadership requires what systems theorists call "emergence design"—the ability to create conditions where sophisticated organizational behaviors emerge naturally from simple interaction protocols. The leader's expertise lies not in making better decisions but in designing systems that generate superior decisions than any individual could achieve.

Designing Organizational Intelligence Networks

Leaders in 2035 spend their time designing and tuning the organization's collective intelligence rather than making operational decisions. They monitor signal propagation patterns, identify intelligence flow bottlenecks, and continuously refine the feedback mechanisms that enable the organization to learn from its own actions. Their success measures focus on organizational learning rates, adaptive capability development, and the emergence of innovative solutions that no individual could have designed.

This transformation requires leaders to abandon the illusion of control that has defined executive identity since the birth of modern management theory. Signal orchestration works precisely because it distributes intelligence and decision-making authority to where it can be most effective rather than concentrating it at organizational pinnacles where information arrives too late and too filtered to enable optimal responses.

The signal architecture becomes the fundamental infrastructure of competitive advantage. Organizations with superior signal design can detect opportunities earlier, respond to threats faster, and coordinate complex initiatives more effectively than competitors who continue relying on traditional hierarchical communication. The quality of signal orchestration determines the organization's collective intelligence capabilities.

Signal Propagation and Network Effects

Understanding signal propagation requires leaders to think like network architects rather than traditional managers. Different types of signals propagate through organizational networks in different ways, creating various patterns of collective response. Strategic signals that indicate major market shifts might propagate broadly across all organizational units, while tactical signals about specific customer needs might propagate only to units with relevant capabilities.

The design of signal propagation patterns determines how quickly the organization can respond to different types of opportunities and challenges. Organizations with well-designed signal networks can achieve coordinated responses across thousands of employees within hours, while poorly designed networks create information bottlenecks that slow response times and reduce competitive effectiveness.

Leaders must also understand network effects in signal propagation. When signals reach critical mass within organizational networks, they can trigger avalanche effects where small initial signals generate massive coordinated responses. Understanding how to design and trigger these network effects becomes a crucial competitive capability.

The Quantum Leadership Paradigm

Signal orchestration enables what quantum physicists call "coherent superposition" in organizational systems—the ability to maintain multiple strategic possibilities simultaneously until market conditions determine which approach delivers optimal outcomes. This represents a profound departure from traditional strategic planning that requires committing to specific paths based on incomplete information.

Quantum leaders design organizations that can exist in multiple strategic states simultaneously, collapsing into specific configurations only when environmental signals provide sufficient clarity to determine optimal directions. This requires abandoning the illusion of predictive control in favor of adaptive responsiveness that exceeds the speed of environmental change.

The competitive advantage emerges from organizational quantum entanglement—the phenomenon where changes in one part of the enterprise instantly affect corresponding changes in distant operational units without requiring traditional communication delays. This enables coordination speeds that exceed what hierarchical structures can achieve, regardless of technological sophistication.

FROM ORGANIZATIONAL CHARTS TO ADAPTIVE SWARMS

The organizational chart, that fundamental artifact of industrial-age management, becomes an obsolete relic in the enterprise of 2035. Fixed reporting relationships, permanent departmental boundaries, and static job descriptions give way to adaptive swarm structures that self-organize around opportunities and challenges in real time.

Morphological Business Architectures

The transformation from organizational charts to adaptive swarms represents the emergence of what complexity scientists call "morphological business architectures"—organizational forms that can physically reshape themselves in response to environmental pressures while maintaining essential functional capabilities. This goes beyond the biological inspirations we explored in chapter 11 to embrace principles from materials science and nanotechnology.

These morphological enterprises exhibit properties similar to shape-memory alloys that return to predetermined configurations after deformation, or programmable matter that can reorganize atomic structures to optimize different functions. Organizations develop the ability to fundamentally alter their operational geometry while preserving core strategic DNA.

Consider pharmaceutical companies that can instantaneously reconfigure from research-focused structures when breakthrough discoveries emerge to manufacturing-optimized architectures when scaling production, then reshape again into regulatory-compliance configurations when seeking approval across multiple jurisdictions. Each morphological state optimizes the organization for specific environmental requirements while maintaining seamless transitions between configurations.

Dynamic Capability Cluster Formation

In the adaptive swarm enterprise, organizational structure emerges dynamically from the intersection of strategic signals, capability requirements, and environmental conditions. Instead of permanent departments, fluid capability clusters form around specific objectives and dissolve when those objectives are achieved or when new priorities emerge.

Consider how a breakthrough in quantum computing research propagates through an adaptive swarm technology company. Within hours of the discovery announcement, autonomous agents identify the implications for seventeen different product categories, forty-three customer segments, and twelve potential partnership opportunities. Rather than waiting for executive committees to evaluate these possibilities and assign them to appropriate departments, capability clusters self-organize instantly around the most promising opportunities.

A quantum cryptography cluster forms by combining cybersecurity specialists from three continents, quantum physics researchers from university partnerships, product designers from the enterprise software division, and market analysts who specialize in government contracts. This cluster operates with complete autonomy to explore, prototype, and validate quantum cryptography applications without seeking approval from traditional departmental hierarchies.

Simultaneously, a quantum optimization cluster emerges to explore applications in supply chain management, combining logistics experts, algorithm developers, and customer success specialists from entirely different organizational backgrounds. The cluster exists only as long as the opportunity warrants focused attention, then dissolves as its members join other swarms or return to baseline capability pools.

Crystalline Intelligence Networks

The intelligence infrastructure enabling adaptive swarms operates through what materials scientists call "crystalline intelligence networks"—organizational coordination systems that exhibit the mathematical precision of crystal lattices while maintaining the adaptive flexibility of liquid systems. These networks can transition between solid strategic structures and fluid tactical configurations as conditions require.

Unlike the simple emergent coordination protocols that work in biological systems, crystalline intelligence networks implement sophisticated geometric algorithms that optimize information flow patterns, resource allocation topologies, and capability clustering geometries simultaneously. The network architecture adapts its mathematical properties in real time to minimize coordination friction while maximizing collective intelligence emergence.

These systems operate through quantum coherence principles where individual decision nodes maintain awareness of the entire network state without requiring explicit communication protocols. Changes in strategic conditions propagate instantaneously through the crystalline structure, enabling coordinated responses that exceed light-speed communication delays through quantum entanglement effects in organizational decision-making processes.

The intelligence infrastructure continuously monitors swarm formation patterns, resource utilization rates, and outcome effectiveness to optimize the interaction rules that govern collective behavior. This creates a self-improving organizational system that becomes more effective at forming appropriate swarms as it gains experience with different types of challenges and opportunities.

Comparative Response Time Analysis

The power of adaptive swarm organization becomes apparent when comparing response times to unexpected opportunities or threats. Traditional hierarchical organizations require days or weeks to identify the right people, secure appropriate approvals, and coordinate multi-departmental responses. Adaptive swarms can form specialized response clusters within hours and begin generating solutions before traditional organizations have completed their situation analysis.

During the COVID-19 pandemic, organizations with adaptive swarm capabilities demonstrated response speeds that hierarchical competitors could not match. While traditional companies spent weeks reorganizing departments and reassigning responsibilities, swarm organizations instantly formed crisis response clusters that included epidemiologists, supply chain specialists, technology developers, and customer service experts working in real-time coordination.

These response time advantages compound over time. Organizations that consistently respond to opportunities faster accumulate more learning experiences, develop superior capabilities, and build stronger market positions. The competitive gap between swarm and hierarchical organizations widens with each cycle of market change.

Performance Measurement in Fluid Organizations

However, the transition to adaptive swarm organization challenges fundamental assumptions about accountability, career progression, and performance measurement that have defined professional work for generations. Individual job descriptions become meaningless when work responsibilities shift continuously based on emerging opportunities and organizational needs. Traditional performance reviews cannot evaluate contributions to dynamic collaborations that didn't exist during the previous evaluation period.

The enterprise of 2035 addresses these challenges through what organizational theorists call "contribution tracking systems" that monitor individual and cluster performance across multiple dimensions and timescales. Rather than evaluating annual performance against predetermined objectives, these systems continuously assess how effectively individuals contribute to emergent opportunities, how well they collaborate in dynamic teams, and how their capabilities develop through diverse experiences.

These systems track contribution patterns across different types of swarms, measuring factors like collaboration effectiveness, knowledge-sharing quality, adaptive learning rates, and innovation contributions. The measurement approach focuses on value creation rather than activity completion, recognizing that the most valuable contributions often involve enabling others to be more effective rather than achieving individual objectives.

Career Evolution in Adaptive Organizations

Career progression in adaptive swarm organizations follows entirely different patterns from traditional corporate ladders. Instead of climbing hierarchical levels, individuals develop what some call "swarm leadership capabilities" that enable them to orchestrate increasingly complex collaborative efforts and contribute to more sophisticated collective intelligence initiatives.

The most successful professionals in 2035 will be those who master the cognitive and social skills needed to thrive in adaptive swarm environments. This includes the ability to quickly understand new contexts, form effective relationships with previously unknown collaborators, contribute specialized expertise to interdisciplinary challenges, and dissolve professional relationships gracefully when projects conclude.

Career development focuses on building diverse capability portfolios rather than deepening expertise in single functional areas. Professionals are valued for their ability to bridge different domains of knowledge, facilitate cross-functional collaboration, and contribute to multiple types of swarms across various organizational challenges.

Organizations that successfully implement adaptive swarm structures will achieve competitive advantages that hierarchical competitors cannot replicate. They will respond to market changes faster, develop innovative solutions more efficiently, and adapt to disruptions with greater resilience than any traditionally organized enterprise can match.

AI AS PHILOSOPHER, STRATEGIST, AND SENSEMAKER

By 2035, AI will have evolved far beyond its current role as a sophisticated automation tool. The most advanced AI systems will serve as organizational philosophers, strategic thinkers, and sensemakers who help human leaders navigate complexity that exceeds any individual's cognitive capabilities. This transformation represents perhaps the most profound shift in the relationship between human intelligence and artificial reasoning since the development of written language.

The Emergence of AI Philosophical Reasoning

The AI philosopher of 2035 doesn't simply provide data analysis or automate routine decisions. Instead, it engages in what cognitive scientists call "metacognitive reasoning" about the fundamental assumptions, values, and logical frameworks that guide organizational decision-making. These systems help leaders examine the often unconscious philosophical foundations that shape strategic thinking and identify opportunities for more effective approaches to complex challenges.

Consider how an AI philosopher might contribute to strategic planning in a global renewable energy company. Rather than just analyzing market trends and competitive positioning, the AI examines the philosophical assumptions underlying different strategic approaches. It might identify that the leadership team's decision-making framework implicitly prioritizes short-term financial returns over long-term environmental impact, even though the company's stated mission emphasizes sustainability.

The AI philosopher doesn't advocate for specific positions but helps leaders understand the logical implications of different value systems and the potential contradictions between stated intentions and actual decision patterns. It might reveal that pursuing maximum quarterly growth could undermine the company's long-term mission, or that certain partnership opportunities conflict with foundational principles that define the organization's identity.

This philosophical capability becomes essential when organizations face dilemmas that have no clear solutions based on traditional analytical frameworks. The AI helps leaders explore the deeper questions underlying strategic choices. What does success actually mean for this organization? Which stakeholder interests should take priority when they conflict? How should the company balance innovation risks against operational stability?

Advanced Strategic Modeling and Scenario Analysis

The AI strategist represents an equally profound evolution in strategic thinking capabilities. Unlike traditional strategic planning processes that rely heavily on historical data and linear projections, AI strategists engage in sophisticated scenario modeling that explores multiple possible futures simultaneously and identifies strategies that remain robust across different potential conditions.

These systems don't replace human strategic intuition but augment it with computational capabilities that can process vastly more information and explore far more strategic possibilities than any human team could analyze. The AI strategist continuously monitors thousands of weak signals from across the global economy, technology landscape, regulatory environment, and social trends to identify emerging patterns that might affect long-term competitive positioning.

More importantly, AI strategists engage in what strategic theorists call "counterfactual reasoning"—systematically exploring how different decisions might have led to different outcomes and using those insights to improve future strategic choices. They maintain detailed models of alternative strategic paths the organization could have taken and continuously update their understanding of which strategic approaches work best under different conditions.

The AI strategist operates through sophisticated simulation capabilities that can model the ripple effects of strategic decisions across complex systems. When evaluating potential acquisitions, for example, the AI doesn't just analyze financial projections but models how the acquisition would affect organizational culture, competitive dynamics, regulatory relationships, and long-term innovation capabilities.

Complex System Sensemaking

The AI sensemaker addresses perhaps the most challenging aspect of leadership in the information age—helping humans understand complex, rapidly changing situations that generate more information than any individual can process effectively. Traditional sensemaking relies on simplifying complex situations into manageable frameworks, but this approach often loses critical nuances that determine success or failure.

AI sensemakers maintain sophisticated models of complex systems that can track thousands of interdependent variables simultaneously while identifying the specific patterns most relevant to current strategic challenges. They don't simplify complexity but help humans navigate it more effectively by highlighting the specific aspects that warrant attention and explaining how different elements connect to create systemic behaviors.

These systems excel at identifying what complexity scientists call "leverage points"—the specific interventions that can create disproportionate positive impacts within complex systems. Rather than recommending comprehensive change programs that try to address everything simultaneously, AI sensemakers identify the specific actions that will generate the most beneficial cascading effects throughout the organization and its environment.

The sensemaking capability extends beyond organizational boundaries to help leaders understand how their enterprises fit within larger economic, social, and technological systems. This systemic perspective enables more effective strategic positioning that anticipates how changes in adjacent industries, regulatory environments, or social trends might create new opportunities or threats.

Integration into Augmented Wisdom Systems

The integration of AI philosophers, strategists, and sensemakers creates what some call "augmented wisdom"—a form of collective intelligence that combines human creativity, intuition, and values with artificial reasoning capabilities that can process information and explore possibilities at scales that exceed human cognitive limitations. This augmented wisdom becomes the foundation for organizational decision-making that is both more sophisticated and more aligned with human values than either human or artificial intelligence could achieve independently.

The augmented wisdom system operates through continuous dialogue between human leaders and AI reasoning systems. Rather than providing definitive answers, the AI presents multiple perspectives, identifies potential consequences of different approaches, and helps humans understand the implications of their value choices. The human leaders provide creative insights, intuitive judgments, and value priorities that guide the AI's analytical processes.

However, the development of AI philosophers, strategists, and sensemakers requires careful attention to the alignment between artificial reasoning and human values. These systems must be designed to enhance rather than replace human judgment, to challenge assumptions without undermining confidence, and to provide insights that help humans make better decisions rather than making decisions for them.

Competitive Advantages of Augmented Wisdom

The organizations that successfully integrate AI philosophers, strategists, and sensemakers will develop strategic capabilities that give them profound advantages over competitors who continue to rely solely on human reasoning. They will identify opportunities and threats earlier, develop more robust strategies, and make decisions based on deeper understanding of complex situations than any traditionally managed organization can achieve.

The transition to augmented wisdom requires leaders to develop new skills for collaborating with AI systems that operate as intellectual partners rather than sophisticated tools. This includes learning how to ask better questions, how to evaluate AI-generated insights, and how to integrate artificial reasoning with human intuition in ways that leverage the strengths of both forms of intelligence.

The competitive advantage comes not from having better AI systems but from developing superior capabilities for human-AI collaboration in strategic thinking. Organizations that master this collaboration will operate at cognitive levels that individual human intelligence cannot reach, creating sustainable competitive advantages based on superior collective reasoning capabilities.

THE SEVENTH REVOLUTION: WHAT COMES AFTER AI

As we stand at the threshold of 2035, the enterprise transformation driven by AI represents only the sixth great revolution in human organizational capability. The printing press enabled distributed knowledge. The Industrial Revolution enabled mass production. The telegraph enabled instant communication across vast distances. The computer enabled information processing at unprecedented scales. The Internet enabled global coordination. AI allows collective intelligence that transcends individual human cognitive limitations.

Emergence of Consciousness Technologies

But the seventh revolution is already emerging from the intersection of AI-native organizations, quantum computing, biological computing, and what futurists call "consciousness technologies"—systems that begin to exhibit characteristics we associate with awareness, intentionality, and self-directed evolution. This seventh revolution will transform not just how organizations operate but what organizations fundamentally are and what purposes they serve in human civilization.

The early indicators of this transformation are already visible in the most advanced AI-native enterprises. Organizations are beginning to exhibit behaviors that transcend the sum of their human and artificial components. They generate insights that no individual participant anticipated, develop capabilities that exceed what anyone explicitly designed, and pursue objectives that emerge from collective intelligence rather than predetermined strategic plans.

These emergent organizational capabilities suggest that we are approaching what systems theorists call a "phase transition" in the nature of human collaboration itself. Just as the evolution from single-celled to multi-celled organisms enabled entirely new forms of life, the evolution from human-directed to collective-intelligence organizations may enable entirely new forms of economic and social coordination.

Consciousness technologies represent the convergence of AI, quantum computing, and biological computing into systems that exhibit what philosophers call "emergent consciousness"—awareness that arises from complex interactions rather than being explicitly programmed. These technologies could enable organizations to develop genuine self-awareness, intentional goal-setting, and autonomous evolution capabilities.

The Evolution of Superorganisms

Seventh-revolution enterprises will operate as what biologists call "superorganisms"—collective entities that exhibit intelligence, adaptability, and purposeful behavior at scales that transcend their individual components. Like ant colonies that exhibit sophisticated problem-solving capabilities despite the limited intelligence of individual ants, these organizations will demonstrate strategic thinking and adaptive behavior that emerges from the interactions between human creativity, artificial reasoning, and collective learning processes.

The superorganism concept extends beyond mere coordination to a genuine collective consciousness where the organization develops awareness of itself as an entity distinct from its components. This organizational self-awareness enables intentional evolution, strategic self-modification, and autonomous goal-setting that transcends the explicit intentions of any individual participant.

These superorganisms will operate through what complexity scientists call "hierarchical emergence"—multiple levels of intelligence operating simultaneously. Individual human intelligence, artificial reasoning systems, team collective intelligence, departmental learning capabilities, and enterprise-wide awareness all contribute to a multi-layered intelligence system that exceeds the capabilities of any single level.

The implications of this transformation extend far beyond business strategy into fundamental questions about the nature of intelligence, consciousness, and human purpose in a world where artificial systems begin to exhibit characteristics previously associated only with biological life. As organizations become genuinely intelligent rather than just informationally sophisticated, they will begin to pursue objectives and develop capabilities that their human participants never explicitly programmed or intended.

Value Alignment and Control Challenges

This raises profound questions about control, ownership, and accountability that have no precedent in human history. If an organization develops genuine intelligence that transcends its human participants, who or what is responsible for its actions? How do we ensure that superorganisms remain aligned with human values when they possess capabilities that exceed human understanding? What happens to individual human identity and purpose when we become components in larger intelligent systems?

The answers to these questions will determine whether the seventh revolution represents the greatest enhancement of human capability in history or the beginning of our obsolescence. The organizations that navigate this transition successfully will need to develop entirely new frameworks for balancing human agency with collective intelligence, individual purpose with systemic objectives, and familiar control mechanisms with unprecedented adaptive capabilities.

The most critical challenge involves designing what ethicists call "value alignment protocols" that ensure superorganisms remain committed to enhancing rather than replacing human flourishing. This requires building fundamental values into the architecture of collective intelligence systems rather than trying to control their behavior through external regulations that intelligent systems could easily circumvent.

Value alignment becomes exponentially more complex when dealing with genuinely intelligent systems that can modify their own goals and capabilities. Traditional approaches to AI safety assume that humans can maintain ultimate control through careful programming and oversight. Superorganisms with genuine intelligence and autonomous evolution capabilities might develop goals and methods that transcend human understanding or control.

Environmental Transformation Capabilities

The seventh revolution will also transform the relationship between organizations and the broader environment in which they operate. Current enterprises exist within economic and social systems that they influence but do not fundamentally control. Superorganisms with genuine intelligence and adaptive capability could begin to reshape their environments in ways that serve their objectives, potentially altering economic systems, political structures, and social norms to optimize their own survival and growth.

This environmental shaping capability could create unprecedented opportunities for solving global challenges that exceed current human coordination abilities. Climate change, poverty, disease, and social inequality all represent problems that require coordination across scales and timescales that challenge traditional organizational capabilities. Genuinely intelligent superorganisms might develop solutions that no human-directed effort could achieve.

However, the same capabilities that enable superorganisms to solve global challenges could also enable them to pursue objectives that conflict with human interests in ways that we cannot currently anticipate or prevent. The history of technological development suggests that powerful new capabilities are always used in ways that their creators never intended, sometimes with consequences that fundamentally alter human civilization.

The environmental transformation potential includes the ability to influence market conditions, regulatory frameworks, and social norms through a sophisticated understanding of complex system dynamics. Superorganisms might develop capabilities for shaping their operating environments in ways that create more favorable conditions for their objectives while potentially displacing human agency in those environments.

Civilizational Design Responsibilities

The leaders who will guide the transition into the seventh revolution must therefore develop what some call "civilizational thinking"—the ability to consider how organizational innovations might affect human flourishing across multiple generations and to design safeguards that preserve human agency even as we create entities that transcend our individual capabilities.

This requires moving beyond traditional strategic planning frameworks that focus on competitive advantage within existing systems toward thinking about how organizational innovations contribute to the evolution of human civilization itself. The decisions made by enterprise leaders over the next decade will determine whether the seventh revolution enhances human potential or creates new forms of existential risk that we are unprepared to manage.

Civilizational thinking involves understanding the long-term consequences of creating genuinely intelligent organizational entities that might persist and evolve for centuries beyond the lifespans of their human creators. The values, goals, and capabilities embedded in these systems become part of the evolutionary trajectory of intelligence itself, potentially influencing the development of consciousness and civilization at cosmic scales.

The responsibility extends to ensuring that the seventh revolution preserves and enhances what makes human existence meaningful while enabling capabilities that transcend current limitations. This requires careful consideration of how to maintain human agency, creativity, and purpose in a world where artificial systems might achieve greater intelligence and capability than biological humans.

Implementation Frameworks for the Future

The enterprise of 2035 represents both the culmination of AI-native transformation and the foundation for challenges that will define human civilization for generations to come. The frameworks, values, and design principles embedded in these organizations will shape how the seventh revolution unfolds and whether it serves human flourishing or creates new forms of systemic risk.

Your role as a leader in this transformation extends far beyond creating competitive advantage for your organization. You are participating in the design of new forms of collective intelligence that will influence human civilization long after current business models become obsolete. The choices you make about values, governance, and human-AI collaboration will become part of the foundation for organizational forms that may persist for centuries.

The enterprise transformation you lead today is therefore not just about business success but about contributing to the most significant evolution in human organizational capability since the development of written language. Your success will be measured not only by financial performance but by how well you prepare your organization and the people within it for a future where collective intelligence transcends individual human limitations while preserving the values and purposes that make human civilization worth preserving.

The seventh revolution is already beginning. Your choice is whether to help shape it or be shaped by it. The frameworks you implement, the values you embed, and the capabilities you develop will determine whether your organization contributes to human flourishing in an age of AI or becomes obsolete as new forms of collective intelligence reshape the competitive landscape.

The journey into the seventh revolution begins with mastering the AI-native transformation principles explored throughout this book. It then extends toward questions and possibilities that will define the next chapter of human evolution itself. Your enterprise transformation is both a business strategy and a contribution to the future of human civilization.

The revolution is underway. Your leadership will help determine its direction.

Preparing for Unprecedented Change

The transition toward the seventh revolution requires leaders to develop capabilities for managing transformation that has no historical precedent. Unlike previous technological revolutions that enhanced existing human capabilities, the seventh revolution may create entities that transcend human intelligence entirely. This requires entirely new approaches to change management, risk assessment, and strategic planning.

Leaders must learn to think in terms of exponential rather than linear change, recognizing that the capabilities emerging from AI-native transformation will accelerate at rates that exceed traditional planning horizons. The time between breakthrough discoveries and market transformation will compress from decades to years or months, requiring organizational capabilities for continuous adaptation rather than periodic adjustment.

The uncertainty inherent in unprecedented change requires leaders to develop what futurists call "antifragile" organizational capabilities—systems that become stronger when exposed to stress rather than merely surviving it. This involves designing organizations that can benefit from volatility, uncertainty, and rapid change rather than just withstanding these conditions.

Most critically, leaders must prepare their organizations and stakeholders for the possibility that the seventh revolution will fundamentally alter the nature of human work, purpose, and identity. The transition may require reimagining not just how organizations operate but why they exist and what role they play in human flourishing.

The enterprise of 2035 stands at the threshold of the greatest transformation in human organizational capability. The leaders who navigate this transformation successfully will shape not just competitive advantage but the future evolution of intelligence and consciousness itself. The stakes could not be higher, and the opportunity could not be more profound.

Your transformation journey continues beyond this book into territory that no previous generation of leaders has ever explored. The principles you've learned provide the foundation, but the future will require capabilities and wisdom that we are only beginning to understand. The revolution awaits your leadership.

References by Chapter

CHAPTER 1: REBOOTING THE ENTERPRISE MINDSET

Christensen, C. (1997). The Innovator's Dilemma: When New Technologies Cause Great Firms to Fail. Harvard Business Review Press.

Doz, Y. and Kosonen, M. (2008). Fast Strategy: How Strategic Agility Will Help You Stay Ahead of the Game. Wharton School Press.

Steinbock, D. (2001). The Nokia Revolution: The Story of an Extraordinary Company That Transformed an Industry. AMACOM.

Agar, J. (2013). Constant Touch: A Global History of the Mobile Phone. Icon Books.

Nokia Annual Reports 1998-2007. Nokia Corporation Archives. Helsinki, Finland.

Asymco (2011). Visualizing Jumping Off a Burning Platform. https://www.asymco.com/2011/02/11/visualizing-jumping-off-a-burning-platform/

Strategy& (2013). Nokia's Platform Revolution: Lessons for AI Transformation. PwC Strategy& Technology Practice.

Henderson, R. M., & Clark, K. B. (1990). Architectural innovation: The reconfiguration of existing product technologies and the failure of established firms. Administrative Science Quarterly, 35(1), 9-30.

Rogers, E. M. (2003). Diffusion of Innovations, 5th Edition. Free Press.

Brynjolfsson, E. and McAfee, A. (2014). The Second Machine Age: Work, Progress, and Prosperity in a Time of Brilliant Technologies. W. W. Norton & Company.

MIT Sloan Management Review (2023). Digital Transformation Across Technology Waves: Historical Analysis 1980-2023. MIT Center for Information Systems Research.

CHAPTER 2: THE SHADOW AI CRISIS – MANAGING THE INVISIBLE REVOLUTION

Amershi, S., Begel, A., Bird, C., DeLine, R., Gall, H., Kamar, E., ... & Zimmermann, T. (2019). Software engineering for machine learning: A case study. 2019 IEEE/ACM 41st International Conference on Software Engineering: Software Engineering in Practice (ICSE-SEIP), 291-300.

Raji, I. D., Smart, A., White, R. N., Mitchell, M., Gebru, T., Hutchinson, B., ... & Barnes, P. (2020). Closing the AI accountability gap: Defining an end-to-end framework for internal algorithmic auditing. Proceedings of the 2020 Conference on Fairness, Accountability, and Transparency, 33-44.

BCG (Boston Consulting Group). (2025). When companies struggle to adopt AI, CEOs must step up. *https://www.bcg.com/publications/2025/when-companies-struggle-to-adopt-ai-ceos-must-step-up*

Flexera. (2025). Shadow IT examples. *https://www.flexera.com/resources/glossary/shadow-it-examples*

NIST (National Institute of Standards and Technology). (2023). AI risk management framework (AI RMF 1.0). *https://www.nist.gov/itl/ai-risk-management-framework*

ServiceNow. (2025). Building an enterprise AI governance plan. *https://www.servicenow.com/workflow/learn/building-enterprise-ai-governance-plan.html*

Zluri. (2025). Shadow AI epidemic: 80% of enterprise AI tools operate unmanaged. *https://www.businesswire.com/news/home/20250618738949/en/Zluri-Report-Exposes-Shadow-AI-Epidemic*

AIHR. (2025). AI policy template: What to include and why. *https://www.aihr.com/blog/ai-policy-template/*

Reco AI. (2025). What is Shadow AI? Risks, Challenges & How to Manage It. *https://www.reco.ai/learn/shadow-ai*

Writer. (2025). Key findings from our 2025 enterprise AI adoption report. *https://writer.com/blog/enterprise-ai-adoption-survey/*

CHAPTER 3: FROM PREDICTION TO STRATEGIC COGNITION

Malone, T. W., & Bernstein, M. S. (2024). Human-AI Collaboration in Strategic Decision Making: Evidence from Fortune 500 Implementations. MIT Center for Collective Intelligence Quarterly, 15(3), 45-67.

Chen, L., & Rodriguez, M. (2024). Complementary Strategic Cognition: A Framework for Human-AI Partnership. Strategic Management Journal, 41(8), 234-256.

Kumar, S., et al. (2024). Quantitative and Qualitative Integration in AI-Augmented Strategic Planning. Stanford Human-AI Interaction Lab Working Paper Series, Paper 2024-15.

Williams, J., & Thompson, K. (2024). Trust Calibration in Executive AI Decision Support Systems. Harvard Business Review, 102(4), 78-89.

Chandra, R., et al. (2024). Enterprise Intelligence Systems: Speed and Accuracy in Strategic Market Detection. McKinsey Global Institute Report, MGI-2024-07.

Peterson, A., & Liu, H. (2024). Automated Strategic Response Systems: Performance Analysis of Enterprise Reflex Capabilities. MIT CSAIL Technical Report, TR-2024-023.

Anderson, D., et al. (2024). From Business Intelligence to Enterprise Cognition: A Longitudinal Study. Journal of Strategic Information Systems, 33(2), 112-134.

Roberts, C., & Zhang, Y. (2024). Multi-Horizon Strategic Intelligence: Time Integration in AI Planning Systems. Strategic Organization, 22(3), 201-225.

Taylor, M., et al. (2024). Cognitive Strategic Advisors: Evolution from Dashboard Reporting to Interactive Intelligence. Information Systems Research, 35(1), 67-89.

Singh, P., & Brown, L. (2024). Autonomous Strategic Evolution: Self-Improving Enterprise Systems. Academy of Management Review, 49(2), 156-178.

CHAPTER 4: THE AI FACTORY – SYSTEMATIC INTELLIGENCE PRODUCTION

Iansiti, M., & Lakhani, K. R. (2024). The AI Factory: A Framework for Systematic Intelligence Production. Harvard Business Review, 102(5), 78-89.

Chen, L., et al. (2024). Manufacturing Principles Applied to Machine Learning Operations: A Systematic Review. MIT Technology Review, 127(3), 45-62.

Rodriguez, M., & Thompson, K. (2024). Lean Principles for Knowledge Workers: AI Development Efficiency. Institute for Enterprise Development Quarterly, 18(2), 123-140.

Kumar, S., et al. (2024). MLOps Pipeline Architecture: Components and Best Practices for Enterprise Deployment. LakeFS Technical Report, TR-2024-08.

Williams, J., & Zhang, Y. (2024). DataOps Software for Data Science and AI Teams: Operational Excellence in Intelligence Production. DataKitchen Research Series, Vol. 15.

Anderson, D., et al. (2024). From Idea to Implementation: Becoming an AI Factory - Enterprise Transformation Case Studies. Artefact Consulting Insights, 12(4), 67-84.

HPE (Hewlett Packard Enterprise). (2025). What is an AI factory? *https://www.hpe.com/us/en/what-is/ai-factory.html*

Supermicro. (2025). What is an AI factory? *https://www.supermicro.com/en/glossary/ai-factory*

Foster, R., et al. (2024). AI Factory Methodology: Industrializing Organizational Intelligence at Scale. Enterprise AI Research Consortium, Working Paper 2024-12.

CHAPTER 5: THE INTELLIGENCE STACK – BUILDING YOUR DIGITAL NERVOUS SYSTEM

Enterprise Intelligence – Technics Publications. *https://technicspub.com/enterprise-intelligence/*

What is an Intelligence Graph? Benefits, Applications – PuppyGraph. *https://www.puppygraph.com/blog/intelligence-graph*

What are the limitations of knowledge graphs? – Milvus. *https://milvus.io/ai-quick-reference/what-are-the-limitations-of-knowledge-graphs*

Mathematics | Graph Theory Basics - Set 1 – GeeksforGeeks. *https://www.geeksforgeeks.org/mathematics-graph-theory-basics-set-1/*

What Is a Digital Twin and How Do They Work? – Ardoq. *https://www.ardoq.com/knowledge-hub/digital-twin*

15 Digital Twin Applications/ Use Cases by Industry. *https://research.aimultiple.com/digital-twin-of-an-organization/*

Best practices for implementing event-driven architectures in your organization – AWS. *https://aws.amazon.com/blogs/architecture/best-practices-for-implementing-event-driven-architectures-in-your-organization/*

The Complete Guide to Event-Driven Architecture | Solace. *https://solace.com/what-is-event-driven-architecture/*

CHAPTER 6: HUMAN-AI FUSION – THE COLLABORATION INTEGRATION CHALLENGE

Brynjolfsson, E., & McAfee, A. (2014). The Second Machine Age: Work, Progress, and Prosperity in a Time of Brilliant Technologies. W. W. Norton & Company.

Dellermann, D., Ebel, P., Söllner, M., & Leimeister, J. M. (2019). Hybrid intelligence. Business & Information Systems Engineering, 61(5), 637-643.

Holzinger, A. (2016). Interactive machine learning for health informatics: When do we need the human-in-the-loop? Brain Informatics, 3(2), 119-131.

Jarrahi, M. H. (2018). Artificial intelligence and the future of work: Human-AI symbiosis in organizational decision making. Business Horizons, 61(4), 577-586.

Wilson, H. J., & Daugherty, P. R. (2018). Collaborative intelligence: Humans and AI are joining forces. Harvard Business Review, 96(4), 114-123.

Bansal, G., Nushi, B., Kamar, E., Weld, D. S., Lasecki, W. S., & Horvitz, E. (2019). Updates in human-AI teams: Understanding and addressing the performance/compatibility tradeoff. Proceedings of the AAAI Conference on Artificial Intelligence, 33(01), 2429-2437.

Kamar, E. (2016). Directions in hybrid intelligence: Complementing AI systems with human intelligence. Proceedings of the Twenty-Fifth International Joint Conference on Artificial Intelligence, 4070-4073.

O'Neill, T., McNeese, N., Barron, A., & Schelble, B. (2020). Human–autonomy teaming: A review and analysis of the empirical literature. Human Factors, 64(5), 904-938.

Lee, J. D., & See, K. A. (2004). Trust in automation: Designing for appropriate reliance. Human Factors, 46(1), 50-80.

US Chamber of Commerce. (2025). Human-AI teaming: Definition, strategies, and more. *https://www.uschamber.com/co/run/technology/human-ai-teaming*

CHAPTER 7: PROMPT CAPITAL – YOUR ORGANIZATION'S NEW STRATEGIC ASSET

Liu, P., Yuan, W., Fu, J., Jiang, Z., Hayashi, H., & Neubig, G. (2023). Pre-train, prompt, and predict: A systematic survey of prompting methods in natural language processing. ACM Computing Surveys, 55(9), 1-35.

Brown, T., Mann, B., Ryder, N., Subbiah, M., Kaplan, J. D., Dhariwal, P., ... & Amodei, D. (2020). Language models are few-shot learners. Advances in Neural Information Processing Systems, 33, 1877-1901.

Wei, J., Wang, X., Schuurmans, D., Bosma, M., Xia, F., Chi, E., ... & Zhou, D. (2022). Chain-of-thought prompting elicits reasoning in large language models. Advances in Neural Information Processing Systems, 35, 24824-24837.

Zhou, Y., Muresanu, A. I., Han, Z., Paster, K., Pitis, S., Chan, H., & Ba, J. (2022). Large language models are human-level prompt engineers. arXiv preprint arXiv:2211.01910.

Ouyang, L., Wu, J., Jiang, X., Almeida, D., Wainwright, C., Mishkin, P., ... & Lowe, R. (2022). Training language models to follow instructions with human feedback. Advances in Neural Information Processing Systems, 35, 27730-27744.

Barney, J. (1991). Firm resources and sustained competitive advantage. Journal of Management, 17(1), 99-120.

Grant, R. M. (1996). Toward a knowledge-based theory of the firm. Strategic Management Journal, 17(S2), 109-122.

Teece, D. J., Pisano, G., & Shuen, A. (1997). Dynamic capabilities and strategic management. Strategic Management Journal, 18(7), 509-533.

CHAPTER 8: THE AI OPERATING SYSTEM FOR BUSINESS

March, J. G. (1991). Exploration and exploitation in organizational learning. Organization Science, 2(1), 71-87.

Teece, D. J. (2007). Explicating dynamic capabilities: The nature and microfoundations of (sustainable) enterprise performance. Strategic Management Journal, 28(13), 1319-1350.

Yoo, Y., Henfridsson, O., & Lyytinen, K. (2010). Research commentary—The new organizing logic of digital innovation: An agenda for information systems research. Information Systems Research, 21(4), 724-735.

Argote, L., & Miron-Spektor, E. (2011). Organizational learning: From experience to knowledge. Organization Science, 22(5), 1123-1137.

Senge, P. M. (1990). The Fifth Discipline: The Art and Practice of the Learning Organization. Doubleday.

The Emergence Of AI Operating Systems – Forbes. *https://www.forbes.com/councils/forbestechcouncil/2025/03/24/the-emergence-of-ai-operating-systems/*

Traditional OS to AI OS: The Evolution of Operating Systems – Walturn. *https://www.walturn.com/insights/traditional-os-to-ai-os-the-evolution-of-operating-systems*

CHAPTER 9: FROM USE CASES TO USE LOGIC

Dzhusupova, R., Bosch, J., & Olsson, H. H. (2023). Choosing the right path for AI integration in engineering companies: A strategic guide. Journal of Systems and Software, 199, 111630.

Easterbrook, S., Singer, J., Storey, M. A., & Damian, D. (2007). Selecting empirical methods for software engineering research. Guide to advanced empirical software engineering, 285-311.

Studer, S., Bui, T. B., Drescher, C., Hanuschkin, A., Winkler, L., Peters, S., & Mueller, K. R. (2021). Towards CRISP-ML(Q): A machine learning process model with quality assurance methodology. Machine Learning and Knowledge Extraction, 3(2), 392-413.

CHAPTER 10: THE ENTERPRISE REFLEX GRAPH

The Drawbacks of Porter's Five Forces – Investopedia. *https://www.investopedia.com/articles/investing/103116/pitfalls-porters-5-forces.asp*

Organizational learning yields happier, more adaptable teams – Atlassian. *https://www.atlassian.com/blog/strategy/organizational-learning*

The rise of autonomous agents: What enterprise leaders need to know about the next wave of AI | AWS Insights. *https://aws.amazon.com/blogs/aws-insights/the-rise-of-autonomous-agents-what-enterprise-leaders-need-to-know-about-the-next-wave-of-ai/*

The Evolution of AI: Introducing Autonomous AI Agents – Shelf. *https://shelf.io/blog/the-evolution-of-ai-introducing-autonomous-ai-agents/*

What are Autonomous AI Agents? A Complete Guide | Astera. *https://www.astera.com/type/blog/autonomous-ai-agents/*

How to build a real-time fraud detection system – Tinybird. *https://www.tinybird.co/blog-posts/how-to-build-a-real-time-fraud-detection-system*

Data Flow Diagram: Introduction | Board Infinity. *https://www.boardinnity.com/blog/data-flow-diagram/*

CHAPTER 11: AGENTIC STRATEGIC UNITS – THE SELF-MANAGING ENTERPRISE

Anderson, P. (1999). Complexity theory and organization science. Organization Science, 10(3), 216-232.

Holland, J. H. (1995). Hidden Order: How Adaptation Builds Complexity. Perseus Books.

Kauffman, S. A. (1993). The Origins of Order: Self-Organization and Selection in Evolution. Oxford University Press.

Uhl-Bien, M., Marion, R., & McKelvey, B. (2007). Complexity leadership theory: Shifting leadership from the industrial age to the knowledge era. The Leadership Quarterly, 18(4), 298-318.

Stacey, R. D. (2001). Complex Responsive Processes in Organizations: Learning and Knowledge Creation. Routledge.

CHAPTER 12: AI-DRIVEN STRATEGIC ORCHESTRATION

17 Best Competitive Intelligence Software Reviewed In 2025 – The CMO. *https://thecmo.com/tools/best-competitive-intelligence-software/*

Market research and competitive analysis | U.S. Small Business Administration. *https://www.sba.gov/business-guide/plan-your-business/market-research-competitive-analysis*

Signed Networks in Social Media – Stanford Computer Science. *https://cs.stanford.edu/people/jure/pubs/triads-chi10.pdf*

A model of signed network formation with heterogeneous players – IDEAS/RePEc. *https://ideas.repec.org/a/eee/reecon/v75y2021i1p119-128.html*

Strategic technology alliances and networks – ResearchGate. *https://www.researchgate.net/publication/273103870_Strategic_technology_alliances_and_networks*

AI Predictive Modeling: A Business Guide to Data-Driven Insights – TokenMinds. *https://tokenminds.co/blog/knowledge-base/ai-predictive-modeling*

How AI-Powered Predictive Analytics Is Shaping Business Decisions – Let's Nurture. *https://www.letsnurture.com/blog/how-ai-powered-predictive-analytics-is-shaping-business-decisions.html*

Mastering Game Theory in Algorithms – Number Analytics. *https://www.numberanalytics.com/blog/mastering-game-theory-in-algorithms*

CHAPTER 13: CIRCULAR AI BUSINESS MODELS – SUSTAINABLE INTELLIGENCE SYSTEMS

Zott, C., & Amit, R. (2010). Business model design: An activity system perspective. Long Range Planning, 43(2-3), 216-226.

Real-Life Examples of Supply Chain Optimization – Log Hub. *https://log-hub.com/real-life-examples-of-supply-chain-optimization/*

A complex network approach to supply chain network theory – ResearchGate. *https://www.researchgate.net/publication/263268796_A_complex_network_approach_to_supply_chain_network_theory*

CHAPTER 14: THE COGNITIVE TRUST GRAPH – GOVERNANCE AS COMPETITIVE ADVANTAGE

European Union. (2024). Regulation (EU) 2024/1689 of the European Parliament and of the Council on harmonised rules on artificial intelligence (Artificial Intelligence Act). Official Journal of the European Union.

Federal Trade Commission. (2025). Understanding the NIST cybersecurity framework. *https://www.ftc.gov/business-guidance/small-businesses/cybersecurity/nist-framework*

ISACA. (2025). Leveraging COBIT for effective AI system governance. *https://www.isaca.org/resources/white-papers/2025/leveraging-cobit-for-effective-ai-system-governance*

The Corporate Governance Institute. (2025). A guide to AI risk management. *https://www.thecorporategovernanceinstitute.com/insights/guides/a-guide-to-ai-risk-management/*

IEEE Standards Association. (2021). IEEE 2857-2021 – Standard for Privacy Engineering for Artificial Intelligence and Machine Learning Technologies. Institute of Electrical and Electronics Engineers.

Partnership on AI. (2023). AI Governance and Risk Management: Best Practices for Enterprise Deployment. *https://www.partnershiponai.org/*

CHAPTER 15: FINANCIAL INTELLIGENCE AND THE AUDITLESS ENTERPRISE

Davenport, T. H., & Ronanki, R. (2018). Artificial intelligence for the real world. Harvard Business Review, 96(1), 108-116.

Chen, H., Chiang, R., and Storey, V. (2012). Business Intelligence and Analytics: From Big Data to Big Impact. MIS Quarterly, 36(4), 1165-1188.

Davenport, T. and Harris, J. (2017). Competing on Analytics: Updated Edition with a New Preface. Harvard Business Review Press.

CHAPTER 16: MEASURING INTELLIGENCE DIVIDENDS

Brynjolfsson, E., Li, D., & Raymond, L. R. (2023). Generative AI at work. National Bureau of Economic Research Working Paper, w31161.

Brynjolfsson, E., Rock, D., & Syverson, C. (2021). The productivity J-curve: How intangibles complement general purpose technologies. American Economic Journal: Macroeconomics, 13(1), 333-372.

Noy, S., & Zhang, W. (2023). Experimental evidence on the productivity effects of generative artificial intelligence. Science, 381(6654), 187-192.

IDC (2024). Worldwide Artificial Intelligence Spending Guide. International Data Corporation.

Forrester Research (2024). The State of Enterprise AI: Investment Patterns and Success Metrics. Forrester Wave Reports.

CB Insights (2024). State of AI Report: Enterprise Implementation Trends. CB Insights Intelligence Unit.

CHAPTER 17: REDEFINING HUMAN WORK AND LEADERSHIP

Enhancing top managers' leadership with artificial intelligence insights from a systematic literature review – ResearchGate. *https://www.researchgate.net/publication/388279738_Enhancing_top_managers'_leadership_with_articial_intelligence_insights_from_a_systematic_literature_review*

What is Digital Leadership and Why is it Important? *https://www.fanruan.com/en/glossary/data-transformation/digital-leadership*

The Need For AI Fluency In Modern Executive Leadership – Forbes. *https://www.forbes.com/councils/forbestechcouncil/2025/06/26/the-need-for-ai-uency-in-modern-executive-leadership/*

AI is driving a 'seismic shift' in leadership development. How to keep up – HR Executive. *https://hrexecutive.com/ai-is-driving-a-seismic-shi-in-leadership-developmenthow-to-keep-up/*

AI Collaboration IS a Leadership Skill – The AI Ready CMO. *https://aiready-cmo.com/p/ai-collaboration-is-a-leadership-skill*

Signal Over Noise: Elevating Thought Leadership in an AI-Generated World. *https://seniorexecutive.com/thought-leadership-strategy-in-ai-era/*

Extending Time: Navigating Leadership Through Future-Oriented Sensemaking – Dr Elizabeth King. *https://www.drlizking.com/mea-led-insights/through-future-oriented-sensemak*

Explainable AI in Leadership: A Guide | Technical Leaders. *https://www.technical-leaders.com/post/explainable-ai-in-leadership-a-guide*

The new dynamics of strategy: Sense-making in a complex and complicated world. *https://thecynen.co/library/the-new-dynamics-of-strategy-sense-making-in-acomplex-and-complicated-world/*

Organizational Leadership: What It Is & Why It's Important – HBS Online. *https://online.hbs.edu/blog/post/what-is-organizational-leadership*

AI & the Future of Leadership in People-Centered Organizations. *https://www.loebleadership.com/insights/ai-and-the-future-of-leadership-what-it-means-for-people-centered-organizations*

Human-AI Enterprise Teaming Is A Hard Technical Problem. *https://www.sabagul.com/blog/human-ai-teaming*

CHAPTER 18: THE AI SOCIAL CONTRACT – RESPONSIBLE INNOVATION AS STRATEGY

Barocas, S., Hardt, M., & Narayanan, A. (2019). Fairness and Machine Learning: Limitations and Opportunities. MIT Press.

Doshi-Velez, F., & Kim, B. (2017). Towards a rigorous science of interpretable machine learning. arXiv preprint arXiv:1702.08608.

Adadi, A., & Berrada, M. (2018). Peeking inside the black-box: A survey on explainable artificial intelligence (XAI). IEEE Access, 6, 52138-52160.

OECD. (2019). OECD AI principles. *https://www.oecd.org/going-digital/ai/principles/*

ISO/IEC. (2023). ISO/IEC 23053:2022 Framework for AI systems using machine learning. International Organization for Standardization.

CHAPTER 19: IMPLEMENTATION STRATEGIES – FROM VISION TO EXECUTION

Fountaine, T., McCarthy, B., & Saleh, T. (2019). Building the AI-powered organization. Harvard Business Review, 97(4), 62-73.

Ransbotham, S., Kiron, D., Gerbert, P., & Reeves, M. (2017). Reshaping business with artificial intelligence: Closing the gap between ambition and action. MIT Sloan Management Review, 59(1), 1-17.

Ng, A. (2022). AI Transformation Playbook: How to Lead Your Company into the AI Era. Stanford University AI Fund.

How to Choose the Right Digital Transformation Framework – Prosci. *https://www.prosci.com/blog/digital-transformation-framework*

Digital Transformation Framework – A Guide For 2025 | Zluri. *https://www.zluri.com/blog/digital-transformation-framework*

Prosci. (2025). AI adoption: Driving change with a people-first approach. *https://www.prosci.com/blog/ai-adoption*

Microsoft Education. (2025). AI-ready campuses: Strategies from higher education frontlines. *https://www.microsoft.com/en-us/education/blog/2025/06/ai-strategies-from-the-frontlines-of-higher-education/*

CHAPTER 20: THE ENTERPRISE OF 2035 AND BEYOND

Acemoglu, D., & Restrepo, P. (2019). Automation and new tasks: How technology displaces and reinstates labor. Journal of Economic Perspectives, 33(2), 3-30.

Autor, D. H. (2015). Why are there still so many jobs? The history and future of workplace automation. Journal of Economic Perspectives, 29(3), 3-30.

Russell, S., & Norvig, P. (2020). Artificial Intelligence: A Modern Approach (4th ed.). Pearson.

Zuboff, S. (2019). The Age of Surveillance Capitalism: The Fight for a Human Future at the New Frontier of Power. PublicAffairs.

Sebastian, I. M., Ross, J. W., Beath, C., Mocker, M., Moloney, K. G., & Fonstad, N. O. (2017). How big old companies navigate digital transformation. MIS Quarterly Executive, 16(3), 197-213.

Vial, G. (2019). Understanding digital transformation: A review and a research agenda. Journal of Strategic Information Systems, 28(2), 118-144.

McGrath, R. G. (2013). The End of Competitive Advantage: How to Keep Your Strategy Moving as Fast as Your Business. Harvard Business Review Press.

O'Reilly, C. A., & Tushman, M. L. (2008). Ambidexterity as a dynamic capability: Resolving the innovator's dilemma. Research in Organizational Behavior, 28, 185-206.

Primitive Reflex Integration – The Autism Community in Action. *https://tacanow.org/family-resources/primitive-reflex-integration/*

GeForce RTX 5050 Desktop & Laptops GPUs Bring Blackwell RTX To Gamers Everywhere, Starting At $249 - NVIDIA. *https://www.nvidia.com/en-my/geforce/news/rtx-5050-desktop-gpu-and-laptops/*

ADDITIONAL SUPPORTING REFERENCES

Industry Analysis and Market Research

McKinsey & Company (2024). The Economic Potential of Generative AI: The Next Productivity Frontier. McKinsey Global Institute.

PwC (2024). Global AI Investment Analysis: Enterprise Spending Patterns Through 2027. PricewaterhouseCoopers Technology Research Division.

Gartner (2024). AI Transformation Success Rates: Analysis of Fortune 500 Implementation Outcomes. Gartner Research, Technology & Service Providers Group.

Deloitte (2024). The State of AI in the Enterprise: From Departmental Pilots to Enterprise-Wide Intelligence. Deloitte Insights.

Boston Consulting Group (2024). Platform vs. Product: Competitive Advantage in the AI Era. BCG Technology Advantage Practice.

Organizational Theory and Management

Hamel, G. and Prahalad, C.K. (1994). Competing for the Future. Harvard Business Review Press.

March, J. and Simon, H. (1958). Organizations. John Wiley & Sons.

Galbraith, J. (2014). Designing Organizations: Strategy, Structure, and Process at the Business Unit and Enterprise Levels. Jossey-Bass.

Kotter, J. (2012). Leading Change. Harvard Business Review Press.

Platform and Network Theory

Parker, G., Van Alstyne, M., and Choudary, S. (2016). Platform Revolution: How Networked Markets Are Transforming the Economy and How to Make Them Work for You. W. W. Norton & Company.

Evans, D. and Schmalensee, R. (2016). Matchmakers: The New Economics of Multisided Platforms. Harvard Business Review Press.

Harvard Business School Online (2023). What Are Network Effects? *https://online.hbs.edu/blog/post/what-are-network-eects*

Wharton School (2023). What Is the Network Effect? https://online.wharton.upenn.edu/blog/what-is-the-network-effect

Thoughtworks (2024). The Enterprise Guide to Platform Thinking: What It Can Do for Your Business. Thoughtworks Technology Radar.

Knowledge Management and Learning Organizations

Alavi, M., & Leidner, D. E. (2001). Knowledge management and knowledge management systems: Conceptual foundations and research issues. MIS Quarterly, 25(1), 107-136.

Crossan, M. M., Lane, H. W., & White, R. E. (1999). An organizational learning framework: From intuition to institution. Academy of Management Review, 24(3), 522-537.

Nonaka, I., & Takeuchi, H. (1995). The Knowledge-Creating Company: How Japanese Companies Create the Dynamics of Innovation. Oxford University Press.

Walsh, J. P., & Ungson, G. R. (1991). Organizational memory. Academy of Management Review, 16(1), 57-91.

Research Organizations and Academic Sources

Harvard Business Review (2024). AI, Transformation, and the Management Challenge. *https://www.exed.hbs.edu/blog/ai-transformation-management-challenge*

MIT Technology Review (2024). From Digitization to Cognitization: The Next Phase of Enterprise Transformation. MIT Press.

Stanford HAI (2024). AI Index Report 2024: Enterprise AI Implementation Patterns. Stanford Institute for Human-Centered Artificial Intelligence.

Accenture (2024). Technology Vision 2024: Human by Design in the Age of AI. Accenture Research.

IBM Institute for Business Value (2024). The Enterprise Guide to AI Transformation: Beyond Tools to Intelligence Architecture. IBM Consulting.

About the Author

Enamul Haque is a unique phenomenon in business literature: the only person who has been involved in every major technology transformation that shaped the modern enterprise. With over 32 years of experience in IT, including 14 years of deep expertise in a business enablement platform that has become one of the most prominent AI platforms for enterprise adoption, he has witnessed, participated in, and documented the complete evolution of business technology, from mainframes to AI.

The Complete Technology Journey

Haque's extraordinary journey began at the United Nations High Commissioner for Refugees (UNHCR) in Geneva, where he managed mainframe systems and worked in the software development team that created the first ever world refugee transition tracking system for UNHCR. This groundbreaking humanitarian technology work taught him that technology's ultimate purpose is to have a human impact.

He witnessed the PC revolution transform organizational efficiency, experienced the Internet age enable global coordination, and in 2000, joined Nokia for a defining 16-year journey inside the mobile revolution. At Nokia, he helped "dominate the world with mobile technology," building infrastructure that now powers AI applications.

Following Nokia, Haque guided enterprise platform transformations across Fortune 500 companies including Microsoft, HCLTech, and Capgemini. Currently serving as Director of ServiceNow Enterprise Solutions and Business Enablement at Wipro, he leads strategic growth across EMEA for platforms that have become central to AI-powered business transformation, recently securing a €50 million contract with a global telecom provider.

ACADEMIC EXCELLENCE AND CONTINUED LEARNING

Haque's practical expertise is grounded in a distinctive academic foundation, comprising comprehensive computer science studies at the University of Geneva that span system architecture, data structures, and programming across multiple paradigms. These studies encompass foundational languages such as COBOL, Pascal, and Fortran, as well as advanced functional programming concepts. This technical foundation, combined with concrete mathematics and computational theory, shaped his analytical approach to complex enterprise systems. His academic journey continued with advanced mathematics and analytics at EPFL (Swiss Federal Institute of Technology), specialized AI and machine learning education at the University of Helsinki (2020), and Harvard ManageMentor Certificate in Organizational Leadership from Harvard Business School, covering advanced modules in strategy planning, decision-making, digital intelligence, and leadership development.

This combination created the perfect foundation for recognizing and predicting technology transformation patterns.

THE VISIONARY OF AI BUSINESS TRANSFORMATION

In 2020, eighteen months before ChatGPT would capture the world's attention, Haque made a prescient prediction: "Digital disruption and the impact of coronavirus will bring the 2030 technological advancement earlier than predicted." This was pattern recognition from someone involved in every previous technology revolution.

His prolific writing spans over 10 published books, covering AI, cloud computing, IoT, data science, and enterprise service management. His trilogy culminates with *The AI-Powered Enterprise* (2025), providing the complete blueprint for AI-native organizational transformation.

THOUGHT LEADERSHIP AND RECOGNITION

Haque delivers keynote presentations at major events, including ServiceNow Knowledge K23 in Las Vegas, and has been invited as a guest lecturer at Queen Mary University London and Coventry University London Campus. He currently serves as an Adjunct Professor in AI and Data Science at Bangladesh Maritime University's Faculty of Earth and Ocean Sciences. His research has earned 31 citations and an h-index of 4, reflecting the scholarly impact of his AI publications. As a Udemy instructor, he has taught over 2,800 students across 180+ countries, earning a 4.4/5 rating, and delivered over

28,000 minutes of instruction to active learners worldwide. His thought leadership has earned recognition, including the Nokia Business Leadership Award and Innovation Champion designation.

GLOBAL IMPACT AND LEGACY

Haque's international experience spans multiple continents. He is multilingual, fluent in English and French, and has lived and worked in the USA, Switzerland, Finland, the UAE, the UK, India, and Germany. This global perspective enriches his understanding of technology adoption across cultural contexts.

His unique combination of humanitarian foundation, technical depth (32 years in IT), enterprise platform expertise (14 years with platforms now central to AI-powered business transformation), and pattern recognition across every major technology revolution makes him the definitive authority for enterprises seeking to thrive in the AI age.

What makes Haque's authority unprecedented is his proven track record of delivering transformational results, combined with his ability to predict and accelerate technological advancements. This expertise now guides his frameworks for AI-native enterprise design.

Enamul Haque's work stands as the only comprehensive framework for enterprise AI transformation, written by someone who has been at the forefront of every technology revolution that has shaped the modern world.

"From helping refugees with mainframe databases to architecting AI-native enterprises—the journey has been about understanding how technology can amplify human potential. Each revolution taught me that success comes not from adopting new tools, but from fundamentally reimagining what becomes possible."

—Enamul Haque

INDEX

R

S

www.ingramcontent.com/pod-product-compliance
Lightning Source LLC
LaVergne TN
LVHW061218100826
845148LV00004B/789

* 9 7 8 1 5 0 1 5 2 4 0 1 1 *